MW00637447

Other McGraw-Hill Books of Interest

DAVIDSON • *Consumer Electronics Troubleshooting and Repair Handbook*
JURGEN • *Digital Consumer Electronics Handbook*
LUTHER & INGLIS • *Video Engineering*, 3/e
POHLMANN • *Principles of Digital Audio*, 4/e
ROBIN & POULIN • *Digital Television Fundamentals*
SOLARI • *Digital Video and Audio Compression*
SYMES • *Video Compression*
TAYLOR • *DVD Demystified*
WHITAKER • *DTV*, 2/e
WHITAKER • *Standard Handbook of Video Engineering*, 3/e

DTV Survival Guide

Jim Boston

McGraw-Hill

New York   San Francisco   Washington, D.C.   Auckland   Bogotá
Caracas   Lisbon   London   Madrid   Mexico City   Milan
Montreal   New Delhi   San Juan   Singapore
Sydney   Tokyo   Toronto

Library of Congress Cataloging-in-Publication Data

Boston, Jim.
    DTV survival guide / Jim Boston.
        p.   cm.
    Includes bibliographical references and index.
    ISBN 0-07-135061-6
    1. Digital television. I. Title.
  TK6678B67   2000
  621.388—dc21   99-054049
                                  CIP

McGraw-Hill

*A Division of The **McGraw·Hill** Companies*

Copyright © 2000 by The McGraw-Hill Companies, Inc. All rights reserved. Printed in the United States of America. Except as permitted under the United States Copyright Act of 1976, no part of this publication may be reproduced or distributed in any form or by any means, or stored in a data base or retrieval system, without the prior written permission of the publisher.

1 2 3 4 5 6 7 8 9 0 DOC/DOC 0 6 5 4 3 2 1 0

ISBN 0-07-135061-6

The sponsoring editor for this book was Stephen S. Chapman, the editing supervisor was Stephen M. Smith, and the production supervisor was Pamela A. Pelton. It was set in Century Schoolbook by Kim Sheran of McGraw-Hill's Hightstown, N.J., Professional Book Group composition unit.

Printed and bound by R. R. Donnelley & Sons Company.

McGraw-Hill books are available at special quantity discounts to use as premiums and sales promotions, or for use in corporate training programs. For more information, please write to the Director of Special Sales, Professional Publishing, McGraw-Hill, Two Penn Plaza, New York, NY 10121-2298. Or contact your local bookstore.

 This book is printed on recycled, acid-free paper containing a minimum of 50% recycled de-inked fiber.

Information contained in this work has been obtained by The McGraw-Hill Companies, Inc. ("McGraw-Hill") from sources believed to be reliable. However, neither McGraw-Hill nor its authors guarantee the accuracy or completeness of any information published herein and neither McGraw-Hill nor its authors shall be responsible for any errors, omissions, or damages arising out of use of this information. This work is published with the understanding that McGraw-Hill and its authors are supplying information but are not attempting to render engineering or other professional services. If such services are required, the assistance of an appropriate professional should be sought.

In memory of my father, Barney,
who had the same love/hate relationship
with television that I have.

Contents

Foreword

The advent of digital television (DTV) in the United States has probably created more controversy in the last 10 years than all other broadcast technical issues since Guglielmo Marconi sent his first wireless message. Even the term *digital* has resulted in a bunkerlike mentality among proponents of the concept as they each espouse their particular solutions. And in the case of the DTV standard, politics, not technology, has been the primary cause of delays in implementation.

The source of today's U.S. DTV system actually traces back to late 1988. The Japanese broadcast network NHK had just demonstrated an advanced television technology with greatly improved images. While many people were impressed with the technology, many more were aghast that America didn't have its own version of the new television system. The result was political hysteria.

Congress was soon screaming for action. With great pontification and chest pounding, legislators told the FCC to do something. The FCC quickly looked for cover, and Commissioner Patricia Diaz-Dennis said, "We need to ship this out of here." And that's exactly what the good bureaucrats did by appointing a committee. The FCC created the Advisory Committee on Advanced Television Service (ACATS). It turned out that this became the first really productive step toward developing a U.S. digital TV system.

By 1993, as many as a dozen companies were working on their own versions of high-definition television (HDTV). During this time, ACATS was overseeing the official testing of each proposed system, but the process was fraught with both technical and political minefields. The procedure was expensive and time-consuming for everyone. Worst of all, no winning solution was emerging among the various proposals. The committee decided it would be best to select the best technologies from among all the systems, and combine them. In other words, to join forces. The question remained, though: Was that politically possible?

After much wrangling and arm twisting by ACATS, seven entities, General Instrument, Sarnoff, MIT, Philips, Zenith, Thomson, and AT&T, finally agreed in May 1993 to join together and co-develop the American DTV system. This system became known as the *Grand Alliance*. While many thought America was now on the fast track to digital, the computer industry had other plans. Feeling left out of the process, these companies collectively stymied any adoption of a U.S. DTV standard for the next three years.

Finally, compromises were negotiated among all the players, and on Christmas Eve 1996 the Federal Communications Commission approved what became known as the Advanced Television Systems Committee (ATSC) standard A/65 for digital broadcasting in the United States. After more than 13 years, the broadcast industry finally had a DTV standard.

Now, just a few years after that important act, there are almost 100 DTV stations on the air. While it may have taken longer than many wanted for the DTV standard to get here, that it did is proof that the system works. Best of all, we've only scratched the surface of what digital television can ultimately mean to our daily lives. Stay tuned.

Brad Dick
Editor
Broadcast Engineering *magazine*

Acknowledgments

Writing my first book has been a wonderful and horrible experience that above all has been enlightening. As with most projects it took more than one person to complete it, but in this case only one person will receive credit. I therefore want to thank the people who helped me with this project who would otherwise not be recognized.

First I would like to thank my wife, Joanne, who asked the really tough question, What exactly *is* DTV? The first three chapters are an attempt to answer that. I would also like to thank my daughters, Jenny and Shelly, who put up with my inattentiveness while I wrote this, and my son David, who encouraged me to see this through.

There are a number of other people who built the foundation for this book: Jerry Whitaker, who convinced me that I could actually do this; David Lingenfelter, one of the most tenacious and best engineers this business has, who answered the questions I asked and with whom I had discussions that pushed my knowledge forward; Frank Foge, who has forgotten more about RF than I will ever know and who shed light on RF in a logical way; Bill Beeman and Jim Evers, who not only managed television station KICU to success, but fostered an almost lablike atmosphere where experimentation and learning was celebrated; my lifelong friend Ted Shimkowski, who waded through the manuscript when it was in a chicken scratch state and provided much needed input; Jim Baird, with whom I discussed where technology might be taking the industry; John Neundorfer, who reviewed the first three chapters; and finally Stephen Smith at McGraw-Hill, who guided and helped me during the metamorphosis of my text into the book in your hands.

Jim Boston

The March to
Digital Television

1.1 What Is Digital Television?

DTV, or digital television, is a "ready or not, here it comes" proposition. This is true not only for consumers but also for television professionals. Part of the universal confusion surrounding DTV is that it represents different things to different people. To most consumers DTV means sharper pictures and *extremely* expensive receivers. To most broadcasters it means uncertainty and large capital outlays with very little initial return on investment. Even people in the electronics industry think DTV means only compressed bit streams. However, digital television is not entirely new. It has been serving niche markets in the television universe for more than 20 years. Anyone who subscribes to Direct TV ™ or to one of the other broadcast satellite services has already been receiving DTV. The set-top box that decodes the digital signal from the satellite receiver feeds a television signal that is usable by your current set. Some cable companies now have digital channels on their systems. Again, you need a set-top box to receive these signals. The set-top box will become much more prevalent in the next few years because boxes that receive local DTV signals and convert them to signals your current set can display will become available. Digital television is nothing new to most television engineers.

1.1.1 National Television System Committee (NTSC)

First, a few additional terms and a little history are necessary to an understanding of DTV. NTSC, which is the television format that broadcasters transmit today, was the acronym for the National Television System Committee. The committee was formed by the Federal Communications Commission (FCC) back in the 1940s, and its purpose was to decide on the technical attributes of the television signal. An inside joke among broadcasters, especially our European counterparts (who use a different and, what many claim, a better standard), is that NTSC means "never twice the same color." The NTSC originally specified how black-and-white video, with its accompanying audio or sound, would be transmitted over the airwaves. Later, the committee was reconvened to determine how color information could be added to the signal. NTSC is also called *analog television* by some.

NTSC is considered a *standard-definition* (SD) television signal. Many other signals, including many DTV signals, are also SD television. Confused? You're not alone. Many equate DTV with *high-definition* (HD) television only. This is understandable because DTV grew out of the initial Japanese development of HD. In the early 1970s a number of Japanese organizations started experimenting with analog HD television. We should expand upon analog versus digital television at this point. *Analog television* (digital television will be introduced in Sec. 1.2.3) is a signal that is continuous in nature, with a varying voltage whose value indicates the picture's brightness and whether it is time to start a new scan line or a whole new picture (Fig. 1.1).

Later, when color was added, an additional frequency was embedded into this varying voltage. Its amplitude indicated how saturated the color was, and its phase determined the color (Fig. 1.2).

The problem with this approach is that the television receivers and processing equipment used in television stations can't completely separate the color signal from the black-and-white picture. You see the result when a gray tweed suit on someone develops a rainbow pattern. To eliminate this problem, some *videotape machines* (VTRs) are used in professional television applications that keep the chroma, or color information, separate from the luminance, or black-and-white information. This is what is known as *component television*. But before it can be broadcast to your home, the chroma and luminance must be recombined. This is called *composite television*. NTSC is analog composite television. Early Japanese HD was analog component television, but with a lot more picture information than the television we are used to.

1.1.2 Advisory Committee on Advanced Television Service (ACATS)

In the early 1980s the head of engineering at one of the major U.S. networks claimed that the Japanese had invented HD television to spark a rise in TV set sales, which had flattened out at the time. The Japanese kept plugging away. In 1987, 58 broadcasting organizations petitioned the FCC to initiate proceedings to explore the issues arising from the possible introduction of advanced television technologies. The broadcasters were concerned about the possible impact that HD might have on them. At the time, it was believed that high-definition television (HDTV) could not be broadcast in a standard television channel, which is 6 MHz wide. The FCC created the Advisory Committee on Advanced Television Service, known as ACATS. Its mandate was to determine the form that HDTV should take. Initially, ACATS was only looking at analog high-definition television. At one time there were 23 analog

Figure 1.1 An NTSC signal is displayed on a video monitor that can be time-shifted to show the horizontal blanking interval. Below the video monitor is a waveform monitor that is displaying the varying voltages that occur during the horizontal interval.

proposals on the table, and the technology was heavily dominated by the Japanese. In 1990 U.S. and European concerns performed an "end run" around the Japanese and their analog HDTV. Four digital television transmission systems were proposed by consortiums consisting of AT&T (now Lucent), Sarnoff Research Center (formerly RCA's research lab), General Instrument, Massachusetts Institute of Technology (MIT), Philips, Thomson, and Zenith. ACATS decided that the new U.S. television transmission system would be digital. The Japanese were out. However, in 1993 ACATS tested the

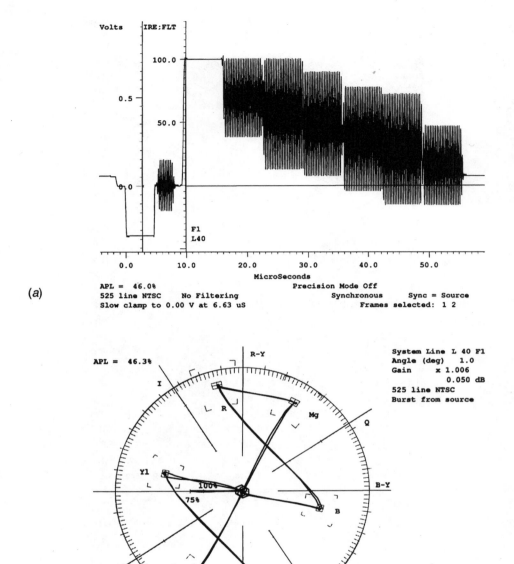

Figure 1.2 (*a*) Voltage versus time display as seen on a waveform monitor. Shown here is the signal generated each horizontal line by a "color bars" test signal, specifically the end of the previous horizontal line followed by a negative-going pulse (horizontal sync), the color reference signal (color burst), and then the video. Notice that the first part of the video is simply a straight line. This is the "white" bar in the test signal. The rest of the signal is additional bars of various color hues and saturation along with descending values of luminance. (*b*) Vectorscope display. This device filters out only color information and represents the color energy (voltage) in a vector display. This device displays color saturation and hue. The phase of the color signal determines the color or hue. The farther away the signal is from the center, the greater the color energy or saturation. The horizontal spike to the left is the color reference signal, which is also known as the color burst. This reference signal occurs at the beginning of every horizontal line.

competing systems and found no clear winner, so it asked the competing parties to work together and compromise to create a system that used the best
attributes of each proponent. The competitors came together and became
known in the industry as the Grand Alliance. They developed the new television transmission scheme, which the FCC eventually accepted, and which we
call DTV.

1.1.3 Joint Committee on Intersociety Coordination (JCIC)

However, no standard is an entity unto itself. There were—and are—a number
of international standards committees involved with the creation of the DTV
system. Just because we have developed a transmission standard that televises
digital data into the home, that doesn't mean we have decided how that data
should be organized. The following groups became involved: the Electronics
Industry Association (EIA), the Institute of Electrical and Electronics Engineers
(IEEE), the National Association of Broadcasters (NAB), the National Cable
Television Association (NCTA), and the Society of Motion Picture and Television
Engineers (SMPTE). An intergroup coordination committee was formed and
was called, what else, the Joint Committee on Intersociety Coordination (JCIC).
This body, which would advise the FCC as to their members' interest and con-
cerns, created the Advanced Television System Committee (ATSC). The ATSC
split itself into six specialized areas of interest, with the goal of nailing down a
complete system. The system for turning video pictures into data streams was
taken to another totally separate group, the Motion Pictures Expert Group
(MPEG). The system that MPEG developed has been accepted as an international standard by the International Standards Organization (ISO), but more on
MPEG later. The system for taking audio and creating a data stream was a system created by Dolby Labs, and it is called AC-3. Wrapping the various data
streams into one complete data stream was also taken from MPEG, with slight
modifications. ATSC adopted the Grand Alliance transmission system for getting the digital data to the home.

1.2 SDTV versus HDTV

Before advanced television turned into a digital system, it was thought that
two television channels would be needed to broadcast HD—one channel for the
SD signal and the second channel to carry the extra information needed for
high definition. When the digital approach was adopted with the MPEG system
included, it became apparent that one HD program could be transmitted in a
television channel. Soon engineers inside the television industry began to realize how expensive it was going to be to add a DTV channel while still operating an old NTSC channel. In addition, to make matters worse, HD equipment
is much more expensive than SD equipment. There would be few viewers during the first few years, which would mean a dearth of advertising revenue.

Sentiment began to grow to *not* use all the data bits for a single HD picture, but instead to divide the bit stream up and send multiple SD pictures. This is currently not sitting too well with Congress because many of the members thought the country was getting HD as DTV. There are many issues within the television industry now that border on being philosophical, and they quickly become almost religious discussions when broached. HD versus SD is one of those issues. Many claim that HD will not be needed in newscasts and, in fact, many news anchors would not want more facial details sent into their viewers homes. Many professional studio cameras actually have circuitry that softens the details of skin tones. Technical arguments have been made that most studios do not have sets with enough spatial content to warrant HD inside. Most experts agree that HD shines when used for sporting events or just about anywhere outdoors. Another area where it stands out is when viewing a movie or program that was originally made on 35-mm film. Television production people with film backgrounds still claim that 35 mm is the best method of capturing HD pictures. In 1997, 70 percent of prime-time programming for CBS was shot on film. CBS has said that 16-mm film conversion to HD doesn't work. However, television production people with video backgrounds counter that if what is being shot is going to end up on television, then a television camera is the best acquisition tool. This is another of those philosophical arguments.

There are those who argue that no one can prove what program stream in the DTV signal is the most important. Are multiple SD educational programs on an educational station less important than one HD educational program? Is one football game in HD better than letting the viewer watch the same game and choosing between multiple SD angles? Is an SD program with supplemental data less worthy than the same program in HD without any additional data? We have already mentioned the SD versus HD debate, but in Sec. 1.2.5 we will begin to define what constitutes an SD and a HD picture.

1.2.1   Multimedia

Some in the computer industry are now developing multimedia presentations called *tele-presence*. Tele-presence sends multiple video and audio streams that allow the viewer to interact with the story and to actually select alternate viewing locations or, in some cases, to actually select a different story line. Terms like "freedom," "location" or "position," and "gravity" are used to define the viewers' interaction with the presentation. The amount of gravity means how much or how little freedom the viewer has or, in other words, how tied the viewer is to a particular view or position. A panoramic show that allows the viewer to change view direction but not location or position has a single story line with high gravity on position and low gravity on viewpoint. Traditional television is simply a single story line with gravity so high that the viewer can only view from the perspective broadcast to them. Shows which allow full freedom have low gravity for both position and viewpoint. Shows which broadcast multiple camera angles or views have high gravity with multiple story lines. Although this technology was initially used in software applications like games and virtual tours, it is intended for use on DVD discs and over the Internet. With DTV able to send multiple data streams it could find a home with DTV.

1.2.2 Studio quality

As mentioned earlier, the NTSC television you watch now is SD. The FCC has mandated that each DTV channel must contain at least one SD program equal in quality to today's single NTSC channel. Many would flinch at that thought. Most people have never seen what the NTSC signal looks like before it leaves the television station. More recently, newcomers that see NTSC inside the station have asked if the picture they are viewing is HDTV. "Studio" quality NTSC is actually not too bad. In most cases it shouldn't be. In this day of $200 VCRs, $600 camcorders, and $100 receivers, television stations can still spend $40,000 for VTRs, $100,000 for studio cameras, and $10,000 for video monitors. Why so high? Ever hear the saying that "20 percent of something usually requires 80 percent of the resources"? The same holds true for professional television equipment. In order to get significantly better performance than that provided by your home equipment, a significant engineering effort is needed for products that sell very few units. Why bother with such high quality? Besides the fact that many advertisers demand it, every time a video signal passes through a cable or piece of equipment, it is slightly degraded. It is not uncommon for a particular program to experience many trips on and off videotape, plus a couple of satellite and microwave hops, along with additional passage through switching and processing equipment on its way to the television station's transmitter, each time taking a slight hit in quality.

1.2.3 Digital's lineage

Because of continual quality degradation, many television stations have started their migration to digital technology. It can be safely said that, today, no NTSC video reaches your home that hasn't been "digital" at some point in its life. Over 20 years ago the processing equipment in VTRs converted analog video that came off the tape into digital data for processing. The subsystem of the VTR that did this was known as a *time-base corrector* (TBC). The digitally processed video was then converted back to analog at the output. The technology in the TBC was expanded to allow video signals arriving from outside the television station, such as the networks (ABC, CBS, Fox, NBC, UPN, WB, etc.), to be synchronized to all the video in the station. Video that is not locked in phase to each other can't be mixed together or have locally produced graphics laid over it. Next, the technology was expanded to create systems that allow full-screen video to be shrunk back and flown around the screen, or raster. Separate shots of a news anchor or a reporter in the field are done using these techniques. Now, some professional VTRs and cameras can internally process the video and audio as digital data. Most television stations have these digital "islands," which are pockets of digital equipment surrounded by analog equipment. However, few facilities are shipping the video, and even fewer the audio, around the television station as digital data.

The main advantage of making video and audio into digital data streams is that once the analog video and audio become what are essentially number sequences, they will not be degraded at all, unless something in the system is broken. The quality will remain the same. However, the problem has been that, up until now, the digital video had to be converted back to analog to be transmitted. What the DTV revolution is bringing to the broadcaster that is totally new is digital techniques for transmission. These techniques are sorely needed because it is the trip from the broadcaster's antenna to your home, either straight from the antenna to yours or via a detour through a cable system, that causes the most grief. The NTSC pictures that some think are HD when viewed at the television station will surely not be mistaken for HD by the time they get to your TV receiver. Noisy pictures, ghosting (multiple images), and interference from other sources are the most common complaints. Once again, many people who see SD on a DTV receiver think they are looking at HD.

1.2.4 Cliff effect

Like all things digital, it generally either works or doesn't. What this means is that digital signals are usable right up to a point where they stop working altogether. There is no warning or gradual degradation, no increase in visible symptoms; it just quits. Component digital video (some bits describe black-and-white values and some describe chroma values) can be sent down a coax only so far before losses overcome the signal and render it unusable. Generally, SD digital video can travel approximately 1000 ft down high-quality coax before the signal is degraded to a point where it can't be received without errors occurring. Errors occur when the receiving device can no longer decode all digital bits correctly. When errors start, adding only another 20 ft of coax will usually render the signal totally unusable. DTV broadcasting has the same steep rise in error rates. Those testing DTV have witnessed the received DTV signal going from perfectly fine to unusable over a distance as small as 300 ft as you reach the edge of a television station's service area. Of course, if you go behind a hill or a building, just like NTSC, it could get a lot worse in a few feet. Signal strength, the amount of radio frequency (RF) energy available at a receiving antenna, can vary in strength by over a 3000:1 ratio. That means that some people close to the station's antenna might receive a signal 3000 times as strong as someone out near the edge of a television station's reception area. Another dramatic way to describe the cliff's slope is that once you are at the edge, a change in the signal as small as 0.07, or 7 percent, can cause the signal to go from usable to unusable. The good news is that DTV appears to be very robust. We have seen a number of cases where the NTSC signal is not watchable, while the DTV signal is still perfect (Fig. 1.3).

Figure 1.3 Here a received DTV signal is just over the cliff. Notice the video blocks that are out of place. Also notice that there is no increase in noise, no ghosting, just "blockiness." Some encoders and decoders can create block patterns that appear to be noiselike.

1.2.5 Defining HD

Back to HD versus SD. If SD is going to look so much better as DTV, how much better is HD DTV going to look? Well in one flavor of HD (yes, there are a number) the picture content is 6 times more than SD. In another flavor the picture content is only about 3 times greater than SD. Another philosophical argument in the industry is over what exactly is HD. Most agree that more picture content is a requirement. Is 3 times more picture content enough or should it be 6 times? As it turns out, HD is more complicated than that. Two additional ingredients in the HD recipe are how many pictures we present each second and how those pictures are displayed. Thomas Edison discovered that if you present still pictures at a fast enough rate, you create the illusion of motion. Hence, the motion picture. The standard frame rate settled out to be 24 frames/s, which means that every second 24 still pictures are flashed on the movie screen, and our eye and brain make them into a continuous moving picture. Television sends individual pictures to impart a sense of motion as well. Because the frequency of alternating-current (ac) power in this country is 60 Hz, it was easiest in television's infancy to base the rate at which television presented pictures to also be 60 Hz/s.

1.2.6   Resolution

Like any two-dimensional picture, a television picture has a horizontal and a vertical component. The way we draw this picture is one horizontal line at a time. We start at the top of the screen and scan one line from the left all the way to the right of the screen. This is done by an electron beam inside the picture tube. In black-and-white receivers the electron beam is made more intense for bright areas along the scan line than it is for dark areas. For color receivers the process is much more complex. Three scan lines, one for each of the primary colors (green, red, and blue), are swept across the screen. If all is working correctly, each beam will only land on its color phosphors on the face of the tube. These phosphors are what allows this scanning technique to work. Once they have been scanned and made to glow by the electron beam, they will continue to glow long enough so that our eyes (which aren't fast enough) do not notice them dim until the beam finally gets back to this line again. Once the first line has been scanned, the beam is turned off, and the beam-aiming (deflection) circuitry positions the beam back on the left side of the screen, slightly below where it started the first line. The second line is then scanned. This is done 238 more times, at which time the beam is at the bottom of the screen. The beam is turned off and brought back up to the top of the screen to start the process again. Each complete "painting" of the screen is called a *field*. The lines scanned during the first field have enough space left between them that another set of lines could be scanned or interlaced between them. That is exactly what happens in the next field. Both scans together are called a *frame*. These 480 active scan lines, which took two fields to create, is one NTSC picture. Each field happens in 1/60th of a second; therefore, one frame happens in 1/30th of a second. In addition to the 480 lines that make up the picture, there are 45 lines over the two fields in the vertical interval for a total of 525 lines/frame. These lines were needed in television's infancy to allow the television set to stop scanning vertically at the bottom of the picture and return (or retrace) to the top and start a new scan field.

In Fig. 1.4b each pulse (or dot along the bottom) represents the start of a horizontal line. These are called *sync pulses*. The first four and one-half horizontal lines at the left represent the last four and one-half active scan lines of a field. Then, notice the six half lines. These are called *equalizing pulses*. Next come six wide pulses followed by another six equalizing pulses. Together, these are called the *vertical interval*. The equalizing pulses and wide vertical interval pulses were needed by early television sets so the scan circuitry could stay in lock and to indicate when to start the vertical scan retrace. The blank horizontal lines after the vertical interval were originally there to allow time for the television set's deflection circuitry (which controls horizontal and vertical positioning of the electron beam or scan) to settle after the retrace. Now, these lines often carry data such as closed captioning, ghost-canceling signals, program information (for both broadcaster and viewer use), and test signals. The first three lines that are not blank are actually test signals used by broadcasters for quality control. The short pulses right

Figure 1.4a   Video monitor able to be time-shifted to display the vertical interval.

after the sync pulses at midscale (amplitudewise) are the color reference (also known as color burst) information. These are the horizontal pulses referred to in Fig. 1.2. Since there are two fields in a frame and the total number of lines is an odd number (525), one field has an additional horizontal line. This is achieved by starting and ending one field with half lines. These half lines ensure that the proper interlace will take place. *Interlace* occurs when the scan lines of one field fall between the scan lines of the next field. More on this in Sec. 1.2.10.

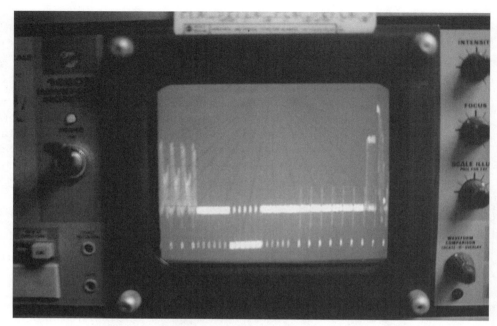

Figure 1.4b Waveform monitor below displays the varying voltages that comprise the vertical interval.

This means that if a camera shot a tall ladder, we could, in theory, see 240 rungs of that ladder. No, not 480, because to separately see each of the ladder's rungs, we would need one line for a rung, and at least one line of the space between the rungs. We would say that NTSC has 240 vertical lines of resolution. Now many things limit that resolution. One is how well the television set can interlace the scan lines for each field into a frame. This issue will resurface again very soon in our discussion.

What about horizontal resolution? Horizontal resolution is even less straightforward. Initially, horizontal resolution was mainly based on the bandwidth of the system. *Bandwidth* is a measure of how wide a range of frequencies can be sent. Your telephone sounds "tinny" because only 3000 Hz of voice, or audio, information can get through. This frequency range has been kept low to keep phone rates from rising higher. Modems play many modulation tricks to get data throughput much higher. Your AM (amplitude-modulation) radio sounds slightly better because back when the FCC set up AM broadcasting, each AM station was given only 10 kHz of spectrum, which equates to an audio bandwidth of only 5 kHz, as the audio information is duplicated on the upper and lower part of the signal (called dual sideband). Although the signal could be made to occupy only the upper or lower (called single sideband) side, it would result in much more expensive receivers. FM (frequency modulation) was given much more bandwidth, and thus it can broadcast audio frequencies up to 15 kHz, which is near the upper hearing limit of most people. That is why most music is on FM, and AM has gravitated to news and talk shows, where the fidelity doesn't have to be as good. So television must have very wide bandwidth compared to other RF services to have good horizontal resolution.

1.2.7   Bandwidth

When it came time to allocate spectrum for NTSC television, it was decided that a 6-MHz-wide television channel would be enough. However, that channel had to hold not only the video or picture but also the aural or sound. Television stations are really two broadcasting stations in one. The picture, or visual half, uses the same transmission technique as AM radio, except that a bandwidth of just under 5 MHz is available for pictures, unlike the 5 kHz for AM radio. The aural or audio part of the signal is FM just like any FM radio station. Like FM radio, 150 kHz of the NTSC television channel is allocated to the audio. In AM the amplitude of the carrier is modulated (varied) to convey information to the receiver, and in FM it is the frequency that is modulated. Now why doesn't it take 10 MHz to transmit 5 MHz of video you might ask? Because television is not transmitted double sideband. It uses a technique called *vestigial sideband,* which means that the entire upper sideband is transmitted and only part of the lower. The gains at various frequencies are controlled so that any frequencies below the carrier are combined to make the response across the whole channel flat. This means that, in theory, all frequencies across the channel are treated equally (Fig. 1.5).

So now that we've explained how much bandwidth is available for video in the NTSC television channel, we can get back to horizontal resolution. A signal with a frequency of 5 MHz would have a period of 200 ns (0.0000002 s). This period would be represented by a sine wave with a positive and negative excursion. It takes approximately 52 μs (0.000052 s) for the visible portion of a horizontal line to be swept across the screen, which means that 260 occurrences of our 5-MHz sine wave could occur across the screen. If instead

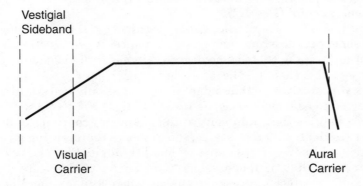

Figure 1.5   Response across NTSC channel.

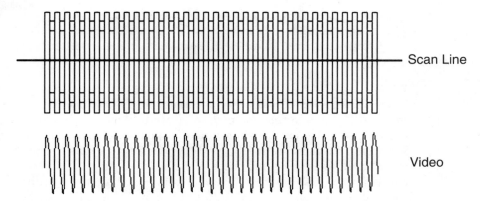

Figure 1.6   Picket fence scanned by camera and resultant electronic signal.

of displaying a vertical ladder on the screen, we shot a picket fence extending horizontally across the screen, we would find that the maximum pickets that could be displayed would be 260 (Fig. 1.6). The reason is that because each picket, and the corresponding space between pickets, could be represented by our 5-MHz sine wave. Of course, not all picture content is at 5 MHz; picture content can be a cacophony of many frequencies from 0 to 5 MHz.

1.2.8   Maintaining quality

The NTSC television signal we receive today has a theoretical resolution of 260 (horizontal) by 240 (vertical). But, and that is a big "but," that resolution is only obtained if the NTSC television station you are watching has the best equipment with no engineering or equipment problems, the transmission process and your receiving antenna/cable system are perfect, and finally your set is in perfect working order. Care to place your bets? Every time an analog signal passes through cable, or a piece of equipment, some small nip is taken off the high-frequency information, and all the lower frequencies are scrambled a small amount (Fig. 1.7).

This situation is where the digital technology that has slowly spread through some broadcasting facilities has helped. A television station that is digital, or at least digital to some degree, will generally be able to deliver better quality to its transmitter because digital technology is based on binary values. Binary means two, so that a digital signal has only two states, high or low (also referred to as true or false, or 1 and 0) (Fig. 1.8).

Thus, as long as the equipment in the chain can continue to differentiate between two widely different states, the numbers representing pieces of video information will never change, not by one literal bit (Fig. 1.9). Working with these number values is much easier than trying to duplicate a constantly varying voltage which represents analog video. But as mentioned earlier, even if the signal is digital in the television station, it must be converted back

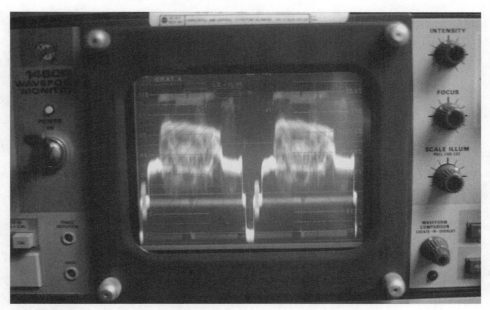

Figure 1.7 Received NTSC video waveform.

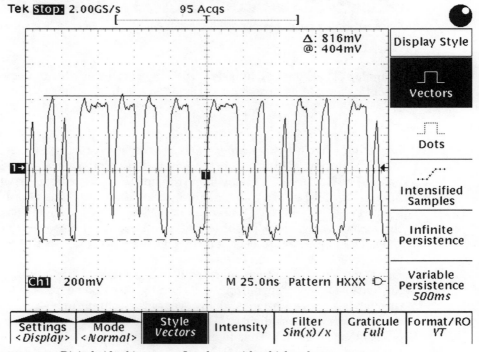

Figure 1.8 Digital video bit stream. Levels are either high or low.

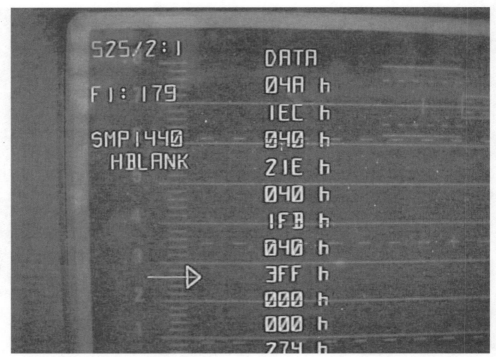

Figure 1.9 Number sequence representing digital video. These values represent the last few samples of the video in a horizontal line.

to analog NTSC for transmission, with the result that the analog transmission/reception problems already mentioned are still possible.

1.2.9   Initial DTV

Restating what was mentioned earlier, at least to start, the biggest part of the DTV revolution is going to be the digital transmission of television to your home. The link from the television transmitter to your home is what is changing. In Fig. 1.10 notice that the amplitude values take one of only eight possible levels. It would take 3 binary bits (2^3) to represent eight possible values; thus, conversely, eight possible levels represent 3 bits for each DTV sample time. The values found between the eight levels represent the transition time between actual value sample times.

Many television stations merely take their analog NTSC video and convert it to digital at the transmitter and nothing else. Some pass digital SD video through the entire station, and have very little content in NTSC at all. A few others will actually pass digital HD video. Initially, 5 percent of television facilities did some form of HD. That figure will probably approach 20 percent early into the transition. The global HDTV *common image interface* (CIF) adopted by the

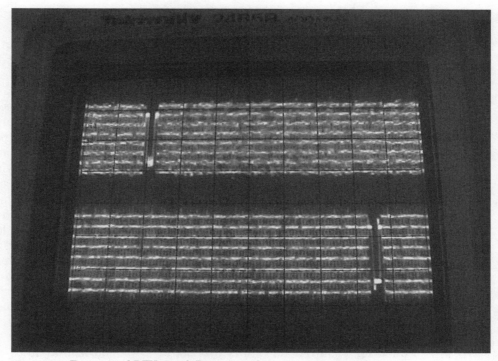

Figure 1.10   Transmitted DTV signal. Preprocessed (top), postprocessed (bottom).

International Telecommunications Union (ITU) is 1080 horizontal lines. ATSC only considers 1080i and 720p (i = interlaced, p = progressive) as HD, but the European Telecommunications Standards Institute considers 480p to be HD.

1.2.10 Progressive scanning

ABC and Fox insist that HD have progressive scanning. CBS and NBC aren't opposed to progressive scanning but they want 1080 lines. The technology, along with bandwidth considerations, means that, today, 1080 must be interlaced. A possible compromise, first advanced by the International Teleproductions Society (ITS) and discussed within SMPTE, is 1080p at 24 frames. This is even a lower data rate than 1080i. It also conforms to the ITU's CIF. Since most network shows are still shot on film, and because it is within 4 percent of the 25-frame/s rate used in Europe, it would seem to be a good compromise. The problem is that no one supplies 1080@24p equipment yet. Even 720p doesn't have all the necessary infrastructure yet. ABC originally used a supercomputer to transfer film to its 720p format. Another problem that occurs with 1080p at 24 frames/s is in displays that *actually* display at only 24 frames/s. The fatigue from the display flicker with such a low scan rate will hinder production monitoring at the 24-frame/s rate. 1080p24 will require monitors that display at a much higher rate even though the actual frame scan rate is only at 24 frames/s.

We covered the resolution part of what makes an HD picture, but only partly. When we talk of horizontal and vertical resolution, we are referring to *spatial resolution,* that is, resolution in two-dimensional space. There is another type of resolution called *temporal,* or resolution that occurs over time. As already mentioned, NTSC provides new complete pictures at 30 times/s, which is the frame rate, but remember that it takes two scans, or fields, to make up that frame. It is the interlacing of those two fields that results in the frame. There are proponents who claim that 60 frames/s is a better, maybe higher, definition picture, but interlace was originally adopted to limit the bandwidth needed. Intuitively, it should be obvious that to send all the scan lines on each vertical sweep should double the information or frequency bandwidth required. That is one reason why interlace was adopted early on. Another reason is that the phosphors used in early picture tubes, or CRTs, didn't glow long enough between scans if all the horizontal lines were sent each scan but at a rate that didn't increase the bandwidth. That is, one vertical sweep would take 1/30th of a second instead of 1/60th. This problem is what is generally called *flicker.*

The computer people are big proponents of sending all the horizontal lines each vertical scan. This is called *progressive scanning,* and is what your computer monitor does. However, to eliminate flicker, vertical scanning is usually done at 72 times/s instead of 60, which results in a very wide band of frequencies required for this video because 480 horizontal lines are sent 72 times/s. This makes your computer monitor display 28,800 active horizontal lines/s versus half that for NTSC video. However, since it only has to be sent a few feet and not through a television channel, it doesn't matter.

Computer people who feel production, or broadcast, facilities should do 480p, or 720p, because it is computer friendly and because cameras in that format are available have not grasped the entire scope of even a small facility. While TV people seem to add complexity in their minds when it comes to computers, most computer people simplify the infrastructure in a television plant to cameras, maybe routers, transmitters, and monitors. And most computer people think all storage is on NT servers. To be fair, 720p switchers and VTRs are now available. But computer types have claimed that the broadcasters' problem is due to long-standing regulation by the FCC so that the future path is only charted by managers who have no grasp of new technology. Some of that is undoubtedly true. However, creating specifications for the whole DTV pipeline based on what is easy for a PC to display will most likely optimize PC display at the expense of real-time television display. This fact might be true only because the computer industry is adamant that optimizing the bit stream for display on a television receiver will stunt PC use for DTV.

NHK had pressed Panasonic to stop 720p demos and production in 1998. But the U.S. Department of Defense's (DOD) complaints were forwarded to Japan, which stopped NHK's efforts. The National Imagery and Mapping Agency (NIMA), a part of the Pentagon, has embraced 720p.

So the argument goes, is it better to refresh all the lines each vertical scan and eat up bandwidth that way (the progressive scanning approach) or is it better to send only half the lines each vertical scan and use the bandwidth for spatial (horizontal and vertical) resolution? Temporal resolution, the number of complete pictures we send a second, and interlace versus progressive scanning are linked together.

Besides pushing for progressive scanning, the PC camp is worried about interactivity. The Advanced TV Enhancement Forum, a coalition formed by Microsoft, Intel, Disney, Direct TV, CNN, and others, recently agreed on a standard for enhanced TV content. This standard applies to over-the-air (terrestrial) cable and satellite transmission for the enhancement of interactive TV.

1.2.11   Aspect ratios

There is one other issue clouding the SD/HD discussion: aspect ratios. The television screen you watch today is one-third wider than it is tall. In other words, for every 3 in high it is 4 in wide. Hence, the term *4:3 aspect ratio* is used to describe the dimensions of the screen. This was the original film format, but cinematographers have realized that our peripheral vision extends wider than taller, so they started making films with wider aspect ratios. Many films are shot with aspect ratios as wide as 21:9. With the advent of HDTV, it was decided that television should have a wider aspect ratio also; 16:9 was settled on after early pioneer NHK adopted it for their HDTV experiments. HD formats 720p and 1080i are both 16:9. Once the aspect ratio is settled on horizontal, pixel count and vertical line counts fall out. The H/V ratios for the two HD formats are as follows:

1920 pixels ÷ 16 = 120   120 × 9 = 1080 lines
1280 pixels ÷ 16 = 80   80 × 9 = 720 lines

Many program producers are starting to acquire SD material in 16:9 format also. Many professional cameras offer options that allow acquisition in 16:9. Some mistakenly believe that any 16:9 programming they see must be HD, just as all DTV programming is HD. Neither is true.

1.3   Why Compression?

SD versus HD is only a part of the DTV equation. Digital television ushered another transforming technology into prominence—compression—more exactly, lossy compression. As already stated, digital has only two states. When digital changes states, it does so abruptly. When it goes from one state to the next, it does it almost instantaneously. What's instantaneous? A few nanoseconds, which results in lots of sharp edges. What does it take to make lots of sharp edges? Very high frequencies. This means that any coax carrying digital signals better be able to handle extremely high frequencies. How high? Up into the gigahertz range. As these high frequencies are rolled off, or attenuated, the nice

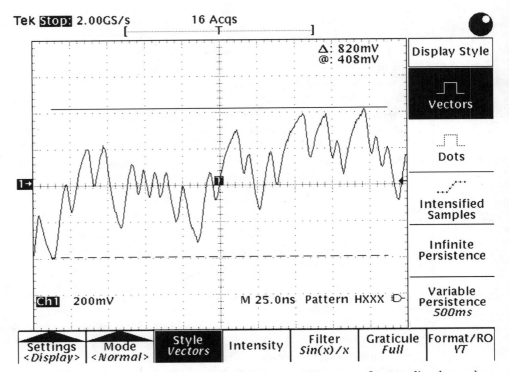

Figure 1.11 Digital signal that has lost its high-frequency components after traveling down a long coax cable.

sharp edges that should be at the transitions disappear. The edges become pro-longed slopes so that the receiving device has a hard time determining the state of the incoming signal. Although coax that carries only NTSC needs a few megahertz, for SD digital that coax should carry at least 500 MHz or all bets are off. In fact, if the coax was only capable of carrying 500 MHz, the signal could not travel very far down that cable, maybe 100 ft maximum. If you want to send the signal down 1000 ft of coax, the coax has to be capable of passing frequencies above 1 GHz. With HD digital it gets far worse. To be able to send the digital signal through a few feet of coax, you need frequencies over 2 GHz. To get it over 100 ft, you're looking at frequencies over 3 GHz (Fig. 1.11).

These frequencies are considered baseband frequencies, which means that the signal is not riding on a carrier. If baseband digital is modulated onto a RF carrier, the baseband frequencies end up as sidebands. If a 3-GHz carrier is chosen to modulate a digital baseband signal onto, the sidebands, in theory, would use up most of the usable RF spectrum. Shortwave, AM, very high frequency (VHF) TV, FM, ultrahigh-frequency (UHF) TV, and some satellite channels would all be occupied by this one signal. Of course, no transmitter or receiver could be built that would be able to handle this bandwidth. Direct transmission of digital signals just isn't going to work.

1.3.1 High bit rates

In analog AM and FM NTSC transmission, the sharpness of the edges is limited so that the transmitted information stays inside its allocated channel. The information in digital signals gets so high because of the number of bits that must be carried. With SD digital-component video, there are 1440 digital samples or number values per horizontal line. This number sequence is comprised of 10 bits, and these bits are sent serially, one after another. First, the 10 bits of the first sample, then the next 10 bits of the second sample, etc., for a total of 14,400 bits of video per horizontal line. Besides the video bits, there are approximately one-third more bits sent that aren't part of the picture but are used for synchronization and identification, and to carry audio data along with the video. Over the course of 1 s it works out to be 270 million bits. That's 1 bit every 3.7 ns, which doesn't leave much time for the edges, or transitions, to occur when needed. As a very general rule of thumb, the minimum frequency needed to keep the edges sharp enough is $1\frac{1}{2}$ times the bit rate. That means that, for SD digital, component frequencies in the range of 405 MHz are the minimum. By the way, digital composite is not used much because it has many of the same artifacts that analog composite NTSC has, only in digital form. Its only real benefit is that its bit rate is lower, 144 million bits/s. Plus, since it is digital, if no processing is done, the quality will remain constant.

So, we can pass baseband digital down reasonable length cables, but not through the air. Another problem is that the frequencies are too high for reasonably priced videotape machines to handle. Also, if we want to record them straight onto computer hard disks, we will use up 27 Mbytes of storage every second (overlooking the fact that SD digital video is generally 10 bits and a byte is 8 bits). A 9-GHz drive would be filled up in under 6 min. Compression to the rescue! It was discovered that if we put each picture in memory and broke it up into blocks, we could use techniques to throw away some information that wouldn't be missed too much. We do this by taking the block out of the time domain and into the frequency domain. What that means is that instead of having values that describe each pixel in the block, we come up with values that represent which combination of frequencies would be needed to regenerate the block in the space and time domain. Now the values, or coefficients, generated are numerous, but we usually find that only a few are of any magnitude, with the rest being very small in value. If we throw the ones with small values away, we can describe a block with far fewer values then before. Now those values that are thrown away will cause a slight degradation of the picture but, as it turns out, not by much. This is known as *lossy compression*.

1.3.2 Joint Picture Experts Group (JPEG)

The compression you perform on your computer, such as when you use a program like PKZIP™, is lossless. This means nothing is lost or thrown away. Lossy compression won't work on software because it would corrupt the program and make it useless. After the lossy compression is completed, we can also use lossless compression techniques on the strings of frequency coefficients generated. This lossy, followed by lossless compression, process is exactly what a group known as the Joint Picture Experts Group developed. This compression scheme is known as *JPEG*. Each picture or frame is compressed; it worked well for still photographs and greatly helps with moving ones like television. Compression ratios of 4:1 can be achieved with essentially no degradation at all. Compression ratios of 10:1 yield results that are not objectionable to most people, but as you approach 20:1 the results start to look frayed. At 10:1, we've gotten our bit rate down to 27 Mbits/s for SD digital. Now our disk drive can hold almost 1 h of video. At 2:1 compression, videotape technology can handle the resulting bit rate at a reasonable price.

1.3.3   Motion Pictures Expert Group (MPEG)

It was soon realized that JPEG compression didn't quite take the process far enough. JPEG performed only spatial compression, that is, two-dimensional compression. But motion pictures, including television pictures, often do not change much over time. Essentially, the same picture is sent over and over again, with small amounts of change. A group came together known as—yes, you guessed it—the Motion Pictures Expert Group. It developed a way to perform *temporal compression,* or compression over time. This process requires a lot of memory. A series of pictures are stored, and each is analyzed to determine what is different. A signal which represents only the change is developed. That differential picture is then compressed just like a normal JPEG picture. It also has what is known as motion vectors, which represent the direction and magnitude of objects that are moving in the picture. To reset the picture differential process, a normal JPEG frame, compressed only with reference to itself, is sent as a sort of anchor frame. Then a series of the different pictures are sent before another anchor frame is sent. MPEG compression, therefore, uses spatial and temporal compression. Compression ratios of 50:1 yield reasonable results. MPEG allows even HD to be brought down to reasonable bit rates. One common flavor of HD has a bit rate of 1.5 Gbits. MPEG allows a reasonable rendition at 19 Mbits (recent developments have put this bit rate as low as 12 Mbits/s). Now, if some of the same modulation techniques are applied that let us use 56K modems over 3K-bandwidth phone lines are applied, we can fit one HD picture into a 6-MHz-wide television channel. Or several, say five, SD pictures into the same channel.

   MPEG compression is not transparent. You will see some artifacts—some bad, some good. What could be good? Well in the process of cutting up the picture for compression, a lot of edges in the picture become very sharp and pronounced. People like this because it makes everything look sharper and, in some cases, harsher. This sometimes goes against what the program's producer had in mind. Viewers liked the Trinitron™ picture tube introduced by Sony over 20 years ago because it tended to make things appear sharper than they really were. The bad aspects are generally what are known as motion artifacts. Fast action and motion can sometimes swamp the compression circuitry and cause sample blocks to be wrong or out of place. This is called *blockiness*. The interesting thing is that if you could analyze the motion artifacts in slow motion, the ones you would find most objectionable are not usually the ones you find irritating at normal speed. In many instances of compression, motionless video is extremely clear, but when objects move, blocklike features blur the picture. Clear stationary pictures that morph into unrecognizable forms when motion starts might be disconcerting to a viewer who is used to his or her brain discarding action that is too fast to follow, not the broadcaster doing it.

1.3.4   Native formats

ATSC has a 70-member Top-Down Implementation Subcommittee whose goal was to identify all the typical systems and interfaces that could exist at a station moving to digital, depending on its implementation scenario. A large chunk of the panel's work dealt with possible conversion scenarios, and the impact these would have on quality. No one is yet quite sure how many conversions are OK. The Top-Down Committee's final report examines four different implementation scenarios for plant native formats: NTSC, combination of analog and digital, SMPTE 292M (baseband HD; see Chap. 6), and MPEG/DV. Many stations are considering a plant native format since so many different formats are available. All incoming material will be converted to that format. It is still undecided as to who will specify a plant native format: networks, advertisers, program suppliers? Also, keep in mind that receivers convert to their own native format. Audio is a whole other issue. Dolby Labs states that there is no native format for audio. You can't take six channels to two and then back to six with anywhere near the same result. Experiments have also shown that conversion from interlaced to progressive is more likely to degrade the signal than conversion of progressive to interlace. Therefore, some have argued that a progressive scan native format is better.

Other groups are wrestling with what the television station will look like in the future. A group called the National Institute of Standards and Technology (NIST) was formed by broadcast and computer interests to tackle problems encountered in implementing MPEG and computer technologies in the television plant. This group is looking at technologies quite foreign to most broadcasters on which to build the plant's technical infrastructure.

1.4   The Agreement to Disagree

So, at this point, we have a number of groups, each of whom has a stake in the DTV game, but most of them can't agree on the rules of the game. The computer industry wants interactivity and lots of data delivery to progressive scan PCs. They have no initial interest in HD. The PC folks suggest a phased-in approach to HD after their SD/data goals are met. But some members of Congress believe the second television channel handed to the broadcasters was for HD, not SD or data delivery. Cable isn't happy about carrying the additional DTV signals.

Besides the PC folks, broadcasters, cable, and Congress, four other groups will affect the rollout of DTV: the content providers such as production houses and studios, advertisers, professional equipment manufacturers, and consumer electronics manufacturers.

If the content providers decide that HD isn't worth the effort or investment, HD will not happen. An additional issue that is being worked out is fee charges, if any, needed for airing programs on two separate channels during the NTSC/DTV transition phase. Advertisers have to decide if they will foot the bill for HD, and when they will consider the broadcaster's additional DTV channel worthy of payment for advertising on it. Most of the early DTV broadcasters are doing little more than simulcasting their NTSC video and audio on the DTV channel. These stations are not charging their advertisers anything for the additional channel. The professional equipment manufacturers have to decide which DTV formats they will support. Whereas 1080i@60 equipment seems to be fairly well supported already, 1080p@24 and 720p@30 equipment is still in development. It is still unclear as to whether any of the other formats, such as 480p@60 or 480p@30, will have much in the way of product offerings. Finally, the TV set manufacturers will decide how to display the various formats. They will also decide on how cable and terrestrial broadcasters will interface with their products.

CBS says the biggest issues facing DTV programming are

1. A 59.94- versus 60-Hz field rate.
2. Number of concatenations of audio compression that can occur.
3. The number of audio channels.
4. Which tape formats to use.

DTV issues fall into three categories:

1. *Production.*   Content creation, choice of "native" DTV production format, distribution from the producer through the network to the affiliate.
2. *Transmission.*   Towers, antennas, transmitters, studio-to-transmitter links (STLs), DTV transmission format, ATSC encoders.
3. *Display.*   Choice of DTV display size, choice of native DTV display format, set-top boxes (STBs).

For consumers to want DTV there must be program content, good picture portrayal, and great sound. DTV has created a decoupling of production, transmission, and display formats. Different sets of winners and losers will evolve, based on how the issues just mentioned play out. Cable wins if DTV fails, especially if they are trying to implement some sort of advanced television. The PC industry derives no benefit from DTV since there are Internet alternatives to broadcasting. Early adopters will pave the way for the rest of us. They will pay a premium to use, or have access to, the technology first. The feature set will improve with each new iteration of product and the price will drop. This is not fair, but necessary.

1.4.1 Nothing's fair

Little in the broadcast industry has ever been fair. Historically, UHF stations have always had disadvantages when compared to their VHF counterparts (VHF and UHF stand for the band of frequencies of the broadcast stations). The VHF band is the frequencies between 30 and 300 MHz. Television channels do not occupy the whole band. Channel 2, which is the lowest VHF channel, has its lower channel edge at 52 MHz. The channels go up from there. There is actually a small gap between channels 4 and 5 (4 MHz), which is why you can find these channels adjacent to each other in the same or nearby markets.

Early television-receiver technology prevented the placement of channels adjacent to one another in the same market. Adjacent channel separation for VHF channels has been 60 mi. That means that channel 7 in San Francisco and channel 8 in Monterey, or Detroit and Cleveland, can exist because each station's transmit site is at least 60 mi apart. For UHF the separation is 55 mi. The FM radio band, aircraft, and some public safety services are located between channels 6 and 7. That is why broadcast facilities with channels 6 and 7 can be in the same or nearby markets. The FM radio channels are, therefore, in the VHF band as well. However, to avoid confusion, the radio stations that are found between 88 and 108 MHz are called FM after their method of modulation (frequency modulation). The VHF television channels start up again with channel 7 at 174 MHz and go to the top end of channel 13 at 216 MHz. Channels 2 through 6 are referred to as low-band VHF, and 7 through 13 are known as high-band VHF. There is a big gap between channel 13 (VHF) and 14 (UHF). The low end of UHF, channel 14, is at 470 MHz. The UHF television band runs uninterrupted up to channel 83, whose upper-channel edge is at 890 MHz.

Depending on your geographical location in the country (the FCC divides the country into three zones) and whether the station is UHF or not, broadcasters using the same frequency must never be closer than 155 mi from each other (northeast U.S. UHF), although spacing of up to 220 mi can be required (southeast VHF). As mentioned previously, early television receivers limited the combination of channels that could exist in a geographical area. In the UHF band you would think that you could have broadcast stations on every other channel, but that was not possible when channel assignments to the various markets were first handed out. The tuners, or front ends as they are called, in early sets could not filter out unwanted energy or signals from other stations or the harmonics and intermodulation from those other channels. Harmonics are basically multiples of frequencies in a signal. *Intermodulation,* or intermod for short, occurs when two separate signals mix together to create additional signals. This usually happens in the receiver because its tuner can't filter out, or throw away, all the unwanted signals and only keep the desired one. Additionally, sometimes when high-power broadcast transmitters are near one another, intermod can be created at the transmit end.

Except for adjacent channels in VHF this wasn't a major problem in the VHF band because the band was fairly narrow, barely over 150 MHz wide, with an 86-MHz hole between channels 6 and 7. However, the UHF band is 420 MHz wide. So besides the 55-mi adjacent channel taboo, a number of other taboos exist. Channels that were within five channels of each other had to be at least 20 mi from each other or intermod in the receiver could result. Additionally, channels that were 7, 8, 14, and 15 channels apart also presented reception problems to most early receivers. Channels that were 7, 14, and 15 channels apart had to have separations of 60, 60, and 75 mi, respectively, which is why there are not as many UHF stations as you might think. Television channels are a precious and, thus, often contested commodity.

1.4.2   Problems handling UHF channels

Cable systems have greatly added to the mix of programming available to the home, and we will examine them later in this chapter. However, early cable systems were limited as to the amount of programming capacity they could add. When cable began in earnest in the 1970s, the adjacent channel taboos were greatly suppressed by the receiver technology of the day. So channels could be packed next to one another on a cable system. The problem that the cable systems had was bandwidth. Most early cable systems had upper bandwidths that stopped at 300 MHz. The infrastructure of the cable system, mainly the amplifiers, splitters, and set-top boxes, was the limiting factor with regard to the size of the information pipe for a given cable system. Although early coaxial cable could pass higher frequencies, it had greater loss (especially at high frequencies), which required more amplifiers, and remember that each one slightly degrades performance to push the signal a given distance. A top end of 300 MHz covered the VHF but not the UHF band.

To add additional channels to the cable system, many systems down-converted the UHF channels 14 to 22 to the frequencies above the FM band and below the band where channel 7 begins (120 to 170 MHz). These frequencies were called midband VHF because they were between the low- and high-band VHF channels. UHF channels 23 through 36 were moved to the frequency range directly above the high-band VHF frequencies (216 to 300 MHz). These frequencies were called superband VHF. This shuffling of channels required cable customers to have set-top boxes from the cable company to view these channels. In the 1980s equipment became available to rebuild systems to allow top-end bandwidths of over 500 MHz, but many systems had bandwidths that topped out at 400 MHz. This still allowed over 50 television channels to be carried. Now, many systems are being rebuilt with fiber-optic cable. Thus, the roll-off due to copper (coaxial cable) will be eliminated. In addition, television sets that are "cable ready" are now universally available. A cable-ready set has two channel modes. The first one is the historical terrestrial television channel mapping. The second one tunes the channels based on the cable channel–shuffling scheme just described.

As you now see, UHF stations are at much higher frequencies than VHF stations, which poses a number of problems. The first problem that most early UHF stations had was that no sets could receive the channels. Even after color sets were in full production, most didn't have the ability to tune UHF. Most early UHF viewers did so with a UHF converter box sitting on top of the set. The set-top box, as it turns out, is not new and will probably make a return engagement with DTV. The lack of viewers wasn't the only problem the pioneer UHF broadcaster faced. Transmitters at higher frequency tend to be much less efficient, which means more incoming power is needed to get the same RF transmit power that you would at lower, that is, VHF, frequencies. Once the RF power is out of the transmitter, it must be sent to the top of what is typically a tall transmit tower to the antenna (Fig. 1.12).

Figure 1.12 UHF antenna side-mounted atop a tower that sits atop a mountain ridge. Television signals, especially UHF, must be transmitted at high power and from high places to maximize coverage.

1.4.3 Radiating RF

Let's digress a moment to talk about towers and antennas. A *tower* is a narrow platform usually made of steel. An *antenna* is a device to radiate RF energy. Confusion exists on the difference between a transmit antenna and a transmit tower. In AM radio the antenna and tower are the same. Due to the low frequencies used with AM (centered at 1 MHz) it is important to create a strong ground-wave signal. Ground-wave signals can go a long way and easily travel over the horizon. So AM antennas are near the ground. The height of these towers is usually related to the wavelength of the transmitted carrier. Thus, AM stations with lower frequencies tend to have taller towers than ones at higher frequencies. To minimize the height for a low-frequency AM broadcaster's tower, a station at 540 kHz might use a one-quarter wavelength height tower, while one at 1540 kHz might use a one-half wavelength height tower. It is important for efficiency that the tower be some fraction of the transmitted carrier's wavelength. Since these towers are also the antennas, the RF power is applied directly to the whole tower. The base of the tower is isolated from the ground, and these towers are fenced off because if you walked up and touched one, it would ruin your day, and maybe your life. Often you see multiple antennas near one another at the AM transmit site. These antennas are called an *array*. They are driven by the AM transmitter such that they amplify the RF transmitted in some directions and minimize it in others to protect other AM stations in other markets. Thus, these antennas are often called *directional arrays*. AM ground waves are best created in wet, highly conductive soil so AM towers are usually in low-lying areas, often near water or swamps.

In contrast, FM and television antennas and towers are separate entities. At high frequencies very little, if any, ground wave is created. The radiated signal is said to travel only in a line of sight. The RF signal can be thought of as being like a street light. To cover the widest area with that light, you want lots of height and a lot of light or power. Therefore, the tower is used as a platform to get the antenna as high as possible. That is why television and FM antennas often sit on top of mountains or tall buildings. If neither is available, towers that are often 1000 ft high are erected, although towers as tall as 2000 ft are not uncommon (Fig. 1.13).

Now let's go back to UHF difficulties. The transmitter at the base of the tower produces RF power, although generally less efficiently for UHF than for VHF frequencies. That power must be sent up to the top of the tower to the antenna. Half the power can be lost in the trip up the tower (Fig. 1.14).

To maximize what is left, television (FM also) transmit antennas don't radiate straight up into space or straight down the tower. They are generally designed to radiate only toward the horizon. If it were possible to see the RF energy being emitted from the antenna, the pattern would look like anything from a fat donut to a thin phonographic record (today the reference should probably be to a CD). Donut patterns are good for close-in coverage, while thin or flat patterns are used to push the signal out toward the horizon. With thin patterns, viewers close in to the transmit antenna might have far worse recep-

Figure 1.13 Bridge carrying RF power (through a transmission line; see Chap. 13) from the transmitter building to the base of the tower, and then up the tower.

tion than viewers much farther out. Fat donut patterns cover near-in viewers well but limit the far reaches of the station's coverage.

VHF signals can reach to the optical horizon and then some. It is said that the radio horizon for VHF is greater than the optical horizon. For UHF the radio horizon is often less than the optical horizon. To make up for the UHF propagation shortfall, the FCC lets some UHF stations effectively radiate power toward the horizon that is 50 times greater than their low-band VHF counterparts. In contrast, high-band VHF broadcasts are only allowed to be a maximum of just over 3 times "hotter" than the VHF low bands. However, in order to generate all this power, UHF stations tend to have much greater power bills than VHF—6 to 10 times is a fairly common disparity. In addition to more power, the UHF station can build higher towers and find higher mountains and buildings on which to place their antennas. Height turns out to be very important in maximizing UHF coverage.

1.4.4   DTV channel assignments

Now with DTV the FCC has assigned a second DTV channel to every TV station. In most markets there isn't any room for additional VHF channels because their saturation prompted the original UHF allocations. It should be noted that, in several markets, the FCC experimented with handing out only UHF and no VHF licenses. During the 1960s, the FCC toyed with the idea of

Figure 1.14 RF transmission line, which appears to the layperson as water pipes, carries the RF energy from the UHF transmitter up to the antenna on top of the tower.

moving all broadcasters to UHF and reclaiming the VHF band. Now the plan is to have all television broadcasters end up on channels between 2 and 51 when NTSC broadcasts go dark. These are referred to as the "core" channels. Many VHF stations have UHF assignments for DTV. When NTSC goes dark, they must hand one channel back to the FCC. It doesn't matter which one; it can be either the UHF or the VHF, with the following exception. If a station has an initial DTV channel assignment above channel 51, that station must move back to the current NTSC channel to continue DTV broadcasts when NTSC ends.

Now it would seem that it is a "no brainer" as to what the VHF NTSC broadcaster with a DTV UHF assignment will do when it is time to choose, but it is not that simple. Many components in a television transmitter are frequency or channel dependent, as is the transmit antenna. The cost to purchase and install the DTV transmitter and antenna can easily be above $3 million. If NTSC is discontinued earlier, say around 2006, then later it might make economic sense to keep the UHF frequency. However, the VHF NTSC broadcaster who is now on a UHF DTV channel will have major incentives to switch back to the original channel. The FCC not only handed out channel assignments but also the maximum allowed power levels. The power level was based on what it would take to duplicate the coverage provided by the station's NTSC signal with the station's DTV signal. A tremendous increase in power would be necessary for VHF coverage to be duplicated by UHF. So in a simple effort to save money on power, a VHF to UHF to VHF movement for such a station might prove advantageous no matter how early the initial DTV equipment might have to be scrapped. The downside is that DTV coverage is sort of like the Oklahoma land rush. The first stations up on a channel will have more protection from interference than later arrivals.

There is a saving grace with DTV modulation. The modulation process and error recovery ability of DTV receivers are such that a DTV signal 20 times weaker than a NTSC signal will provide comparable coverage. This is creating an interesting situation for the few broadcasters that have NTSC and DTV assignments that are both in the VHF band and the few broadcasters going from UHF to VHF assignments. In both cases these broadcasters are receiving power assignments that are extremely low. Whereas an NTSC UHF broadcaster might have had an effective radiated power (ERP) of well over 1 MW, now the ERP for that DTV channel might well be under 10 kW. Low-band VHF DTV power assignments as low as 1 kW have been handed out. Typical maximum low-band VHF ERPs for NTSC have been 100 times that amount. In some markets assigned power levels for DTV stations have differed by a ratio of 500:1. This has happened where NTSC VHF stations have been given DTV UHF assignments while NTSC UHF stations in the same market have been given DTV VHF assignments.

1.4.5 Television transmitters

The fact that some stations will have much lower power bills than their competition creates other fallout. As was already mentioned, UHF transmitters tend to be more inefficient than their VHF counterparts. Less efficiency means that more heat is generated, which means larger air conditioning units, requiring more space. Greater transmission power also means larger transmitters and more RF plumbing to accommodate a larger, more complex system (Fig. 1.15).

Additionally, UHF transmitters must cope with two opposing laws of physics. To handle more power, the mass and size of many components must be made larger to dissipate heat; conversely, higher frequencies require that components be made smaller as wavelength decreases. This is why devices that create high-power UHF RF signals tend to be much less efficient. However, they also require more support systems. Whereas VHF transmitters

Figure 1.15   Newly installed DTV (IOT; see Chap. 13) UHF transmitter.

of all power levels have always tended to be air cooled, high-power UHF transmitters tend to be water cooled as water is a better medium to carry heat away
quickly. This leads to complex plumbing systems, with their tendency to leak,
and large heat-exchanger systems (Fig. 1.16).

High-power UHF transmitters use technology that tends to work at higher voltages. All transmitters can be extremely dangerous systems. Most who
have worked around these systems have first- or second-hand accounts of
people being blown across rooms, out of shoes, or worse by coming in contact
with, or just even being near, these potentials. The high voltages found in
VHF transmitters tend to be under 10 kV, with 7 to 8 kV common. Historically, UHF transmitters tend to run at triple that voltage, around 25 kV.
Now new technology for more efficient high-power UHF amplifiers is running as high as 35 kV (Fig. 1.17). High voltage doesn't like dust, water
leaks, or things that change in conductivity. Additionally, the inductive and
corresponding current spikes that occur on power-up or shutdown tend to
produce fail-ures. Because of the voltage potential, these failures can be
spectacular.

Since VHF transmitters generally run at low power levels, solid-state technology has found wide acceptance. Although all VHF and UHF transmitters
have been solid-state, except for the final few amplifying stages since the early 1960s, now triodes and tetrodes (vacuum tubes) used in VHF final amplification stages are being replaced with solid-state modules. These modules work

Figure 1.16 Heat exchanger used by a UHF transmitter. Water is used to carry heat away from the final amplifier stage. The heat exchanger cools the water for a return trip to the transmitter. If more than one final amp is used, additional heat exchangers may be required.

at fairly low voltages, with 45 V common, but handle a fair amount of current. Lots of arcing potential (high voltage, low current) has been replaced with lots of welding potential (low voltage, high current). Whereas a tube would slowly burn up its filament, and consume its cathode, the solid-state devices do neither. As the tubes age, adjustments have to be performed periodically. Solid-state devices generally require no such tweaking. It is claimed that the most important maintenance procedure for solid-state transmitters is the timely replacement of air filters.

Today even some UHF transmitters are solid-state. The architecture of the solid-state transmitter is the limiting factor in the amount of power a solid-state transmitter can efficiently produce. Each solid-state amplifier module handles only a few dozen watts so there are many of these amplifiers in parallel. It is the combining process of summing the power for the many parallel amplifiers that wastes power. At some point, the combining networks reach a size where their combined power loss from each individual combiner makes the overall transmitter too inefficient (Fig. 1.18).

The major advantage to this architecture is that if a module fails, the output power drops only by the amount that the failed amplifier was contributing. There is no single point of failure. The module can even be pulled "hot," which means while power is still on, and replaced.

Figure 1.17 High-voltage supply (often called a beam supply) for a high-power UHF transmitter. This supply is filled with oil to minimize the chance of arcing.

1.4.6 DTV reception

A lot of DTV will be on UHF, at least initially. As we will see later in the book, it is not yet clear what the cable systems will do with the additional DTV channels. Many are betting that many early DTV viewers will have to use a set-top antenna or put up a roof-top antenna. Following the enactment of the Telecommunications Act of 1996 by Congress, the FCC instituted a rule known as 47 C.F.R. 1.4000, which preempts local ordinances and homeowner agreements from prohibiting the mounting of terrestrial television or satellite

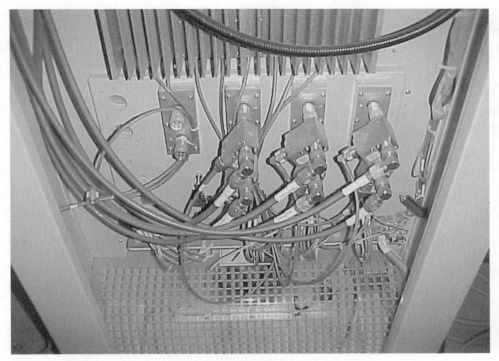

Figure 1.18   RF power combining in a small solid-state transmitter.

receive dishes on dwellings. This is surely a series of court fights waiting to happen. Just as UHF has greater losses and lower efficiencies at the transmit end, the same is true at the receive end. Poorly installed receive systems will hurt UHF channels much more than VHF channels—NTSC or DTV.

Common problems with NTSC reception are a noisy picture caused by low signal levels, "sparklies" caused by electrical noise (which actually affects VHF more than UHF), intermod (which was discussed earlier), and ghosting, which is caused by multipath. Multipath occurs when the signal bounces off natural and man-made objects on its path between the station's transmit antenna and your receive antenna. The received signal could consist of a signal that traveled directly to your receive antenna, along with one or more signals that bounced off buildings, hills, or airplanes to get to your antenna. Because the paths these bounced signals took is longer than the direct path between transmit and receive antennas, they arrive later in time. DTV signals can have the same problems, but you won't know that any exist until they are bad enough to swamp the error-correction circuitry in the DTV receiver.

2

The Players

2.1 The Local Television Station

When television first began, only two choices existed when it came to generating television pictures: the television studio camera or the film chain (Figs. 2.1, 2.2, and 2.3).

Although today most images and sounds are still initially captured by either film or video cameras, they usually end up on videotape or a hard disk. Most local television stations no longer have the ability to play film straight to air. All program material that comes in the door must be on tape, although much arrives via satellite. Programs are even moved from place to place over computer networks, similar to computer files. The process of gathering program material has been coined "acquisition" by the equipment vendors. As just mentioned, this material, or "content" as it is often referred to now, ends up on a hard disk–based "video server" or in a tape library. A growing percentage of local stations use video servers to store and play back the commercials, or "spots" as they are called, during commercial breaks. The same is happening with news stories.

But back to television's beginning. To end up as a television signal the program had to be loaded on a projector and shot into a modified studio camera, which is part of the film chain, or be in front of a camera in a studio. Although television could be taken on the road, which is known as a "remote," or an "outside broadcast" (OB) by the British, this process was difficult and expensive. Early television equipment was very fragile, big, and heavy, and consumed a great deal of power. Remotes were an "iffy" business. In addition, the means of getting the signal back to the station consisted of using the local phone company, most of which were never sure they wanted the hassle that came with the business, or using a microwave system owned by the local station. Videotape recorders, let alone satellite and fiber links, didn't exist. Processing film was a complicated and expensive business, so most local stations didn't want to shoot a lot of their own film. Movie studios were afraid of the new medium, and weren't

Figure 2.1   Early black-and-white studio and field camera.

quick to start producing films intended only for the television stations. As a result, most stations faced the prospect of producing a large share of the daily programming, including commercials, live.

A television studio in the late 1940s and early 1950s could be a bustling place. While one program was going to air at one end of the studio, commercials could be ready in another part of the studio, and the next show's set was being put up at the other end of the studio. Many local television stations didn't, and still don't, have the luxury of having more than one studio. Quick camera moves (the term is called "trucking") from one part of the studio to the other were often necessary. Also, the zoom lens had not yet been invented so that the camera had to be moved (dollied) to change the perspective on a scene. Many in the business lament the adoption of the zoom lens. They argue that zooming is done way too much and that you don't get the same sense of depth as when the camera is dollied to a new position. The film industry still avoids the "zoom" as much as possible.

Another glaring difference between the film and the television studio is the use of lighting. Early television cameras were not very sensitive, and needed a great deal of light to produce acceptable pictures. Television studios could have many times the amount of light found on a comparable film set. A result was a great deal more heat. Performers on a television set often endured temperatures on set that approached 100°F. It took 30 years before the amount of light used on the television set finally dropped because even though the imaging devices (orthicon and vidicon camera tubes) became more sensitive, black-and-white cameras were replaced with color cameras (Fig. 2.4).

2.2   The Television Networks

Early television studios were hot and crowded places and the pace was hard for a local station to maintain—not only from an equipment standpoint but

Figure 2.2 The film chain consisted of one or two film projectors and maybe a slide projector pointed at a specialized television camera through a series of mirrors. (Early television was done with mirrors!) If the show wasn't done live in the studio or shot on film, it could not be broadcast on television.

also from a production and monetary standpoint. It took more space, people, and—given enough shows—more equipment to produce the many hours of live television each day. So radio networks stepped in. The NBC radio network, which was part of RCA, started in 1926. It competed against CBS. David Sarnoff, who ran RCA, had an early interest in television. NBC mounted a television transmit antenna on top of the Empire State Building (1931) even before Fay Wray and King Kong had their tryst. RCA ran a couple of sister radio

Figure 2.3   Room with two film chains and two early VTRs.

Figure 2.4   The light that came through the lens of a color camera was split up among as many as four imagers instead of one (as will be discussed in more detail in Chaps. 8 and 9). A new imaging tube in the late 1960s (the plumbicon) changed the light needed in the studio. The amount of light used dropped by half. Then in the 1980s a softer look for many news shows (primary local programming done today) prompted another 50 percent cut in the amount of light used.

networks, which came to be known as the Red and the Blue networks because of patching and switching color codes used by the phone companies to transport the programming between cities. NBC was forced to spin off one of these networks, which became ABC. NBC and CBS experimented with television throughout the 1930s. NBC took television on the road in 1937 with the first mobile (remote) TV unit. Both CBS and NBC proposed color systems in the 1930s. World War II interrupted experimentation on television, but in 1946 development continued in earnest.

Just as critics of Sarnoff thought that no one would pay money to send messages (advertising) to "no one in particular" (radio broadcasting) in the early 1920s, in the 1940s many thought pictures added no value to radio broadcasts. An old joke in television was (and still is) that video without audio is a mistake, but audio without video is radio. In fact, many early shows offered by the first three television networks (ABC, CBS, and NBC) were extensions of existing radio shows. These first "Big 3" saw the need to fill the programming void in television. These networks realized that if they modeled television after their radio networks, they could provide the programming that the local stations needed. However, as we examined earlier in this book, the bandwidth required for television (video) is many times greater than that required for radio (audio). Trunk cables, which carried telephone voice and radio network audio between cities, could not be used for video. Either AT&T (who owned the only long-distance lines for the local phone companies at the time) had to run coaxial cable from city to city or use a fairly new technology to move the video between cities. This new technology was *microwave*. Microwave, before it was known as a quick way to cook food, was an extremely high frequency radio service for creating wideband frequency links between two points. The frequencies used for microwave were so high that they acted very much like light. Parabolic dishes (looking like metallic snow saucers) would form the microwave transmit signal into an RF beam that was pointed at the receive dish (Fig. 2.5).

The distance between the transmit and receive dishes depended on the height of the tower on which the dishes were mounted. The phone companies generally had their microwave towers 20 to 35 mi apart. A pair of microwave transmit and receive dishes are referred to as *microwave links*. When these dishes are mounted at high elevations, such as on the peaks of mountains, microwave links can be well over 100 mi long. The phone companies built a "network" of microwave links across the country for carrying multitudes of voice telephone calls. Each call had its own frequency or channel. The microwave path could carry hundreds of these voice channels. With towers placed every 20 to 35 mi apart, microwave networks were incrementally built across the country. If someone were willing to buy hundreds of these microwave channels on a particular path, they could be combined to carry video. The fledgling television networks would be able to send their programming from one city to the next. This is exactly what was done.

As with the radio networks, television network programming was produced in New York City. However, because of the expense, only a few microwave paths out of New York were used. One made its way toward Chicago and

Figure 2.5   Microwave dishes mounted on a tower.

another toward the south. The microwave link literally went from city to city, using as few microwave paths as possible. Cities at the end of this network would have signals that had traveled through dozens of microwave links. Some markets (cities) had to erect their own microwave hops to get the network signal from a city along the official microwave network into a spur to a market bypassed by the network backbone. The local stations that carried one of the networks were known as *affiliates*. Some affiliates got their network's signal by receiving the off-air signal of another network affiliate in a nearby market and then rebroadcasting it.

The three networks could now produce shows, mostly still live, and send them to the affiliates for broadcast to their local markets. The networks could also sell time, based on the fact that the programming reached the majority of people in the country. In order to get the affiliates to carry the networks' programming, each network paid its affiliates to carry the network's programming, based on how many households it served. This meant that an affiliate in Boston was paid much more than the one in Boise, Idaho. Also, as is still the case today, the west coast is 3 h behind the east coast, so ABC, CBS, and NBC send programming aired live on the east coast to Los Angeles for delayed airing. Fox, UPN, and WB, reflecting their west coast movie studio origins, originate two separate feeds from Los Angeles, an early one for the east coast and a later one for the west coast. In television's infancy the only way to delay programming was a process known as the *kinescope*. This process took live television and transferred it to film. The film was quickly processed so it could be rebroadcast along the west coast from Los Angeles after the 3-h delay. Kinescopes didn't have much resolution, and shows appeared grainy. It was a good thing that the television receivers of the day had such small screens because of all the picture degradation that incurred between the microwave hops and the kinescope delay.

The networks grew up converting radio shows to copies for television and on game shows. Many of the early shows were actually owned by the sponsors who advertised on the show. The networks sold blocks of time to the sponsors, and the sponsor actually owned and, in many cases, produced the show. The network collected a fee for itself and its affiliates. RCA's Sarnoff told *Fortune* magazine in 1958 that "We're in the same position of a plumber laying pipe. We're not responsible for what goes through the pipe."

The networks soon realized that they could make more money by having the producers of the shows sell the shows directly to the network instead of to the sponsors. The quiz show scandals of the late 1950s prompted the networks to make this switch. The Big 3 networks became the gatekeepers for much of the national programming. In fact, the networks' power in the 1960s led Congress to limit the networks' ability to own most of the shows they aired. This restriction has recently been lifted. The networks had such a tight control on programming that most first-run network programming lost money for the program producers because the networks would not pay enough to recoup the cost of the show. The majority of shows only had a chance at becoming profitable if they went into syndication.

2.2.1 Network affiliates

Most local stations became local affiliates of one of the Big 3 networks. A large percentage of the daytime and prime-time programs came from the network via AT&T. Most were produced live or on film until an audio recorder company in Redwood City, California, invented the videotape recorder. When the Ampex VR-1000 was demonstrated to CBS executives at the 1956 National Association of Broadcasters (NAB) convention, near pandemonium erupted. Ampex secretly recorded the CBS group entering the booth for the demonstration, and when they were seated, played back the tape. Most instantly realized the potential of the VTR. Although they were big, and expensive to buy and maintain, they

Figure 2.6   Ampex VR-1000, the first VTR. The cabinet on the left held most of the electronics.

changed how television shows were produced almost immediately. Shows could be videotaped in advance, mistakes could be reshot, individual commercials would only have to be done once, and the use of the kinescope was over. The networks could videotape their programming for delayed airing on the west coast (Fig. 2.6).

The local affiliated stations greatly benefited from the relationship. They didn't have to buy or produce the majority of their programming, and were paid for "clearing" or airing the programs. Currently, the networks pay approximately $200 million/year to the affiliates as compensation. Over half of the historical value of a station could often be traced to it being an affiliate. The networks would like to lower, and probably eliminate, the payments to the affiliates. While in the 1960s and 1970s many worried that the Big 3 networks would end up controlling too much of what we see, the 1990s have changed the landscape. In the late 1970s the Big 3 controlled over 90 percent of the prime-time audience. Each of the Big 3 networks now averages about 13 percent of the viewers. Fox controls about 11 percent, WB controls 4.7 percent, and UPN 2.7 percent of the viewers. ABC and CBS lost 6 percent of their viewers over the past year. WB gained 9 percent, Fox 2 percent, and UPN lost 41 percent. Advertisers spent $14 billion on airtime at the Big 4 (ABC, CBS, Fox, NBC) last year out of a total of $40 billion for all TV ad time. Although the television advertising pie continues to grow, the major networks' share is remaining relatively flat. Total TV ad revenue in 1983 was only $17 billion. In 1993 it was close to $31 billion. Local TV spots were $4 billion in 1983 and $8 billion in 1993. By comparison, newspaper ads went from $21 to $32 billion in the same period. Only NBC is operating in the black. NBC earned $300 million, but its profits were 40 percent less than the year before. Besides losing viewers, the cost that the networks must pay per show has skyrocketed. While an average episode in the mid-1970s cost less than $200,000 an episode, today a top star alone can command $200,000 an episode. Episodes can run well over $10 million each, and the average cost to produce a show has doubled in the past 5 years. This has forced the Big 3 to run more news and news magazine shows since they own them and they are cheaper to produce.

2.2.2   Program ownership

Due to earlier worries about the Big 3 networks becoming too powerful, the networks were not allowed to own the bulk of their programming until 1995. The financial-syndication rule, as it was known, didn't allow the networks to have a financial interest in the programs they aired, except for news shows. Now that the networks can own the shows, some critics claim that this is tempting the networks to air inferior shows which they own over better shows which they don't own. Fox, as a start-up network, was able to own its shows at the outset. Even with the high prices being paid by the networks for programming, the economics of network program production is such that the company producing the show usually can't get the network to pay enough to air it to make the show profitable during its first run. The producer of the program hopes that the network will keep the show on long enough so that the show can be sold in syndication. What this means is that when enough shows are "on the shelf," the production company can start selling the "old" programs to local television stations to air. These local stations can be network affiliates or independent stations. Today, some are sold to cable networks. In a few recent instances a syndicated show has been sold for airing on both cable networks and local stations. Many shows are aired initially on one network, and end up being aired in syndication by the affiliates of other networks. However, before a station will consider buying a syndicated program for broadcast, enough episodes must be available so that the broadcast entity buying the syndicated show can "strip" the show. This means running the show every night or at least 5 days/week. So the network that is initially running the show must renew the show for enough seasons so that about 100 episodes are produced. Completion of this many episodes takes 3 to 4 years of network airing, and most shows do not have initial runs that long. As a result, program producers lose on most shows, hoping to make it up with a hit every so often.

In an effort to get maximum mileage out of their programming, the big networks are coming up with multiple uses for the same show. ABC is "repurposing" soap operas that are produced in house to not only be used on the broadcast network, but to also be run on cable networks. In several instances networks have found that they can make more money using one of their commercial slots to push a self-produced movie or program, or one produced by an affiliated studio, than by selling the time to an advertiser.

The traditional networks have a problem with demographics. The viewers tend to be older on the traditional Big 3, with CBS sometimes being called the geriatric network. The median viewer age for CBS is 52. ABC and NBC do better in many advertisers' eyes, with an average viewing age around 40. Fox has the thirty-somethings (average viewing age of 33). WB and UPN have even younger viewers. Many cable networks also skim off the younger viewers. Older viewers, and women viewers, command lower advertising rates because they tend to be heavy watchers of TV, which makes them easier to reach. Young adults, especially males, tend to watch fewer television outlets or channels. There might be 1 to 2 million fewer 18- to 34-year-old viewers than a few years ago. A premium is paid to the outlets that have them as viewers. This means that shows with fewer viewers, but viewers with the right demographics, in some cases, command twice the ad rates of shows seen by less-valued viewers.

2.2.3 Network evolution

The network now serves other purposes. The Big 4 networks are evolving into program suppliers for the stations they own and operate because these stations are still lucrative. Warner Brothers and Fox are the top providers of prime-time shows. Fox produces shows for ABC, CBS, and WB. Although the networks' margins are falling, the margins on owned stations remain high. CBS sees its network as a tool to boost local ad sales for their owned and operated stations (O&Os). Disney uses ABC to promote its movies, which means ABC owns content and distribution. GE uses NBC to promote MSNBC, CNBC, and its online service called SNAP. News Corp. uses Fox to create shows that it sells around the world on its cable and satellite services. Since it is difficult to predict where the profits will be made in the future, Fox is deploying its efforts in several different areas—studios, a network, local stations, and cable. Time Warner uses WB and Viacom uses UPN to ensure that the programs they produce reach a national audience. (As this book goes to press, Viacom and CBS are merging, so the fate of UPN is in doubt.) However, the major television networks still reach about 100 million homes, whereas the large cable networks reach only about 75 million homes. The major networks lose share to cable, cable loses share to online services and direct broadcast satellite (DBS), and home video will lose to pay-per-view. As a result, the networks are looking to cut costs. The first time they reduced compensation to the affiliates was in the 1960s after AT&T increased its rates to the networks for distribution. The networks have since totally gone around AT&T by using satellite as the distribution channel to the affiliates.

At the beginning of the 1990s, the Big 3 were getting ready to cut compensation again but Fox intervened. The networks used to look to news shows to bring them the prestige that would, in turn, bring intangibles (goodwill, increased viewing of other shows, increased access to potential advertisers, etc.) to the rest of the network. Now sports programs seem to have displaced news as a loss leader. Fox started prime-time programming as the fourth big network in 1986 but is considered to have become a first-tier player when it acquired NFL broadcast rights in 1994 for $1.6 billion, which temporarily ended CBS's long-standing relationship with the NFL. This started a game of "musical chairs" among many affiliates. The fact that Fox now had the NFC half of the NFL, along with no talk of tampering with affiliate compensation, prompted a number of stations to switch affiliateships. Fox picked up VHF stations in a number of markets where the affiliates were previously UHFs. Although all of the Big 3 were hurt, CBS got the worst of it. In several markets it ended up with UHF affiliates where it previously had VHF.

Fox has been aggressive in building a vertically integrated media concern: Fox owns content (production) and distribution (the network and local television stations). Fox has been acquiring all types of broadcasting assets. For example, Fox paid $1.5 billion for the Family channel. Since Rupert Murdoch (the owner of Fox) has a newspaper background, where news content is gathered, printed, and delivered using assets usually owned by the paper, he would naturally gravitate to using this same type of vertical integration in his television operations. Fox changed a lot of the rules that television had played by. From pushing the programming envelope, to radically upping the cost of sports programming, to scrambling the affiliate lineup, to scooping up many of the top writers for their studio by doubling and tripling the salaries of the top comedy writers, Fox has changed the broadcasting, satellite, and cable landscape.

Fox has also had a hand in pushing the cost of sports television rights through the roof. Last year ABC, ESPN, CBS, and Fox agreed to pay $18 billion for NFL rights over the next 8 years. NBC was squeezed out of this contract by CBS. Fox is losing about $70 million on its NFL contract each year. CBS affiliates gave up some commercial inventory and are actually paying some cash to CBS for its annual $500 million NFL contract. The affiliate help is considered worth about $50 million. Disney (ABC) pays $550 million/year for NFL rights, and $120 million/year for NHL hockey. ABC lost $70 to $80 million on the 1988 Olympics in Calgary and CBS lost more than $500 million on the 1994 Olympics in Lillehammer.

Like many mature businesses, consolidation is the order of the day. Disney bought ABC for $19 billion in 1995. Westinghouse bought CBS for $5 billion in the same year (now Viacom has bought Westinghouse/CBS). Both networks had come under new ownership approximately 10 years earlier. Capital Cities, which owned a group of TV stations, bought the much larger ABC in 1984. Laurence Tisch, whose family owned Loews Theaters along with cigarette and insurance businesses, bought CBS in 1986. ESPN came with the ABC purchase. ESPN was bought by ABC a short time before ABC was bought by "Cap"

Cities. ESPN was founded in 1979 and lost $80 million up to the time it was purchased by ABC for $227 million. When ABC acquired it, it had just turned profitable. ESPN is now the largest cable network, worth more than $3 billion.

After ABC and CBS were bought by other companies, both went through much bloodletting, although CBS's seemed to be much more prolonged than ABC's. That was a decade ago. Unfortunately, the situation for the major networks (ABC, CBS, and NBC) has gotten worse. As already mentioned, 20 years ago the three major networks controlled over 90 percent of the viewer share. When the 1980s downsizing occurred at ABC and CBS, the Big 3 controlled approximately 75 percent of the audience. Today, the Big 3 have slightly under 50 percent. If viewers from Fox are added, the top 4 networks have 59 percent of the viewers. Some think that $1 billion in advertising revenues could shift away from the major networks in the next few years. Cable networks have been consolidating as well. Time Warner bought Turner Broadcasting (including CNN) for $7.5 billion.

2.3   Group Ownership

It is not only the networks that are merging into larger entities; the locally owned station is rapidly becoming endangered as well. Originally, most stations were put on the air by people or companies who were part of the community or market, as television people like to say. Often an existing radio station or newspaper would get the original construction permit from the FCC to build the station. Some companies, namely, the Big 3 networks, owned more than one station early on. Five was the maximum number of stations that was allowed originally. The Big 3 networks naturally wanted to own these stations (O&Os) in the largest markets, so all three had stations in New York, Chicago, and Los Angeles. Other large markets that had one or more O&Os included Philadelphia, Washington, D.C., Detroit, and San Francisco. Slightly smaller markets like St. Louis and Cleveland had O&Os also. In fact, in Cleveland, NBC and the owner of its Philadelphia affiliate KYW squared off. NBC owned a station (WKYC) in Cleveland, and it wanted a station in a larger market. Westinghouse (which now owns CBS) owned KYW in Philadelphia. In 1956 NBC coerced Westinghouse to swap the Philadelphia station for the Cleveland station. KYW's call letters went to Cleveland and WKYC's went to Philadelphia. Nine years later the FCC ruled that the swap had to be reversed.

Over the years the FCC has increased the number of stations that a company can own. It went from 5 to 12 in the 1980s. The station limit had another cap associated with it—those 12 stations couldn't reach more than 25 percent of the U.S. market. The Telecommunications Act of 1996 allowed a single company or group to own any number of stations serving no more than 35 percent of the country. But any of the stations owned that are UHF only have one-half their reach counted toward the cap. The network with the largest number of stations is Fox, which owns 22 stations. Because of the UHF "handicap" clause just mentioned, Fox stations reach about 40 percent of the U.S. market, which means its affiliates are only needed for the other 60 percent. CBS, with 14 stations, reaches 32 percent, Disney's (ABC) 10 stations reach 24 percent, and NBC's 11 stations reach 25 percent. In addition to the Big 4 networks, Paramount reaches 19 percent with its 13 stations. The two Spanish language networks, Telemundo and Univision, own 8 (21 percent) and 12 (20 percent) stations, respectively. A network wannabe, Paxson, owns the most stations, 44, which provide a reach of just over 50 percent. A number of newspaper companies own groups of stations: Tribune (17 stations, 35 percent), Gannet (18 stations, 18 percent), Belo (16 stations, 14 percent), and Scripps-Howard (10 stations, 10 percent). Some owners with only a few stations worry that issues from negotiating network compensation and syndicated programming to borrowing money seem to favor economies of scale.

2.3.1 Stations become valuable

The prices of stations have gone up dramatically over the last few years. Stations are selling for up to 20 times the multiples of their cash flow. Television stations in markets that are well less than 100,000 people have sold for $40 million. Stations in markets less than 500,000 have sold for well over $100 million. In the largest markets individual properties have changed hands for over $500 million. Many may be wringing their hands over the state of the big broadcast networks, but there still seems to be plenty of faith in the local television station. The total value of engineering facilities, including the studio and transmitter, in a large station is seldom worth much above $10 million. Even with the average station looking at approximately a $5 million price tag for its DTV conversion, it is not the cost of engineering DTV that is driving up these prices. The NAB has estimated that the 1576 U.S. TV stations will spend $16 billion over the next 10 years to convert to DTV. As you probably suspect, though, the spending is going to be unevenly distributed among the players. The largest 4 percent of stations in the United States will spend over 50 percent of the total engineering capital over the next year. A few have budgets over $10 million for the year, and a few dozen have budgets over $5 million, but approximately 50 percent of the stations have capital budgets of $500,000 or less next year. Even though the largest stations are spending millions of dollars per year, their capital dollar per viewer can be one-half to one-third of what a small market station is spending per viewer with their budget of a few hundred thousand dollars.

A few years ago the FCC started letting a single company own multiple radio stations in the same market if 30 separately owned stations remained in a top 50 market. This is known as a *dualopoly*. The FCC has recently started to allow television stations to head down the same path. Some stations have also set up local marketing agreements (LMAs). An LMA exists when one station in the market manages and programs a second station in the same market.

DTV will probably increase segmentation of the viewers because of increased program streams from each station. NBC and ABC have already pioneered this approach with Disney's (ABC) ESPN networks and GE's (NBC) CNBC and MSNBC. These additional program outlets for the two companies allow them to capture viewers that would be watching a competitor's station.

2.4 Cable Systems

Cable companies got their start among the ridges of the Appalachians in rural Pennsylvania in the 1950s. Their original service provided acceptable delivery of off-the-air, that is, terrestrial, television stations. In the early 1970s when lower cost, and in many cases lower quality, television cameras and VTRs became available, some cable systems began to offer television channels on their systems that were originated by the cable company itself. This was known as *local origination*. Local origination consisted mainly of community access, government, and movie channels. Some cable systems were more ambitious and provided news and local high school sports. As the broadcast networks had already done over 20 years before, some cable systems set up microwave links that tied separate cable systems together. In this way one cable system could provide local or additional programming to the other systems. Some terrestrial television stations had the same idea. They would arrange to have their signal microwaved to many cable systems so that they could reach markets not possible from their transmit or translator sites alone. In the early 1980s a technology that transformed the industry, satellite transmission, came into being. This technology probably qualified as a "paradigm shift."

2.4.1 Cable networks

Television stations, so to speak, with no terrestrial transmitter arose, but these "stations" had paths to satellite uplink providers. These new broadcasters would buy or lease one or more satellite transponders (channels) and sign up cable systems across the country to install satellite dishes and feed the received signal into the cable system. These operations became known as *cable networks* (Fig. 2.7).

While the big broadcast networks had local stations as affiliates for their program distribution, these new cable networks had cable systems for their distribution. Some of these networks charged cable operators to distribute their signal, for example, HBO. Some simply gave the signal away free and made money off the advertising. Some actually paid the cable systems to carry their programming. Fox has paid cable systems as much as $10/subscriber to carry its Fox News channel. In any event, the concept of the cable network flourished.

During the period between January 1996 and the end of 1998, 109 cable networks were launched. Some are backed by the major networks, existing cable channels, or terrestrial broadcasting companies, but most are not. These new offerings include channels devoted to aging, air and space, alternate lifestyles, antiques, boating, computers, dependencies, fashion, hobbies, infomercials, investments, local newscasts, love, military, museums, outlet malls, parenting,

Figure 2.7 Modern network affiliate with many satellite dishes for feeds into and out of the facility (*courtesy of Sony Systems Integration Center: Benson & Rice*).

pets, real estate, romance, self-improvement, success, theater, and therapy. However, wrestling accounts for many of the most popular shows on the cable networks. At least a few of the cable networks carry a considerable amount of clout in the industry. Even though less than 800,000 watch CNN during the height of prime time on a normal news night, many reporters assume that everyone knows when CNN has reported a story and will back off it. Digital set-top boxes (which accommodate compression) might allow all these cable network start-ups to vie for attention on cable subscriber's screens. Cable and DBS started down the digital highway before the broadcasters. Many existing cable channels, such as Discovery, Fox, HBO, Disney, Lifetime, and Home and Garden, have launched additional cable offerings slated for digital cable channels. Lifetime plans to offer a movie channel for women which cable can multiplex into a digital cable channel. MTV plans to offer seven genres of music with the same purpose, and HBO has similar plans.

Many cable operators won't talk to new cable networks if cable does not own a piece of the action. In the future new cable networks will probably have to have an affiliation with a *multiple system operator* (MSO) like TCI (now AT&T) or Time Warner. Some of the new cable networks have paid lucrative up-front fees to cable systems for carrying their programs. The cable systems are the gateway into most homes today. TCI is the largest MSO with a total of over 14 million subscribers. Time Warner is a close second with over 12 million subscribers. The telephone company U.S. West (in the process of being acquired by Qwest) is a distant third with just over 5 million. The rest, such as Comcast, Cox, and Cablevision, all have less than 5 million each. Some fledgling networks are going completely around cable by garnering space on DBS, or by buying or leasing transponder space on a C-band satellite.

Much of the bandwidth in the 11,000 separate cable systems in the United States today is now tied up carrying cable networks. The space left over for carriage of the local television station has become acute. Two-thirds of the cable systems have no capacity for adding additional channels. As a result, cable systems started to wonder why they needed to carry all the local stations. The National Cable Television Association has stated that they will never accept the idea that every broadcast signal takes priority over any cable network. Cable systems now supply television signals to over 60 percent of the households in the United States. Most of us don't use rooftop or set-top antennas anymore; the majority of us only view what comes in on the coaxial cable that our cable company provides.

2.4.2 Must carry

Local broadcasters became extremely worried as to whether the cable "gate-keepers" would pass their transmitted signal down the cable bottleneck. After much consternation, Congress passed the Cable Act of 1992. This act contained a rule defining "must carry." The must carry rule states that the cable system must carry all local stations in its service area if the local stations request it. This means that if the local station has a certain percentage of viewers in the territory that a cable system serves, the cable system must offer that station to its subscribers. Local stations that opted for must carry were usually UHF stations that carried a shopping channel, a foreign language network, or religious broadcasting.

However, the act also allowed the local television station to play "poker" as a result of another option where the local station could require the cable system to pay the station, through consent agreements, for the right to carry the station's signal. Network affiliates, which usually have respectable ratings, can often get the cable systems to sign these consent agreements. Since cable systems are loath to pay cash to the local stations, the desired local stations that deftly play these hands usually get extra bandwidth from the cable systems in the form of an additional channel. A few powerful local stations have gotten a cable channel, which is used for news or sports programmed by the same local station. However, ABC, or, more exactly, Disney, which owns 10 stations that control 24 percent of the U.S. market, got cable systems to carry ESPN's little-brother network ESPN2 when cable systems didn't want to cough up the cash that ABC asked for to carry the stations owned by ABC/Disney.

This game went against CBS, which didn't have a noncash card to play with the cable companies. CBS had been focused on cost reduction during the 1980s while ABC, Fox, and NBC were building cable programming networks. CBS had no noncash chip to play when its bluff for payment for carrying the CBS O&Os was called. Maybe CBS's focus on the traditional network will be rewarded. The conventional business wisdom of the last decade has been to concentrate on your core competency. CBS did that. In fact, 95 percent of CBS's income is derived from advertising revenue from the television and radio network, along with revenue from its owned and operated stations. The other networks have added cable networks, and their associated subscriber fees, which fluctuate less due to the economy.

2.4.3   Cable's additional DTV burden

Now DTV is threatening to force cable systems to add an additional channel for every local television station. The Telecommunications Act of 1996 states that "ancillary or supplementary" digital TV services should not have must carry rights. Although the FCC has ruled that cable systems must carry the broadcasters' analog channels, it has said nothing about DTV. A member of Congress has asked if the must carry provision was intended to promote localism through any, and all, distribution platforms or is it only necessary when free over-the-air television is threatened. Cable seems to want to carry more channels but not many HD channels. At least not HD channels that the cable companies themselves do not own entirely or partially. SD is easier for cable to do with existing facilities. Since today's compression technology allows six SD channels in one 6-MHz television channel, this equates to a lot more channels with the bandwidth they have today. Their bandwidth today is limited mainly by the coax used to wire their systems; however, many systems are starting to replace their coax with fiber-optic lines to increase bandwidth. Bandwidth, though, is like computer hard-drive space; as larger and larger drives become available, we seem to find more software to fill up the space.

Even the broadcasters understand that this is a fair-sized additional burden. Some of the larger MSOs are adding fiber-optic infrastructure to accommodate this, but many smaller systems, which are still copper-based (coax), and will be for the foreseeable future, are unsure about what to do with DTV. Compression will greatly help these cable systems. Analog services that used to take a full 6-MHz channel bandwidth can now be shrunk to a fraction of that bandwidth. This solution will work for cable networks installing digital overlays on cable systems, but the terrestrial broadcasters expect that the space on a cable system allotted to their new (and second) DTV channel will pass the entire 6-MHz bandwidth. This could be just one HD signal or multiple SD channels. According to the FCC, cable systems cannot degrade a terrestrial broadcaster's DTV signal, but it is still totally unclear as to what that means. The FCC has told the broadcasters that if multiple channels are sent on a single DTV transport stream, at least one program stream out of that channel must be equal in quality to its current NTSC channel. Some cable companies have threatened to down-convert HD to 480p or 480i. It appears that cable companies and broadcasters will strike DTV carriage deals on a case-by-case basis. Bell Atlantic's Ray Smith has said that the broadcasters' "great strength is their ability to create intellectual property. Now, if the broadcasters continue to see themselves as transportation companies, then they will be preempted as these new networks are created." Increasing competition from cable and DBS might require terrestrial broadcasters to do something new and different with the DTV spectrum. Some think broadcasters might start competing for the pay-per-view customer, but broadcasters have no history or infrastructure to sell, bill, and collect for pay-per-view. Broadcasters might end up using a significant portion of their DTV channels' bandwidth to carry data.

Cable doesn't need a big, expensive transmitter consuming lots of power and space or a massive tower on which to place an antenna to send a signal down a cable. However, the broadcaster needs a transmitter to distribute its signal through the air to individual cable headends. These headends are portals to at least 60 percent of their viewers. Although it is true that some broadcasters have direct links to cable headends, through fiber, copper (coax), or microwave, most broadcast television finds its way into your cable system via the terrestrial airwaves. Some have questioned the continued use of large amounts of RF spectrum for fixed, land-based transmitters and receivers. A few years ago many thought that terrestrial television broadcasting would eventually go away, that is, local television stations would give up their transmitters and feed the cable system headends directly via fiber-optic cable or via microwave.

Cable operators committed to digital television are using quadrature amplitude modulation (QAM) to send digital signals down their cables. QAM is the same modulation scheme used by many satellite service providers in Europe and the United States. This modulation method, along with the MPEG stream it carries, is commonly known as digital video broadcasting (DVB). *Business Week* described "modulation" schemes as the methods of getting digital bits onto so-called carrier waves. ATSC uses a different modulation scheme, which is known as 8-VSB (vestigial sideband modulation; see Chap. 10). To satisfy cable's interests, ATSC modified the 8-VSB system into the 16-VSB system. As you might suspect, the 16-VSB system has double the information capacity of 8-VSB. ATSC advocates claim that 16-VSB is compatible with 8-VSB, and that it delivers performance identical to the cable QAM system. However, cable still seems to prefer QAM over VSB, which means, at least initially, consumers may have to choose between cable or over-the-air reception to view DTV. A solution to this might be receivers or set-top boxes that can handle either DVB (QAM) or ATSC (8-VSB) modulation schemes. At this time, though, the only way to receive terrestrial broadcast DTV is via an antenna. Eighty percent of homes are passed by cable TV, and 80 percent of those, or 65 percent, subscribe to cable services.

2.5 Direct Broadcast Satellite (DBS)

Another competitor has emerged over the last 10 years—the direct broadcast satellite (DBS) systems. These companies use Ku-band (a set of frequencies above 10 GHz) satellites to send many television signals straight to your home. These signals are often the same cable networks found on the cable systems. Over 9 million subscribe to DBS in the United States. These transmissions are 100 percent digital, but these digital signals are converted back to analog via the set-top box for display on a regular NTSC set. Direct TV uses three high-power Ku-band satellites to deliver 200 channels. One uses 16 120-W transponders; the other two use eight 240-W transponders. Each transponder averages six program streams (channels) (Fig. 2.8).

With their limited bandwidth most small cable systems will have to use compression to offer enough channel selections to compete with the satellite services. The small systems can't easily increase bandwidth because they don't have enough subscribers over which to amortize the cost. Since TCI (now AT&T) has many small systems, it has its own uplink facility to efficiently deliver cable channels to these systems.

Congress would like to see some competition to cable systems. DBS wants to be able to provide local stations to their customers, and this is called *local into local*. However, the technology doesn't exist yet to allow all local stations to be put up on satellite transponders efficiently because there are simply too many stations. One proposition is to put all television station signals on Ka-band (another frequency band) satellites and spot-beam them back to their home markets. The source for funding this scheme is unclear. As a result, the DBS providers want to select only certain stations from across the country that would represent ABC, CBS, Fox, NBC, PBS, UPN, and WB affiliates everywhere. However, a problem arises with must carry. If DBS provides any local stations, both local stations and cable companies want DBS to live up to the must carry rules. This problem is beginning to be solved by the DBS providers signing marketing agreements with the local phone companies. Some phone companies are starting to realize what the cable folks had already discovered. Building cable systems doesn't necessarily make you knowledgeable about phone nets; the reverse can also be true. A number of phone companies are helping DBS providers integrate the DBS receiver into a combination satellite/terrestrial television receiver that also can be interconnected with the household phone lines. In addition to the phone connection, the phone companies want to add value by providing local support to the DBS subscriber.

Figure 2.8 Master control center of DBS provider (*courtesy of Sony Systems Integration Center: Benson & Rice*).

2.6   The Computer Industry

In 1997 then FCC chairman Reed Hundt predicted that the computer industry would have sold 40 million computers capable of receiving digital TV signals before the TV manufacturers sold their first million. He added, "It's hard to see why the computer industry should care whether digital television comes into the home over the air or through a cable." While many in the broadcast industry and the FCC, which is being prodded by Congress, are hung up on SD versus HD, the computer industry thinks DTV is simply video plus data. The computer camp isn't very enthusiastic about HD because of the processing power needed for the higher bit rate. Their position has been that if broadcasters embrace formats that are "computer" friendly, namely, progressive formats, PCs will be able to greatly speed up the roll-out of DTV. Other computer friendly DTV applications are interactivity and multimedia applications. Terrestrial, satellite, and cable providers of these will enhance the PC's place in the living room. Compaq has stated that "We can bootstrap digital television many times faster than traditional television manufacturers, and we plan to."

Broadcasting is thought of as a one-way process, while telecommunications is considered a two-way process. In addition, telecommunications doesn't have to be symmetrical in the bandwidth allocated in each direction. Oracle is a proponent of not having the consumer store data; video or data will all reside on the Web or at least central servers. The consumer only needs an appliance for access—whether a "network computer" or a "web television." Microsoft takes the opposite approach. The PC becomes increasingly powerful, and most of the control and content remains local. In this case the Web is used only to transport material to the local PC. In both widely differing views, though, data to the user is much higher than the data or requests from the user. Many are working on approaches to use DTV to widen the pipeline to the user while keeping the path from the user at its present size.

Thirty-five million PCs have been installed in homes, which is less than one-third of the number of televisions. Although people with PCs tend to be higher-end consumers, the PC industry desperately wants to gain some of the TV market. Some have envisioned a system with higher-resolution TV and lower-resolution PCs, each having different capabilities. The PC market has centered on the consumer buying new PCs every few years. The television market doesn't enjoy that rate of turnover. This buying pattern shows that people may have different mindsets or the same person may have a different mindset depending on whether he or she is television shopping or PC shopping. To give the Wintel camp time to develop the capacity to handle HD, Microsoft, Compaq, and Intel offered a three-step HD proposal in 1997 dubbed HD0 (480p30), HD1 (480p60), and HD2 (720p24). These steps are all progressive. However, the broadcast industry looked upon them as a delaying tactic. The PC industry has stated that it could build 10 million PCs a year that would be able to receive DTV with only an additional $100 in cost.

The television and PC marketers work under different rules. In the television market the number of features increase as the price increases. In the PC business at the high end of the market, performance, or capacity, increases but not the basic features. Traditional television sets display pictures and sound at the same rate no matter what the cost. However, inexpensive television sets tend to have smaller displays than more expensive ones. On the other hand, low-cost computers today would most likely not handle video or audio as well as more expensive, faster computers. The size of the display, though, can be the same for low- and high-cost computers. DTV receivers will most likely be more like computers than their television ancestors.

2.7   The Web

The latest competitor for viewers is the World Wide Web (WWW). Up to now, viewing the WWW has been done via a computer and a modem connection to an Internet service provider (ISP). Many web sites exist as another distribution channel for manufacturers or as a virtual showroom to educate potential and existing customers. Many web sites provide information and services but no other product. These sites support themselves by placing advertisements on their web pages. Many companies have already sprung up to place ads on the web. It is predicted that advertising on the Web will top $5 billion in the year 2000. Web TV has sold over 400,000 units. The Web and computer industries are counting on the fact that DTV will have "users," not just viewers. There are 80 million Americans with access to the Web (online); these people watch 15 percent less television. Some in the broadcast and television receiver manufacturing industry fear that PCs will change broadcasting in ways that the broadcaster is ill-equipped to cope with and will preclude the necessity of TV sets. However, computers are in far fewer homes than televisions. PC manufacturers fear that the DTV receiver could become the main appliance used to access the Web.

Cable companies are getting into the web connectivity business. AT&T has formed a company called @Home, which will provide turnkey data services for cable companies. As a result, local cable companies can become connected to the web backbone with the help of companies like @Home, and use their cable system to deliver web content to their subscribers. Many upgraded cable systems have as much as 20 percent of their systems wired with fiber. The subscriber would request web content or data over a standard modem connection but would receive the response over the service's high-speed data connection via the cable system. This process is known as an *asymmetrical data subscriber loop* (ADSL), and it means the data pipe to the subscriber is much wider than the one from the subscriber. Since most of the data traffic is to the subscriber, and not from the subscriber, the difference in bandwidth doesn't matter. Intel has invested in @Home, and already several other competitors have come into being. Microsoft has invested in Comcast Corporation and Road Runner cable-modem service. A cable industry research group, Cable Labs, has talked about setting up 10-Mbit/s Ethernet "neighborhoods" serving a few hundred customers in a cluster. Each subscriber would be a node.

2.8   Conclusion

The game is evolving and creating new paths for the players to take. One might think that this upheaval is new to television. Students of television history know that uncertainty has gripped the industry before. In Chap. 3 we will explore the change that has previously been saddled on the industry.

Chapter

3

The Roads That Have
Already Been Taken

3.1 We've Done This Before (Color Television)

Television has undergone a number of transformations. Line and field/frame rates have changed several times, and each new improvement seemed to overwhelm the early viewers of the latest technology employed to enhance the picture. E. B. White, after witnessing TV at the 1939 World's Fair, said, "(Civilization) shall stand or fall by television." The system used was a standard produced by the Radio Manufacturers Association (RMA), and was a 441-line, 30-frame interlaced system. The black-and-white NTSC 525, 30-frame, interlaced standard was introduced in 1943; NTSC was modified in 1953 to carry color information as well.

3.1.1 CBS versus RCA

Color television has been around a lot longer than most people realize. In 1939 CBS demonstrated a color television system that had a spinning color wheel with three filters mounted in front of the camera lens. A similar rotating wheel could be found in the receiver. The two wheels had to be in sync with one another. The color information in the scene being shot was sent sequentially. One field would send the red information, the next field the green information, and the next would send the blue, and then repeat. This "field sequential" system was later adopted for the all electronic "SECAM" system used by the French, Russians, and a number of other countries. Competing against the CBS system was an all-electronic system from RCA. The RCA system used a prism behind the lens of the camera to split the color into red, green, and blue components. These primary colors went to three separate black-and-white cameras mounted in the same housing. The picture tubes phosphors painted on the front of the tube were arranged in the receiver in a triangular pattern of red, green, and blue dots. The RCA system is what we currently use, but in 1949 when both systems were tested by the FCC, the CBS system was more stable. As a result, the CBS system was selected, and CBS started color transmission in 1953. However, the CBS color system was not compatible with existing black-and-white sets, while the RCA system was. The television set manufacturers didn't like the incompatibility and the mechanical aspect of the color process. Few sets were ever manufactured and sold. By 1952 CBS had given up on its system although they had invested $60 million into development of the system. By the end of 1953 the RCA system was approved by the FCC. It is interesting to note that today Texas Instruments has a system with a light that shines through a spinning color wheel and reflects on rapidly pulsed shutters (one per pixel) to generate an HD display.

In the RCA approach color information was conveyed by adding a carrier, a sinusoidal (sine) wave, that is mixed in with the black-and-white picture information. The phase of that sine wave (which is actually referred to as a subcarrier) determines the tint, or color, displayed. When color information was added using the RCA system, some very subtle changes were made. The total number of horizontal lines transmitted each second was reduced by 16. The line rate had been 15,750 lines/s; now it was 15,734. This change was made so that the chroma, or color information, along with an additional "beat" frequency created by the presence of the 3.58-MHz chroma carrier, and the aural carrier at 4.5 MHz, would fall in between energy bands approximately 15 kHz apart across the channel spectrum. Fewer horizontal lines (one-half per frame) meant a slightly smaller field count. Instead of 60 fields, now there are 59.94 fields/s. These numbers were close enough to the original values that black-and-white sets would still lock to them, while minimizing the artifacts from the color information in a picture displayed on a black-and-white only set. If you get close enough to the display on a black-and-white only television receiver (not recommended very often), you can see a dot pattern where the color information would be slowly moving up through the picture. This effect is known as *dot crawl.*

3.1.2   Market acceptance

NTSC color transmission was authorized on January 22, 1954. In March 1954 Westinghouse and RCA offered the first color TV receivers. The cost was $1300, which was about 80 percent of the cost of a new car then. By the mid-1950s color sets cost $500 to $1000. By 1962, only a million sets had been sold. The 1962–1963 season was when NBC committed to 2000 h of color programming. By the beginning of 1964, the number of color sets totaled 3.4 million. The year 1963 was monumental for color TV.* The number of sets sold doubled. The next

---

*The number of years it took some products to reach 85 percent penetration in the marketplace is as follows: microwave oven, 18; radio, 20; color TV, 28; telephone, >70 (more households have color TVs than have telephones). Some products that have not yet attained 85 percent penetration and the number of years they have been available: dishwasher, 50; PC, 24; VCR, 24 (almost at 85 percent); cellular phone, 16.

year the total number of color sets sold almost doubled again. By 1968, the number of color sets totaled 15 million. It was not until 1978, though, that the number of homes with color sets exceeded homes with black-and-white sets. Color-set penetration didn't reach the 85 percent threshold until 1983. That is the DTV receiver threshold for NTSC service being shut down. That's 28 years for color TV sets. However, broadcasters weren't forced to transmit color, and viewers could still receive black-and-white renditions of color programs on their black-and-white sets.

When television transmission started in Britain in 1936, two systems were tested. One was a 240-line, 50-frame/s progressive system; the other was a 405-line, 25-frame/s interlaced system. Both were black-and-white systems. The 405-line system won out and was used until it was replaced by the current 625-line, 30-frame/s interlaced color (colour according to the English) system in 1967. This newer system used a color scheme known as *phase alternate line* (PAL), and it swapped the phase of the color information every other line, which allowed the television receiver to detect any color-phase errors and correct for them, thus eliminating the need for a color-phase (called "tint" on most sets) knob. The 405-line system was discontinued in 1985. Thus, Great Britain had a much more disruptive move to a color system than the United States.

The evolution and roll-out of color television had the support of broadcasters, which today's DTV launch does not have. These broadcasters were the first two television networks, NBC and CBS. RCA was spun off from GE in the early 1920s (RCA was bought back by GE in 1986, although the name lives on as a brand name sold by Thomson Electronics). It started NBC in 1926. In 1928 Sam Paley bought the United Independent Broadcasters radio network. Its new president was Paley's son William, who renamed it the Columbia Broadcasting Company (CBS). David Sarnoff ran RCA, which owned NBC. Sarnoff was a technocrat while Paley was a salesman and broadcasting programmer. After World War II, Sarnoff was eager to launch television, while Paley dreaded the prospect.

3.1.3 Manufacturer and network synergy

Both networks had strong engineering cultures where change was encouraged (Fig. 3.1). NBC (backed by RCA) made television practical. As mentioned earlier, both took an early interest in color. Even though CBS didn't have a manufacturing parent to push it along, over the years CBS offered equipment it had developed to other broadcasters through CBS Labs. In fact, the concept of postproduction was facilitated by CBS's partnership with Memorex in creating CMX. However, the parent of RCA had adopted vertical integration long before the concept became commonly known. Not only did it produce programming, it also sold the receivers on which to view the programming. Additionally, it sold the antennas, towers, and transmitters used to broadcast the signal. It manufactured the microwave systems to get the signal from the television station to the transmitter site. It also sold VTRs, cameras, and most of the other infrastructure needed by a television station. Many years ago a chief engineer

See the Jack Paar Show in Living Color—Monday thru Thursday nights—NBC-TV

Figure 3.1   RCA publication pushing color TV sets, color television programs, and color television cameras and VTRs.

was heard to remark, "If RCA doesn't make it, we don't need it." Many stations still used RCA products to a large extent (80 percent) well into the 1970s. Until the late 1970s only a few companies even tried to compete across the board with RCA; GE and Ampex were those companies. A number of companies competed in niche markets: Norelco in cameras and Grass Valley in switchers were the predominate niche players.

3.2   The Competitive Landscape Changes

In the late 1970s the Japanese became serious competitors. The electronic news-gathering (ENG) craze started (Fig. 3.2). Television news departments discovered that a new tape machine format, the "Umatic," could be coupled with portable hand-held cameras that were becoming available. The predominate camera was the TK-76 from RCA, and the discovered VTR was the VO-3800 from Sony. Sony's broadcast business grew up on supplying ENG gear. The Japanese proved to be much faster to market with new technology than RCA. Although RCA was able to bring large technological leaps to the market every 10 years or so, the Japanese would bring small evolutionary leaps every few years. Another perception that developed between all things American and all things Japanese in the early 1980s is that Japanese products worked out of the box, while American products did not. In 1986 GE bought back RCA and shut down its broadcast operations. Sony is now the closest to filling the void left by RCA in the late 1970s in terms of one-stop shopping. The one area that Sony doesn't cover that RCA did is RF. The other large Japanese competitor is Panasonic. It is interesting to note here that Panasonic has decided to support 720p while Sony has not, at least not yet. This situation is reminiscent of the VHS/Beta wars of the 1980s. Although Beta was slightly better quality than VHS, the market decided on VHS. Sony invented Beta, but eventually found itself building VHS machines.

Figure 3.2   Early ENG camera (TK-76).

3.2.1   Lack of vertical integration

Although RCA could force its broadcasting arm, NBC, to produce shows in color using RCA-supplied equipment to promote the sales of color television receivers, no other entity has had quite that much vertical integration. Three foreign manufacturers come close though: Panasonic, Philips, and Sony. All three manufacture broadcast and consumer television equipment but none has the broadcast distribution channel, namely, a television network, to promote DTV. Sony owns Sony Pictures (formally Columbia Pictures), so it is a content producer with no distribution channel like Disney or Fox. Conversely, neither Disney nor Fox has broadcast or consumer manufacturing operations. Thus, today no entity has leverage over the entire chain to promote DTV.

Broadcasters are being forced toward DTV. To many this means only a DTV transmitter, which makes companies like Zenith, Harris, Acrodyne, and Comark happy. Broadcast equipment vendors, like Sony, Philips, and Tektronix, are trying to move the DTV transformation back to the television station itself by promoting their digital SD and HD solutions. Television set manufacturers are hoping that these vendors succeed, at least in the HD realm, so consumers have compelling reasons to buy DTV receivers. However, the broadcaster is looking at large capital outlays without any payback on the horizon. It is fair to say that while the average broadcaster is not against DTV, the broadcaster just doesn't want to be critically injured by it. Many, though, are hoping for DTV's demise; cable and DBS companies, along with the PC industry, are in that camp.

A rapid roll-out in the United States might help 8-VSB's strength in other countries, and it would help U.S. companies. It would not help DVB's promoters or people who think a third modulation scheme, that is, *coded orthogonal frequency division multiplexing* (COFDM), for DTV in Europe should have been adopted. COFDM is a form of spread spectrum transmission. What does that all mean? Instead of transmitting information 3 bits at a time as 8-VSB does (16-QAM does 4 bits at a time), COFDM transmits many bits at a time but at a slower rate. In essence, each bit is put on its own carrier, all in parallel. Instead of the three 8-VSB bits occupying the whole channel bandwidth, each parallel carrier has a small slice of the bandwidth (the FDM part of COFDM). Because each carrier has a small slice of the overall bandwidth, the rate at which new bits replace the previous bits is slower. Adjacent carriers are 90° out of phase with one another to limit interference with each other (O part of COFDM). The overall bit rate, then, is about the same as the more serial VSB and QAM systems. The advantage to this approach is that errors from multipath are less likely since each bit can be transmitted longer in the parallel than the serial approach.

Most stations are used to selling audience shares and demographics. Now some stations find that they have to sell sets to create an audience. The prime time for selling such sets might not be the traditional prime time. Currently, no HD VCRs are available for stores to feed sets in the store. Many stores are simply feeding SD DVD players into the sets on display. Moreover, consumers aren't even on the same page with Congress. Over 60 percent of people think the key selling point of DTV is more free channels. About one-half think better picture quality will sell DTV. Approximately one-third think DTV's ability to provide Internet content is a selling point. Most viewers claim they don't care about online supplemental information for entertainment programs. Most people who surf the Web view it as a distinctly different activity than viewing television. You usually sit upright when browsing the Web, and it is usually an activity done alone. These are attributes most people don't subscribe to while watching television.

3.3   Receivers Evolve (Fig. 3.3)

A few facts concerning TV sets are in order:

- Customers that see HD are willing to pay a $1200 premium for it.
- Fifteen percent of homes in the United States have home theaters.
- Forty percent of the households that have systems in their homes that cost $1300+ earn less than $20,000 a year.
- Eighteen million households have a TV set that costs more than $2000.
- The average U.S. household owns 2.5 sets, which means there are 250 million sets in this country or about one for every person in the United States.

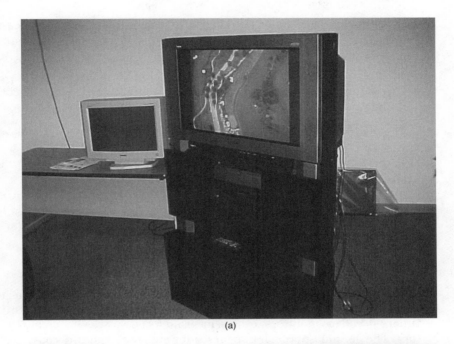

(a)

(b)

Figure 3.3   (*a*) DTV display. (*b*) Two DTV set-top boxes.

- It is interesting to note that even with all the Asian and European company names on the sets that are bought in the United States, half are still produced here. Companies like Sony have large manufacturing and even design operations in the United States.
- Critical mass for DTV sets in the home is said to be 1 million sets. CBS and Thomson think this number will be reached in 2002.

The head of a large station group has stated that the receiver manufacturers can control the ability of the public to actually receive DTV signals off the air. If the manufacturers do not engineer the sets properly, the public might decide that off-air reception is too difficult to cope with and look to cable for DTV distribution. Successful HDTV roll-out requires content and affordable receivers. The PC industry has considered making PC look like a television. Compaq and Gateway tried selling large-screen PCs for the living room, but neither effort did well.

3.3.1 Convergence

Early adopters of DTV will be faced with a confusing choice of options in acquiring a DTV receiver. While some sets are sold as complete integrated tuners-decoders/monitors, others are sold as separate set-top boxes and monitors. The second approach appears safer since the tuner-decoder technology is changing much more rapidly than the displays and compatibility with cable is still evolving. Additionally, a separate monitor would be more likely to be interfaced with other appliances and boxes on the horizon. Displays are evolving also. Sony, Fujitsu, NEC, and Pioneer are offering plasma displays, but they are extremely expensive. Broadcasters can't agree on a transmission format so instead of the TV stations having the expensive box to convert their in-house format to a standard one, each receiver has conversion circuitry to convert any of the 18 DTV formats in use to the set's native display scan.

Convergence means that televisions will start to function like a monitor and PCs will start to have TV functionality. Many members of the PC industry have formed the PC Theater Initiative to determine interconnectivity issues between PCs and monitors. The multimedia "theater" PC, as it is being called, has a fast processor, DVD/CD-ROM drive, modem, TV tuner, USB, and IEEE 1394 (firewire), and is capable of combining PC and TV audio and video. It must have wireless input devices (keyboard, mouse, etc.) so that it can be operated from a couch. This system is a big-screen video-display device with an internal PC, and it will have a single standardized connector between the PC and the monitor. The display will have the functionality of a computer monitor and be the size of a typical television. The monitor will need two different viewing modes—one for graphics and the other for regular television.

3.3.2 Set-top boxes (STBs)

The first generation of STBs lacks conditional access (CA) systems. In a few years broadcasters will be current on the intricacies and billing systems needed for CA, which will make these early boxes obsolete. Most current STBs only handle a narrow set of MPEG2 tools, which limits the types of compressed digital formats they can handle. The set of tools they handle is known as main profile at main level (MP@ML), which means the STB will need to pass any HD it receives on to a television receiver equipped to handle the HD MPEG2 tool set, known as main profile at high level (MP@HL). The interface between the STB and the television is an Institute of Electrical and Electronics Engineers (IEEE) standard known as IEEE 1394 but commonly referred to as "firewire." This high-speed interface might also be needed to send the graphics generated in the STB for program-specific information protocol (PSIP). Some cable systems are planning to buy STBs that pass through 480p, 720p, or 1080i. Cable companies might just pass 8-VSB through to the receiver. Microsoft is trying to position its Windows CE as the operating system for the set-top box.

The FCC has mandated that STBs have separate PCMIA security cards so that consumers can buy their STBs at retail outlets instead of getting them only from their cable company. These PCMIA security cards are being called point of deployment (POD) devices. Use of such devices would allow consumers to get STBs that work with terrestrial or cable modulation schemes and not lock them into their cable system's solution.

There are five possible early evolution paths for STBs:

1. Perform signal processing, decompression, demultiplexing, and security functions.

2. Demod the signal, maybe even both QAM (modulation method used by DVD) and VSB.

3. Demod only DVB, and pass VSB on to the receiver.

4. Detect, demod, and demultiplex the signal while the TV receiver decompresses.

5. Do all of these things and output an analog SD (NTSC) signal for display.

As already mentioned, the DTV receiving system tends to separate receive, demodulation, and decoding from the display. So the display will be either a new HD monitor, a computer-type monitor, some of which can display well over 1000 pixels horizontally, or a set-top box feeding an NTSC receiver much like the VCR does now. Direct-view big-screen displays range in size from 27 to 40 in. Projection televisions range in size from 40 to 80 in. Broadcasters can send pictures that will be displayed one of four ways on the display:

1. 16 × 9 HDTV format—"wide screen"

2. 4 × 3 EDTV—"pan and scan"

3. 4 × 3 movie compress—horizontal "squeeze play"

4. 4 × 3 letter box—reduced vertical resolution

Some networks plan to send DTV media channels along with audio/video programming. Zenith and Philips have developed DTV receiver PC cards.

Commuters could read *The Wall Street Journal* on their laptops. Traffic and train/bus schedules could be broadcast to Palm Pilots. Many are already planning to mix Internet protocol (IP) content into their MPEG streams.

3.4   Fifteen Years of HDTV

Purists used to claim that HDTV should have 30 MHz of bandwidth. This phenomenon is not being demanded by the viewing public, but is being pushed forward by the competitive nature of the business and the fear by many in the business of being "left behind." The late Julius Barnathan, when he was Vice President of Operations & Engineering for ABC, claimed that Japan developed HDTV because the NTSC TV receiver industry was saturated and they wanted a new market.

There were many early proposals for HDTV. One proposal, a system from Farouda Labs, was really just an improvement over NTSC in which an encoding system, which was called Super NTSC, was offered. This system used special adaptive comb filtering to greatly improve chroma handling. Special decoding was not needed at the other end.

The next step in the early HDTV game was a system that NBC proposed. It was called advanced compatible television (ACTV), and was introduced in 1987. This system fit within the 6-MHz bandpass of NTSC. The nuts and bolts of ACTV were that it was a 1050-line per 29.97-frame/s system with an aspect ratio of 16:9. It had a luminance bandwidth of 6.2 MHz and a chrominance bandwidth of 3.75 MHz for the I and 3.25 MHz for the Q subcarrier channel. High-frequency luminance components were inserted in the chroma band (2.0 to 4.2 MHz), producing a 30-Hz complementary flicker that is essentially hidden at normal chroma saturation levels. ACTV used a color subcarrier that was 180° out of phase every other field. Existing NTSC TV receivers would have been able to continue to display a standard 4:3 NTSC picture. The center part of the NTSC-compatible ACTV signal contained the center panel of the original wide-screen image as well as side-panel low frequencies which were time compressed into 1 μs on each side of the active picture. These compressed side panels would be hidden by normal NTSC receivers.

There were also ACTV proposals where a basic system had 410 lines of horizontal resolution and 480 lines of vertical resolution that was fully compatible with NTSC. This was known as ACTV-3. The ACTV people proposed a second "augmentation" channel which would carry the high-definition part of the signal. This second channel would be located in the upper UHF band, would carry additional picture detail, and would provide an ACTV picture with 650 lines of horizontal and 800 lines of vertical resolution. This was known as ACTV-II. ACTV-II would have had twice the chroma bandwidth of ACTV-I or NTSC and a luminance bandwidth of 12.4 MHz. To test the idea of an augmentation channel (which would most likely be on an upper UHF channel) concurrent VHF/UHF broadcast, WUSA in Washington, D.C., conducted a test in 1988 where it presented a simulcast on its normal channel 9 and temporarily on channel 58. The purpose of the test was to check for time and signal strength differences between the two channels. The land mobile interests strongly opposed the augmentation approach because it would effectively double the bandwidth of each television station.

Other compatible HDTV systems were proposed. A 525-line system, with a 6-MHz main channel and another 3-MHz channel for 1125-line HDTV sets, was suggested. A group called the Del Ray group proposed a system where regular NTSC sets would drop 35 lines at the top of the raster and 34 lines at the bottom of the raster; these dropped NTSC lines would carry enhanced video information.

The analog HDTV systems that garnered the most attention were from the Japanese television network NHK, although these systems were not compatible with NTSC sets. NHK now had eight different variations of what it called its "Muse" HDTV system. NHK claimed that these different variations of their basic HDTV system could provide for an evolutionary path toward the full Muse system. Three of these systems fit within the 6-MHz NTSC "window" and were compatible with NTSC sets. The other three Muse variations would have used 9 MHz—a standard 6-MHz NTSC channel and a 3-MHz augmentation channel. These three systems were also compatible with NTSC. A seventh Muse system, known as narrow-Muse, could be transmitted in 6 MHz but required a converter box for NTSC sets. Muse was largely suited for satellite transmissions; its terrestrial tests were highly directional and subject to ghosting.

To cope with all the different competing HDTV/ACTV systems, the industry set up the advanced television test center (ATTC) project to test the various systems. NAB, ABC, CBS, PBS, the Association of Maximum Service Telecasters, and the Association of Independent Television Stations initially funded this project.

Many hope to use HDTV to replace film. As a result, many shooting "video" today are trying to achieve the total film look. Film has a subjective quality that is different than video recorded directly onto videotape. An example of what makes video look like film is the dark areas of a scene. The blacks are much more surpressed in video than in film. Pushing camera gammas (camera transfer function) up to the 60 to 70 percent range helps. For the past 30 years professional television cameras have had circuitry that enhances the edges of objects in the scene, which makes the picture look crisper. This circuitry is called *enhancement* or *detail*. Minimizing camera detail levels by setting the detail threshold at 30 to 40 percent of peak luminance will also help achieve the "film look." This means that areas of low light have little enhancement while highlighted areas have more enhancement. Use of artificial shutters in front of the lens (or different frame update rates through electronic effects devices) to give a 24-frame film rate is yet another technique. The gain of a camera can be increased to give a film emulsion "noise" look. The BBC reduces chroma levels in the camera to get closer to film color-saturation levels. Film also has much less depth of field than video does. Lower lighting levels will create this. Ten years ago Kodak claimed HDTV wouldn't equal 35-mm film quality until CCDs had around 2 million pixels. Today, HD camera CCDs have that level. Ten years ago CCDs had 360,000 (500 lines times 720 horizontal samples) pixels. Today's SD cameras have 600,000 pixels.

Early HD equipment cost 10 times what today's equipment costs. Some early producers of HD programming find that they cannot compete with the new start-up companies who use drastically cheaper, lighter, and less power-hungry equipment. However, the device used to capture what the camera shoots, the CCD imager, has been made smaller in newer cameras. It was reduced to two-thirds its previous size. Many claim the newer HD cameras don't have the same resolution that the older ones had because the same quality lens cannot be used for the smaller imager as for the larger imager in older cameras. This degradation in resolution is generally not due to the cameras themselves, except that smaller imagers use inferior lenses.

Early HD VTRs used bulky reel-to-reel systems with much higher writing speeds, which allowed the video to be recorded without compression. Today's HD VTR is much more compact, and is generally based on tape transports used by their SD VTR cousins. However, the writing speed is lower than that of earlier HD VTRs. Compression is needed to fit HD onto the tape. Some HD VTRs don't even use all the pixels for compression, resulting in a loss of some resolution. On the plus side, though, it cuts down on the amount of compression needed. Many believe that older HD VTRs produce slightly better pictures than the newer models. This leads to the interesting fact that not all HD video with the same 1080i flavor is equal. This inconsistency doesn't affect other HD formats because until recently they didn't exist physically. However, these other formats face the same lens and compression issues that the 1080 does.

At this time, the Europeans aren't pursuing HD broadcasts much. To remain competitive in production, however, Europe will eventually have to adopt HD. The Europeans have also elected to use a totally different transmission scheme—COFDM modulation instead of 8-VSB, which is used in the United States. Many still want to revisit the COFDM versus 8-VSB issue.

To compete in the next century, television organizations will not only need the technical capability but also access to lots of capital and an entrepreneurial drive. Well, that's the freshmen's version of the current state of the television universe. Now that we've touched on why we have the mix of technologies we do, we will move on to look under the hood to see how these components work and come together to create the required infrastructure for the start of this century.

3.5   Trivial Pursuit

- The average American spends one-fourth of his or her time awake watching television.

- Households with DTV at the turn of the century will probably be less than 2 percent.

- The average viewer in the United States expects to pay only $500 for her or his next television set.
- Although consumers have been willing to replace obsolete PCs every few years, it remains to be seen if they will replace televisions and VCRs with the same frequency.
- RCA claims that choice of channels is more compelling to the consumer than high-definition television.
- The cost of early HD receivers will be $5000 to $10,000.
- DTV plug-in boards for computers are now available for well under $1000.
- Thomson thinks DTV set sales will reach 500,000 next year and 2 million by 2002.
- Twenty-five to 30 million conventional television sets are currently sold each year.
- Only a few percent of early adopters are willing to pay over $4000 for a DTV set.
- For more than 50 percent of viewers to consider a DTV set the price must be below $2000.
- Thirty-nine percent of homes have three or more television sets.
- The current receiver manufacturers are Sony, Samsung, RCA, Toshiba, ProScan, Hitachi, Philips, Sharp, Zenith, and Toshiba.
- The inventory of TVs turns over every 8 years.
- Fifty percent of computer owners consider their systems more as entertainment devices than work tools.
- A color TV which cost $500 in 1955 would cost $2500 in today's dollars.
- The average price of a TV today is $325, and it has drastically better video and sound quality.
- Seventy percent of families who set up a home theater spend more time together.
- Eighteen million households (maybe one-fifth) consider themselves early adopters of DTV.
- Thirteen and one-half million households in the United States have a home-theater system.
- Twenty-five percent of all color TVs sold cost over $1300.
- People age 50 and over who own a PC spend 11 h/week working on their computers and 18 h/week watching TV.
- The typical adult consumer watches 20 h of TV a week.
- Forty-seven percent of consumers interested in a TV/PC would expect to pay at least $3000.

- One in four households buys a television in the United States every year. That's 25 million sets.

- Consumers interested in a TV/PC think they would use it for the following: 77 percent would watch TV programming, 73 percent would watch prerecorded videos, 63 percent would record TV programming, 59 percent would use it for word processing, 50 percent would browse the Internet, and 43 percent would send and receive mail.

- In 1961 FCC chairman Newton Minow stated that television is "a vast wasteland."

- Fred Allen, a radio comedian during the transition to TV, quipped: "Why do we call TV a medium? Because it is never well done."

3.6 Bibliography

Beacham, Frank, "Compaq Rejects DTV Standard," *TV Technology,* April 24, 1997, p. 44.
Burger, James M., and Todd Gray, "A DTV Regulatory Primer," *TV Technology,* November 1998, p. 40.
"Cable Channels Stake Their Claims," *Broadcasting & Cable,* January 26, 1998, p. 48.
Forrest, John R., "Convergence in the Industry," *SMPTE Journal,* August 1997, p. 530.
Fredrick, John, "The Approaching Convergence of PC-TV," *Electronic Design,* February 9, 1998, p. 96.
Hoffner, Randy, "TV Standards of the World," *TV Technology,* February 9, 1996, p. 26.
Kapler, Robert, "Cable Has No Space for Digital," *TV Technology,* May 18, 1998, p. 10.
Leopold, George, and Junko Yoshida, "PC Makers Try to Fill DTV Void," *Electronic Engineering Times,* March 24, 1997, p. 1.
Macmillan, Jeffrey, "Folks America Loves to Hate," *U.S. News and World Report,* May 13, 1996, p. 37.
McClellan, Steve, "Can the Big 4 Still Make Big Bucks?" *Broadcasting & Cable,* June 8, 1998, p. 24.
Parshall, Gerald, "The Prophets of Pop Culture," *U.S. News and World Report,* June 1, 1998, p. 56.
Penhume, James, "Digital TV: Where's the Consumer?" *Broadcast Engineering,* December 1998, p. 76.
Schubin, Mark, "HDTV Is Here at Last (Maybe)," *Videography,* December 1998, p. 59.
Silbergleid, Michael, "DTV Era Brings Greater Certainty," *TV Technology,* November 1998, p. 62.
Strassberg, Dan, "HDTV: The Great Picture Isn't the Only Picture," *EDN,* December 17, 1998, p. 48.
West, Don, "Convergence the Hard Way," *Broadcasting & Cable,* April 9, 1997, p. 6.
Yang, Catherine, Neil Gross, and Richard Siklos, "Digital D-Day," *Business Week,* October 26, 1998, p. 144.
Yoshida, Junko, and George Leopold, "Cable Not Ready to Handle Digital TV," *Electronic Engineering Times,* August 31, 1998.
Zdepski, Joel, "Interactivity in TV's Digital Age," *Broadcast Engineering,* December 1998, p. 83.

Going Digital

4.1 Domain Changes

Most video and audio sources do not start out as digital streams. Traditional acquisition devices are analog in nature. Microphones and video cameras have been, and will continue to be, analog in nature. Sound pressure waves create current changes in inductive-type microphones and voltage changes in condenser microphones. These signals are continuous in nature, that is, the changing voltage (or current) has no discontinuities or gaps in its waveform over time. Even with today's "digital" cameras, the devices that capture light and convert it to voltage, the *charged coupled device* (CCD) is still analog. More information on CCDs will be presented in Chap. 9. The CCD is a different type of analog device compared to the microphone. While the microphone is continuous in nature, the CCD is considered discrete (Fig. 4.1).

Over time an analog discrete device develops steps—and even gaps—in its waveform. Each cell in a CCD captures a small amount of light that lands on it from the camera's lens. These cells form an array, and the voltages generated in each cell are marched out in turn. Each cell contains a spatial sample voltage generated by the light that struck the cell. Samples will be discussed in more detail a little later in the chapter.

4.1.1 Composite versus component

Video comes in two flavors—composite and component. In *composite video* the luminance (black-and-white) and the chrominance (chroma or color) information are combined. A color subcarrier is superimposed on the luminance voltage to add the color information. The disadvantage of this process is that luminance and chrominance information mix together, that is, luminance information which happens to have a frequency close to the chroma is interpreted as chroma by the television display. Also, many devices throw out the high-frequency luminance information to facilitate luminance and chroma

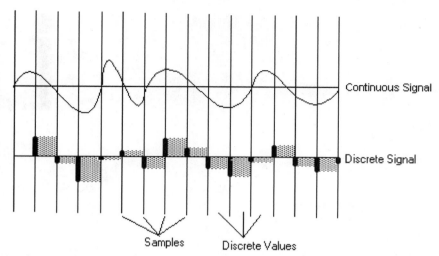

Figure 4.1 Continuous voltage, once sampled, becomes a discrete signal.

separation, resulting in lower resolution. *Component video* was introduced to eliminate these problems. Component video simply keeps the chroma information separate from the luminance information.

4.1.2 Composite encoding

The process of combining the luminance and chroma information into the composite NTSC, or phase alternate line (PAL), signal is called *encoding*. NTSC encoding will be briefly described here. The red, green, and blue signals are combined in separate matrices to form I (in-phase) and Q (quadrature or 90°) signals. These two signals amplitude-modulate a commonly generated 3.58-MHz subcarrier, one of which is 90° behind the other.

The I signal consists of 28 percent of the green signal available (inverted), 6 percent of the red signal available, and 32 percent of the blue signal available (inverted). The Q signal consists of 52 percent of the green signal available (inverted), 21 percent of the red signal available, and 31 percent of the blue signal available.

If these signals are plotted as a vector, their resultants fall at 123° and 33° on the NTSC color wheel, or vectorscope. The I axis falls in the orange/cyan region, while the Q axis falls in the green/magenta region. The I and Q comprise the color, or chroma, information. PAL refers to these two signals as U and V. The gray-scale or luminance information is made up of 59 percent green channel video, 30 percent red channel video, and 11 percent blue channel video. On a waveform monitor the IRE or low-pass mode will show luminance information only, while the high-pass mode will show mostly chroma information.

The human eye sees light in the yellow-green region as twice as bright as red light and almost 6 times brighter than blue. Location (vectorally) of the I and

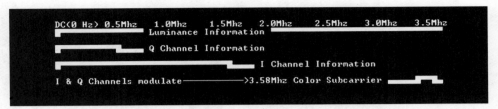

Figure 4.2 For NTSC the luminance bandwidth is much greater than the chroma bandwidth. The two baseband chroma signals that modulate the chroma subcarrier are of different bandwidths, also.

Q axes is based on the eye's inability to detect colors in a scene as the image area decreases and because it has a nonlinear brightness response (Fig. 4.2). As a result, the I and Q channel bandwidths were selected to take advantage of the eye/brain image size versus color-detection characteristic. Q channel attenuation begins at 0.4 MHz and falls rapidly beyond 0.6 MHz. I channel attenuation begins at 1.3 MHz and falls rapidly beyond 1.5 MHz. Therefore, large-size color areas, represented by transmitted frequencies between 0 and 0.5 MHz, are within the bandwidths of both color channels, so true color is displayed on the receiving monitor. Medium-size color areas ($^3/_8$ in on a 21-in screen) are represented by transmitted frequencies between 0.5 and 1.5 MHz. Areas this size can only be perceived as orange, or cyan. Based on this characteristic, they are transmitted only on the 1.5-MHz-wide I (orange-cyan axis) channel. Small-size color areas of $^1/_8$ in square, or smaller, which are represented by frequencies above 1.5 MHz and fall outside the bandwidths of both color channels, are therefore transmitted only as black-and-white detail (Fig. 4.2).

The two major composite standards are NTSC (Fig. 4.3), which is used in the United States, Canada, and Japan, and PAL, which is used in the United Kingdom. In addition to the number of horizontal lines and field rate (which is usually a function of local power frequencies), PAL differs from NTSC in how the embedded chroma is handled. PAL has a higher subcarrier frequency than NTSC, which allows for a higher luminance resolution on most displays. However, the major difference is that the reference color burst is alternated by 90° every other line. This allows the PAL television receiver to correct any phase errors between the color reference and color information in the picture.

4.1.3   Different flavors of component video

Component video can be implemented in a number of ways, the same as for composite. The three implementations of component video are

1. *RGB.*   Red, green, and blue video are sent on separate paths. This format is what most PCs output to computer monitors. Most color cameras output RGB for use in chroma keying. The chroma key is performed by most production switchers or, in some cases, by stand-alone boxes. The chroma keyer is adjusted to create a key signal when one or more of the three color video levels are above the desired threshold. Blue is the most common color used but many facilities also use green. If blue is used as an example, the chroma keyer will

Figure 4.3 Here a composite color-bar test signal can be shown simultaneously on a monitor and a waveform monitor. The waveform monitor indicates both luminance brightness (amplitude) and color information via the carrier frequency superimposed on the luminance stairstep pattern.

replace video above a selected threshold in the blue channel. The original video could be of a person standing in front of a blue screen, and the video that is replacing the high blue areas of the picture could be a weather map, for example.

2. *Y, R-Y, B-Y.* This format is the most common professional television component format. It is also used by most professional analog VTRs. Here a Y or luminance signal is generated from mixing red, green, and blue together. The most common recipe is 30 percent red, 59 percent green, and 11 percent blue. R-Y and B-Y are known as the *color difference signals*. These two values, like I and Q in NTSC, represent two voltages which are 90° apart in phase. When vectorally added together, they present the common vector display on a vectorscope. Similar versions of Y, R-Y, and B-Y are used in Sony's Betacam and Panasonic's MII VTRs. Why the adjective "similar" was used in the last sentence will be explained shortly.

3. *YC.* In this format the chroma signals have already been multiplied together and simply kept separate from the luminance. On the Umatic VTR the dub mode performs this function. S-VHS uses the YC approach throughout. The Umatic, the old consumer Beta, VHS, and S-VHS all use a chroma modulation process called "color under," or heterodyne color (Figs. 4.4 and 4.5).

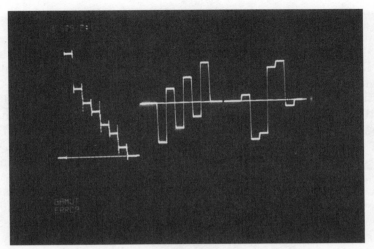

Figure 4.4 Here are the three signals that make up a component color-bar signal in the Y, R-Y, B-Y format. These are all baseband signals that are not used to modulate a chroma carrier (unless the component signal is converted to composite).

Figure 4.5 The drawback to analog component is that three paths are required. Here is a component patch panel. Each patch needs to make three separate connections. Digital component solves this problem by time multiplexing the three signal components.

4.1.4 Color under or heterodyne color

The worn-out retort to "color under" is that it must be from Australia. In actuality the term is derived from the way the color information is laid down on tape. Remember the brief explanation of FM and AM in Chap. 1? Well, the tape formats just mentioned laid luminance information on tape as an FM-type signal with a carrier at a frequency of a few megahertz. The reason FM is used is that the magnetic properties of tape are not linear, so either a high-bias frequency is used and the signal amplitude modulates it (AM) or FM is used. FM is impervious to amplitude linearity problems. Historically, analog VTRs have used the bias method for audio and FM for video. The color information is laid on the tape with the higher luminance information but at a frequency of under 1 MHz. The color signal is under the luminance signal in frequency. While this was a simple solution for inexpensive VTRs, it came at a high price.

The color-under carrier put on the tape was much lower in frequency than the original color subcarrier, which meant much lower color resolution. In addition, since the color information essentially used the bias technique just mentioned, it was affected by nonlinearity and noise. After a few generations (copies of copies) of VHS, the chroma looks hollow and is extremely noisy. A lot of chroma jitter is also apparent. The jitter is caused by a benefit that color under has over composite or component recording. Component and composite VTRs require time-base correctors (TBCs), which used to be external boxes but now are normally built into the VTR.

VTRs are electromechanical devices. Rotating scanners have heads that diagonally traverse the tape, recovering video information in the process. A capstan system pulls the tape using these rotating heads. Scanner servo systems work at keeping the scanner rotating at the right speed and phase. However, these systems are not perfect. The scanner "dithers" around the correct frequency; it continuously speeds up or slows down ever so slightly. This slight speedup or slowdown is known as *velocity error*. The velocity error translates into a slight frequency error in the video or tape. The luminance information error will not greatly change what is seen on most video monitors or television receivers. However, the color information in a composite signal is based on the phase of the color subcarrier, and that subcarrier frequency can be changing by many cycles in frequency. Common velocity errors are between 1 and 5 μs in today's VTRs. That rate will put the chroma phase all over the vector map, rendering useless chroma information. The TBC will remedy this problem by taking video with velocity errors and producing video that is locked to a facilities master reference. TBCs are also needed so that the VTR playback video is in sync with other sources and can be dissolved or keyed over.

Color-under systems bypass the need for a TBC for color lock by using a heterodyne process. The color-under carrier is used to produce a second carrier that is exactly above the original color subcarrier by the frequency of the color-under carrier. For Umatic VTRs the color-under carrier was 688 kHz, and for NTSC color the subcarrier is 3.58 MHz. Thus, the second carrier would be 3.58 MHz + 0.688 MHz = 4.27 MHz. This carrier had the same errors as the color-under carrier. Now if this second carrier had been heterodyned with the first color-under carrier, the difference would be a stable 3.58-MHz color carrier like the original color subcarrier recorded. This is why a VHS machine can be viewed without the need for a TBC. However, the video from a VHS machine will not be locked to the "house," that is, all the other equipment in use. The reason that component VTRs, which don't have the color subcarrier but baseband color-difference signals, need a TBC is that tape velocity errors will result in color-difference signals with incorrect amplitudes or values, resulting in the creation of incorrect colors.

4.1.5 Video levels

Analog levels, both video and audio, have caused much confusion over the years. This confusion has existed since the beginning of television. Original television equipment output video was known as *noncomposite*. No, noncomposite is not component; it is color video that is not quite composite. The term "composite" arose with the advent of color television. Before that time, video was simply known as monochrome. Early television equipment was big and bulky. Every subsystem added to the cost and size of a particular piece of equipment. As part of the EIA (Electronic Industries Association) RS-170 standard, video signals had to have synchronization pulses as part of the signal. From an engineering standpoint it didn't make sense for each individual piece of equipment to have its own circuitry for generating those pulses. As a result, a facility had one master "sync" generator that provided the sync pulses where needed. Video sources straight from cameras tended not to have sync pulses at all; sync pulses were added in the switcher.

These video sources without sync pulses had peak white levels of 1 V while the sync pulses distributed separately had 4-V levels. When both were combined, the resultant video signal was 1 V of video and 0.4 V (one-tenth the original level) of sync pulse. The total value was 1.4 V. If you were to look at the graticule on a waveform monitor, you would now know where the 100-unit video and 40-unit sync scales come from. This 1.4-V level was eventually reduced to 1 V but the 100/40 ratio was maintained. Thus, peak video is now 0.714 V and sync is 0.286 V. The IRE [Institute of Radio Engineers, now known as Institute of Electrical and Electronics Engineers (IEEE)] developed the IRE unit which is one-hundredth of 0.714 V. Europe shied away from these difficult values by rounding these values to 0.7 and 0.3 V. When color was added to the monochrome signal, the term "noncomposite" was coined to denote video with embedded color but no sync or color burst. It wasn't until the late 1970s when companies like RCA developed single-chip sync generators that individual pieces of equipment began to output completely composite signals.

Another video-level confusion factor is "black level setup," which is also known as "pedestal." To prevent early receivers from producing retrace lines, as the television's scanning circuitry performed a retrace to start another line or field, the darkest portions of the video were slightly higher than blanking. The value settled on was 7.5 IRE units. This vestigial practice continues today, which effectively reduces the video amplitude range by 7.5 percent. However, pedestal can be adjusted by an operator just as iris controls peak white levels. The pedestal should be at 7.5 percent but is not always set at that level. As a result, not enough reference black level is present. The United States continues to use setup or pedestal for component signals, whereas most other countries do not. Thus, most VTRs have controls to add or remove setup. To add to the confusion Betacam and MII component VTRs have slightly different peak video and setup values for luminance. The color-difference levels are also different. However, peak values all remain at 1 V. The MII system tends to be closer to the SMPTE/EBU N10 component standard used in Europe than the traditional NTSC values that have been applied to the luminance component channel used in the United States.

4.1.6 Audio levels

Analog audio levels have changed over time as well. The standard reference level was developed by the telephone company, and is known as 0 dBm. The measurement 0 dB is defined as 0.775 V into 600 Ω, resulting in a power level of 1 MW. The analog audio reference used by the broadcast industry is 0 VU. The unit 0 VU has been defined over time and by various manufacturers as +3, +6, or +8 dBm. In addition, a 600-Ω load impedance is not always common. Microphones are often only 110 Ω, and some equipment has a high-impedance input (10 kΩ). As load impedance changes, power levels change as well. A common practice today is to run low-impedance audio into devices set for high impedance, which means that the source impedance is at the traditional 110/600 Ω at the output of a device and the input of the receiving device is at 10 kΩ or higher. Although this configuration results in reflections, runs are short for audio wavelengths, so they can be ignored. There are two benefits however. The signal-to-noise ratio drops by one-half because the received signal is twice that of the terminated amplitude. Therefore, the drive level can be dropped by 3 dB, with a corresponding drop in the signal-to-noise ratio. In the past power was needed to drive audio devices. At that time, impedance matching was important to maximize power transfer. Now voltage, but little current, is needed. Therefore, the maximum applied voltage is desirable while impedance matching is not as important. The second benefit of this approach is that the audio patch panels can be wired in what is known as a half-normal configuration, that is, if a patch chord is inserted in the top jack in the patch field, the path isn't broken—the patch is just bridged across the source, that is, audio monitoring can occur without interrupting the feed. If the patch is inserted in the lower jack, the signal from the top jack is interrupted. The audio monitoring feature is only possible because of the low-feed Z, high-receive Z approach.

4.1.7 Encoding/decoding

Now let's return to composite versus component video. The process of going from composite to component video is known as *decoding,* and the process of going in the opposite direction is known as *encoding*. Decoding is much tougher than encoding. Good decoders are expensive. The best have three-dimensional adaptive comb filters. A comb filter uses the repetitive nature of chroma to extract the chroma. Chroma has the same frequency every horizontal line, and it is also 180° out of phase from one line to the next. A comb filter works in the time domain to extract the chroma (most non-DSP filters work in the frequency domain; see Chap. 8). A one-dimensional filter works only in the horizontal direction (one horizontal line), a two-dimensional filter works over two or three horizontal lines in the vertical direction, and a three-dimensional filter works over multiple fields, or over time. "Adaptive" means that the filter switches between these three modes based on video content.

4.1.8   Analog/digital conversions

The other set of domain changes involves going from analog to digital. As mentioned earlier, both continuous and discrete analog signals can exist. To transform an analog signal to the digital domain it must be discrete in nature. An analog-to-digital (A/D) converter will do that with the use of a sample-and-hold circuit. This circuit lets a sample of the incoming signal through to charge a capacitor at a fixed rate. The sample is let through for a short amount of time. The voltage accumulated on the capacitor is then held at that level until the next sample time. Additional circuitry then converts that voltage into a binary number before the next sample is let in.

Encoding/decoding degrades signals much faster than A/D or digital-to-analog (D/A) conversions. Encoding/decoding can occur in the analog or digital domain. Degradation occurs on the digital side just as it does on the analog side when encoding/decoding. Component VTRs, both analog and digital, are now the most common cause of these domain changes. If a signal starts as NTSC, and does not have to undergo a lot of processing, it is better to keep it in NTSC if it is going to be broadcast as NTSC.

The A/D process is more complex than D/A. More A/Ds are generally required in a plant than D/As. A/Ds are generally locked to the input, unless encoding/decoding is required or has already been done. D/As should be "gen-locked" to the house (locked to the facilities reference sync generator).

4.1.9   Composite digital

Composite and component video can both be found in the digital domain, although composite is on its way out. Early digital VTRs were much cheaper to implement as composite than component because the resulting composite bit rates were lower than component and the composite VTR didn't have to make provisions for decoding NTSC video. With the advances in compression technology, affordable component digital VTRs are now available. Composite digital streams are created by taking composite analog signals and sampling the signal every 67.5 ns, or four times per color subcarrier cycle. This is known as 4Fsc (4 times subcarrier). This process creates approximately 14.4 million samples/s, and is simple and fairly inexpensive. The drawback is that all the sins of analog composite were carried over into the digital domain. Early VTRs created 8-bit samples; now all commonly create 10-bit samples. Early on these samples were transported as parallel bits, but the cable used was big and unwieldy. Parallel digital used 25-pin D connectors, and distances of up to 150 ft (50 m) were allowed. Beyond that, the accompanying clock would begin to skew in time. The cables were bulky and expensive to implement. As a result, the parallel bits from each sample are now lined up one after the other and sent serially. At 10 bits the component serial digital interface (SDI) is 144 Mbits/s.

4.1.10   Component digital and chroma subsampling

Component digital video is more complex. Looking back at the explanation of an encoder in this chapter, we see that color information is less than luminance information, so fewer color samples than luminance samples are taken. In fact, one-half fewer chroma R-Y and B-Y samples are taken. For every four luminance samples two B-Y and two R-Y samples are taken. This process is known as color *subsampling*. The ratio of samples is used to describe the resulting data stream, for example, 4:2:2. If the B-Y and R-Y samples are taken at the same time, they are said to be cosited, but only at every other luminance sample. This sample method is known as SMPTE standard SMPTE 125M, which will be described in more detail in Chap. 6. Other chroma subsampling rates are used. Two common ratios are 4:1:1 and 4:2:0. The ratio 4:1:1 is straightforward; there are four luminance samples for every one chroma set of samples. The ratio 4:2:0 is a bit harder to visualize. In this ratio 4:2:2 sampling occurs on alternate lines, with no chroma sampling on the other lines, just luminance sampling. Thus, the ratio 4:2:2 subsamples chroma in only the horizontal direction, while 4:2:0 subsamples chroma in two directions—horizontal and vertical. These other subsampling methods are found only in compressed video bit streams or as internal formats inside equipment. Only 4:2:2 is used as a non-compressed component digital bit stream between equipment.

With 4:2:2, 720 luminance samples of active video are taken per line, which equates to a sample rate of 13.5 million/s. Each chroma-difference channel has one-half that number, or 6.75 million samples/s. Therefore, B-Y and R-Y together have 13.5 million samples/s. The total number of samples per second is 27 million, which means that if each 27-million-sample/s contained 10 bits, the 4:2:2 serial bit rate would be 270 Mbits/s. That's almost twice the digital-composite (4Fsc) serial bit rate. The SD component digital sampling frequency is the same for the Japanese, Europeans, and Americans—13.5 MHz—and it works in both the 525 and 625 worlds. The same is true for chroma sampling. Sampling rates for HD components, 720P and 1080i, are the same. For component digital no samples are taken in the blanking intervals. Composite sampling samples everything. In Chap. 5 we will look at SD versus HD sample rates. As a point of reference here, many in the industry loosely refer to 4:2:2 (SMPTE 259M) as D1 and 4Fsc (SMPTE 244M) as D2. The reason is because the

SMPTE standard that covers the format at which VTRs record 4:2:2 is called D1 (SMPTE 224-227) and the standard for composite VTRs is referred to as D2 (SMPTE 245-248).

Another sampling scheme is used for keying. It is known as 4:2:2:4, which means 4:2:2 for luminance and color difference and 4:0:0 for the key signal. The resulting multiplexed signal is 40.5 MHz. This is only used internally in devices such as switchers.

The encoding/decoding in the digital domain involves sample-rate conversion. The hardest, and most expensive, conversion process is analog composite to 4:2:2. Next is 4:2:2 to NTSC, and then RGB to 4:2:2. The easiest conversion process is NTSC to 4Fsc.

4.2   Sampling

By looking back at Fig. 4.1, we can see that sampling takes a continuous signal and breaks it into discrete samples. A number of combined issues determine the success of the resultant value from the sample process.

4.2.1   Sampling rate

Remember our picket-fence resolution example from Chap. 1? In order to delineate separate "pickets" along the horizontal length of a fence, we needed to see the space between each one. So if we now want to turn our analog shot of a fence into a digital picture, we will have to have a sample rate that is high enough to capture both the picket and the space between each picket. If we only sample enough times so that we capture only the pickets, our shot of a fence will look like a solid wall, or we will capture only the spaces and see no fence at all. More likely, though, our samples will catch some pickets and some spaces. Our fence will look like a wall with gaps in it. This effect is called *aliasing*. In order to faithfully reproduce our fence digitally, we must sample at a high enough rate to sample each picket and the space between each. We must sample two times for each picket.

Thus, when designing a system to turn analog signals into digital signals, the highest "picket" rate must be known. The picket and its corresponding space would look like a squarewave on a waveform monitor. Obviously, pictures are available without picket fences, but everything that is shot is comprised of groups of frequencies that are summed to describe the object. The higher frequencies represent edges and transitions between objects in a picture, or features within a single object. The high frequencies have to be captured when we convert the analog picture to a digital one for the picture to look as sharp as possible. Therefore, the sample should be done at a rate at least twice the highest desired frequency in the signal. This is known as the *Nyquist rate*, and it can be examined mathematically. As the analog signal is sampled, but before it is converted to digital, in essence a carrier frequency is created at the sample rate. This carrier is amplitude-modulated by the picture frequency components in that sample. Just like a normal AM signal, those modulating picture components end up in the sidebands of the carrier. If the carrier is not at least twice the highest sampled frequency, the sampled baseband frequencies will overlap and remain with the lower sidebands of our carrier. This beat results in aliasing (Fig. 4.6).

Everything involving engineering revolves around trade-offs. We can't simply sample at infinitely high rates because the A/D circuitry gets too expensive and the data stream created becomes extremely large. So we must decide on the highest allowable frequency in the picture. Then a low-pass filter must be installed before the sample circuitry that prevents frequencies that are higher than one-half the sample rate from getting through. In the case of SD component video at 720 luminance samples/line, the highest number of "pickets" in a fence across the entire length of the scene would be 360. It would take a bandwidth of 4.5 MHz to create that fence.

In some cases, such as when converting analog audio to digital, samples at rates higher than the Nyquist rate might be required. The higher the sample, the lower the noise floor of the signal once in the digital domain. Lower sample rates cause ambiguity as to the correct value of the low-order bits. If this ambiguity is noncorrelated, then it appears as noise. If it is correlated, then it appears as distortion. Higher sampling means less noise. How much less? Oversampling at 2 times the Nyquist rate will result in a 6-dB drop in the noise floor. Oversampling at 4 times will result in a 12-dB drop, and so on. Thus, oversampling allows for a higher sampling resolution. An additional advantage of oversampling is that sampling above the Nyquist rate allows for much gentler low-pass filtering to prevent aliasing, resulting in many fewer phase and peaking problems.

4.2.2 Sampling resolution or quantizing

Sample resolution is the other important item to keep in mind when going from the analog to the digital domain. Not only do sampling rates give digital audio and video a noise floor, but so does sampling resolution. Originally, digital video had a sample resolution of 8 bits. The term "bit" probably came from commodity trading where the lowest denomination was one-eighth of a dollar. Another explanation is that it stands for *bi*nary dig*it*. Every sample was converted into a set of eight high or low lines out of those converted. These eight lines (referred to as a byte) could depict one of 256 states:

00000000 = 0	00000111 = 7
00000001 = 1	00001000 = 8
00000010 = 2	00001001 = 9
00000011 = 3	00001010 = 10
00000100 = 4	⋮
00000101 = 5	11111110 = 254
00000110 = 6	11111111 = 255

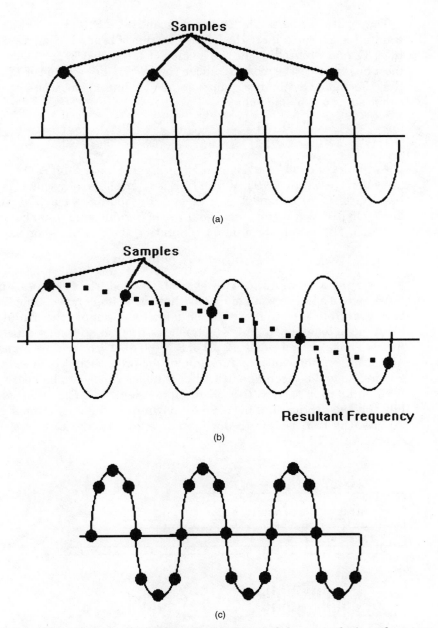

Figure 4.6 (*a*) shows what happens if a signal is sampled once a cycle. A steady-state dc value is generated. (*b*) demonstrates that sampling more than once per cycle, but less than twice per sample, generates a low-frequency alias signal. (*c*) shows sampling at higher than the Nyquist rate.

The rightmost bit is the least significant bit (LSB), and it changes the total count by 1. The next bit to the left has a value of 2, and it changes the total count by 2. The next bit to the left has a value of 4, the next 8, 16, 32, 64, and the leftmost bit, which is the most significant bit (MSB), has a value of 128. This means that the analog luminance component, which has a peak white value of 0.714 V, can assume one of 256 values:

$$\frac{0.714 \text{ V}}{256} = 4.79 \text{ mV quantizing resolution}$$

As analog video is converted to digital values, our 8-bit A/D can't differentiate values less than 4.79 mV. The LSB is often the wrong value. This error acts like noise. One bit out of 8 is essentially adding noise. The voltage log of 256 = 48 dB. As a result, the signal-to-noise value of an 8-bit sampling system is 48 dB. The simple formula for computing the S/N based on sampled bits is

$$\text{S/N} = 6 \times \text{number of bits}$$

This formula shows that, with every bit added, the S/N goes up by 6 dB. A 6-dB increase in the voltage means that the voltage has doubled. So it follows that every bit of sample resolution added will double the sample resolution. Most video today is 10 bits, which means its resolution is 4 times better than 8 bits. That's an additional 12 dB of S/N (60 dB). While 10-bit samples are usually sufficient for video, audio samples for professional television are usually 20 bits. As will be seen later, this limits noise, which can be heard much easier with audio than seen in the video. That means that for audio sampling at 16 bits the least significant bit is 96 dB down from the maximum signal level. At 20 bits, the LSB is 120 dB down, and at 24 bits, the LSB is a mere 144 dB down (Fig. 4.7)!

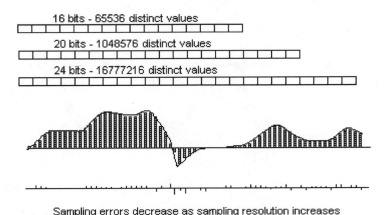

Sampling errors decrease as sampling resolution increases

Figure 4.7 More bits equate to fewer sampling errors. Since the errors are often random, they appear as noise.

4.2.3   Nyquist and the *Z* transform

Designers of digital circuitry typically use a few mathematical tricks to aid in design. It is easier to design in the frequency domain than in the time domain. We will look at this in greater depth in Chap. 8. A set of tools used to ensure that a system meets the Nyquist rules is called the *Z transform*. This transform allows sets of equations to be developed that describe a digital system's transfer function:

$$\text{Transfer function} = \frac{\text{input}}{\text{output}}$$

A *transfer function* is an equation that illustrates or provides a ratio of what will come out of a system versus what was fed in. The transfer function changes over frequency, that is, over the entire spectrum, there will be times when the function equals zero and times when it has infinite values. The *Z* transform can also be used to graphically display how a digital system will react to the frequencies that the digital samples represent. That is what we will look at here. Just as a vectorscope rolls time into a circle, *Z*-transform techniques allow the same type of graphic to be produced. Instead of the circle representing time, it represents frequency (Fig. 4.8).

The point marked 0 on the circle in Fig. 4.8 represents zero frequency. By going counterclockwise around the circle, the frequency increases. Halfway around the circle represents the Nyquist rate. Any response on the circle after

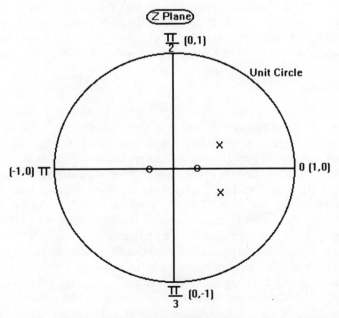

Figure 4.8   *Z* plane used for displaying response via *Z* transforms.

that point will result in aliasing. The relative response is represented by the distance away from the center of the circle. After having gone counterclockwise all the way around the circle, the sampling frequency is reached. Theoretically, it is possible to continue around the circle again, going from the sample rate to twice the sample rate. This process can continue indefinitely but it serves no purpose because the response will be the same with each revolution around the circle. The important thing about the Z-transform display is that it demonstrates that, in digital design, everything is tied to the sample rate. At the most basic level, a designer of a digital processing system doesn't care if the sampling rate is 10 or 100 MHz (as long as the actual circuitry can handle the clock rates). The desired response is relative only to the sample rate.

The 0s and the Xs represent zeros and poles, respectively. A *zero* is where the transfer function equals zero. Zeros tend to make the response break upward. A *pole* is where the transfer function is infinite. Poles tend to make the response break downward. The number of poles and zeros will vary, based on the response desired. Zeros can never outnumber poles or the system will be unstable. They can be on either axis or off the axis, as shown for the poles. When off the axis, they represent complex numbers, which are numbers that not only have magnitude but also a direction, or angle. On the Z plane they have been reduced (through a trigonometric process) to having an X and Y component (also known as the real and imaginary part). Complex numbers are found in Laplace transforms as well (more on this topic will be presented in Chap. 8). Complex numbers in electronics are a result of the fact that it is not only the magnitude that is important but also the phase. The actual response at any point on the circle can be graphically calculated by adding the lengths of all the vectors from each zero to the point on the circle, and then subtracting the lengths of the vectors from each pole to the same spot on the circle.

4.2.4 Signal-to-noise ratios

As mentioned earlier, quantizing errors caused by the A/D process can appear as noise. However, if the errors are correlated with the video or audio being sampled, it will appear as distortion and not noise. In addition, aliasing will result if this distortion happens to be higher than one-half the sampling rate. One way to eliminate the correlated noise from quantizing errors is to introduce dither. Dither decorrelates the errors from the signal, but in the process adds noise. However, the benefit of dither decorrelation can be defeated. For example, if a 20-bit audio signal is passed through a 16-bit box, the least significant bits will be lost along with the dither (Fig. 4.9).

As already discussed, electronics signals can be viewed in either the time or the frequency domain. The same is true with noise. Up to now, noise has been discussed only in the time domain. Occasionally, though, S/N values are given for digital signals. These should not be confused with the noise sampled into the digital signal. How can a digital signal have an S/N value? After all, it's either a

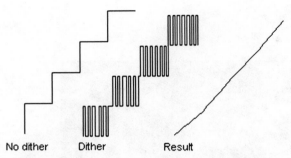

No dither   Dither   Result

Figure 4.9   Dither tends to smooth out sampling errors.

high or a low. This situation is not necessarily the case, as will be seen in Sec. 4.3. One way to determine the distance of a given digital bit stream from the error cliff is to determine the signal-to-noise ratio of the third harmonic of the SDI bit stream. The SMPTE 259M document (more on this topic will be presented in Chap. 6) states in its preamble that the standard applies until the fundamental (135 MHz for 4:2:2 component) has dropped 30 dB in value. Two spectral components have been found whose signal-to-noise values are useful in determining SDI signal health: the bit stream fundamental frequency and the third harmonic (Fig. 4.10).

As can be seen by looking at the frequency spectrum of an SD-SDI stream, the 405-MHz (third harmonic of the fundamental) band is easy to observe. Many think that the 270-MHz component should be used to determine the health of this bit stream. However, Sec. 4.3.2 will show that, in theory, it should not exist and it usually only shows up because of clock crosstalk in SDI transmit and receive integrated circuits (ICs). A number of instruments are commercially available which automatically measure the energy found in the area of the fundamental frequency.

At the output of most SDI drivers, the third harmonic starts approximately 35 dB above the noise floor (versus 45 to 50 dB for the 135-MHz component). After approximately 1000 ft of high-grade coax, it is approximately 8 to 10 dB above the noise floor (Fig. 4.11). As this signal approaches 6 dB above the noise floor, clock recovery becomes unreliable and errors start to occur. The error rate will rapidly increase from rates of one error per day to one error per frame over a range of only 3 dB as we reach this lower limit. In actual tests it has been found that the signal goes from low error rates to unusable error rates over a 2-dB range. This is less than an additional 80 ft of coax. Therefore, any passive path segment that indicates a third harmonic signal–to–noise value of less than 10 dB should be reengineered.

A common method of checking on the health of any digital bit stream is to look at the "eye" pattern. The eye pattern is produced by a waveform monitor that displays only one or two bit cells at a time. A good bit stream will be said to have its eyes wide open while a bad one will have its eyes closed (Fig. 4.12).

Fundamental
135 MHz
↓

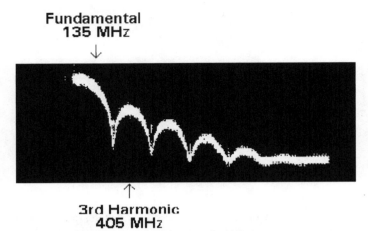

↑
3rd Harmonic
405 MHz

Figure 4.10   Spectrum of 270-Mbit/s digital-component bit stream.

3rd Harmonic
405 MHz
↓

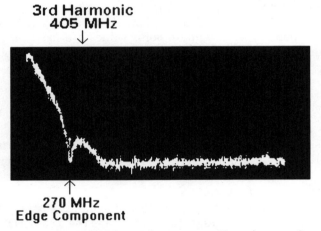

↑
270 MHz
Edge Component

Figure 4.11   270-Mbit/s serial-component video spectrum after
1000 ft of high-grade coax.

It should be understood that the large attenuation factor just mentioned
applies only to the upper frequency components, not the overall height of the
"eye" pattern. The maximum overall eye-pattern amplitude of a signal near
the error cliff might still be close to normal, which is 800 mV, with successive
runs of "ones" at 200 mV. (Chapter 5 will show that digital "ones" create lots
of transitions.) This is only a 12-dB drop from the original 800-mV level. This
is time domain, but in the frequency domain the third harmonic could have
dropped as much as 30 dB (Fig. 4.13).

Pulse rise and fall times and the amplitudes of short-duration pulses will be
worse because of the high-frequency roll-off. Also, the signal will float away

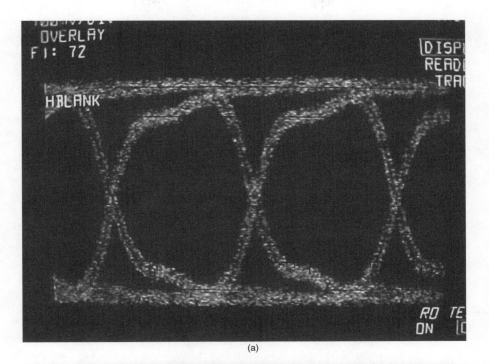

(a)

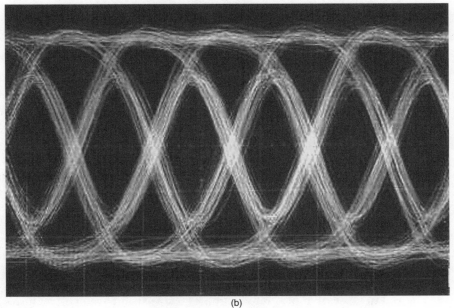

(b)

Figure 4.12 (*a*) This eye pattern is a fairly healthy 270-Mbit/s SD digital video bit stream. Notice how the pattern looks like two eyes. (*b*) This pattern is a healthy 1.5-Gbit/s HD digital video stream.

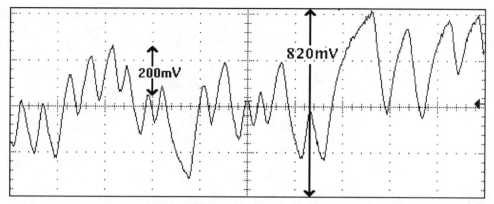

Figure 4.13 Notice that overall peak levels are still approximately 800 mV. However, sequences of high-transition rates, which require high-frequency components, are greatly attenuated.

from the ground, depending on pulse duty cycles because of the large low-frequency and dc components, which do not roll off as quickly.

4.2.5   Transform methods

The simplest way to convert an analog voltage into a digital number sequence is to allow each sample to enable a counter to count. The parallel output of the count is summed and fed back to a comparitor which compares the count value to the input sample voltage. When the two are equal, the count is disabled. The parallel output is the digital value. The problem with this approach is that it tends to be slow. It can only be used at low sampling rates. For SD-SDI video the clock rate for the count would have to be at least 3.5 GHz.

A method related to the count method is called the dual-slope method. The input is integrated over a fixed time. Then, a reference voltage of the opposite polarity is switched in place of the input and it discharges the capacitor until the voltage is back to zero. A reference clock is counted during the second phase; the clock count is the digital value out. A more common method is called subranging. In this method a coarse sample is determined such as the upper 5 bits of a 10-bit video word. This "subranged" value is summed and subtracted from the input. The resulting value undergoes another 5-bit conversion to create the lower 5 bits.

4.3   Bandwidth and the Error Cliff

Digital signals are nonlinear. It is the "high" or "low" state of a serial digital signal, along with its transition time, that determines the state of a data bit cell in a serial digital bit stream. The transition area between the high and low states is undefined when determining the value of an individual bit. In order to maximize the chances of reliable detection, sufficient signal amplitude is needed so noise or receiver inaccuracies do not cause errors.

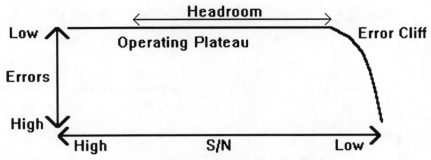

Figure 4.14   Errors stay low as S/N decreases until error cliff is reached.

The transitions, or "edges," between states are just as important. These transitions enable clock recovery from the bit stream in a self-clocking signal such as SMPTE 259M. As Chap. 6 will show, SMPTE 259 takes parallel digital components (SMPTE 125) or digital composites (SMPTE 244) and converts them to parallel bit streams. Without a clock at the receiving end, it is impossible to tell when to check the status of an arriving bit. An algorithm is used to scramble data as it leaves the transmitter to create as many edges as possible to assist the receiver's phase-locked loop (PLL) circuitry in generating a local clock synchronized to the transmit clock.

The error correction and error masking in modern digital equipment ensure that digital signals do not gradually degrade with increasing attenuation in the signal path as analog signals do. Instead, a digital transmission path continues to work perfectly up to the point where it suddenly does not work at all, and this is known as the "cliff effect" (Fig. 4.14).

Serial digital interface (SDI) signals that experience few, or no, errors are somewhere on the operational plateau. Operation stays uneventful until you reach the "error cliff." As the path traverses over the knee of this cliff, errors go rapidly from nonexistent to enough to swamp recovery efforts, making the path unusable. As little as 3 ft extra of coax can be enough to send a signal over the cliff. Many things determine where you are on the operational plateau. We will now examine how to determine exactly where you are on the plateau and how to stay away from the cliff. While almost all the information presented here can be applied to 4Fsc composite SMPTE 244M signals, or most other data bit streams for that matter, the discussion here will now center on component signals.

Although the way in which the SDI signal is used is "digital" in nature, many "analog" attributes of the signal can be used to predict the closeness of a particular digital path to the error cliff.

4.3.1 SDI channel coding

The transmission scheme shown in Fig. 4.15, which is also known as channel coding, is known as nonreturn to zero inverted (NRZI). This approach means

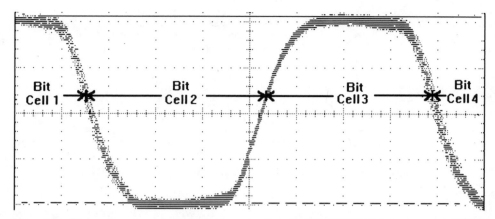

Figure 4.15 The portion shown is of three successive 1s in an SMPTE 259M data stream. A SMPTE 259M data stream changes state at the start of each bit cell if the bit cell has a data value of 1. The three transitions shown here indicate that the transition changes at the start of each bit cell.

that the receiver doesn't need to worry about the polarity (high or low) of the incoming bit stream. This approach also yields a constant high or low if a string of zeros is encountered, hence the use of a bit-scrambling algorithm. The peak-to-peak value of this signal should be 0.8 V, and the rise time, or transition time, should be between 0.75 and 1.5 ns. If the transmission path that this signal took had infinite bandwidth and no phase or group delay, it would be a perfect square wave. No transmission path is ideal, however.

4.3.2 SDI bandwidth

The small snapshot of data seen here has a fundamental frequency of 135 MHz because data bit cells occur at a rate of 270 MHz. Thus

$$Y = \quad 13.50 \text{ million samples/s}$$

$$B\text{-}Y = \quad 6.75 \text{ million samples/s}$$

$$R\text{-}Y = \quad 6.75 \text{ million samples/s}$$

$$\overline{ 27.00 \text{ million samples/s}}$$

$$\times \quad 10 \text{ bits/sample}$$

$$\overline{ 270.00 \text{ Mbits/s}}$$

Approximately 208 Mbits/s are actually devoted to video data in 525-line systems. The rest are used for ancillary data such as audio. Two bit cells in succession constitute the positive and negative half of a square wave with a frequency of 270 MHz/2, or 135 MHz. That 135-MHz square wave has a 135-MHz sine wave as its fundamental (Fig. 4.16).

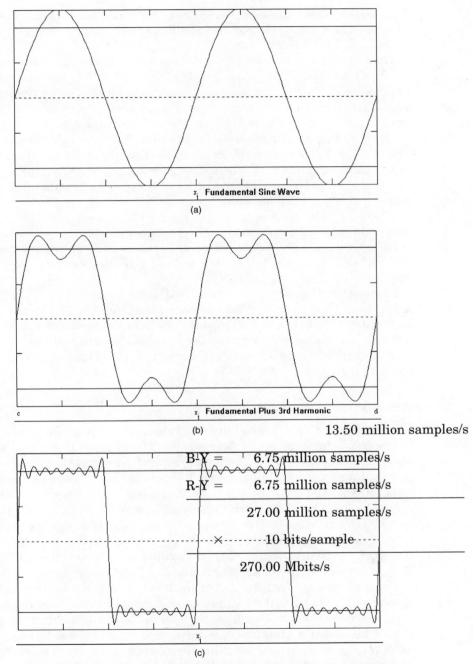

(a)

(b)

$$Y = \qquad\qquad\qquad\qquad 13.50 \text{ million samples/s}$$

$$B\text{-}Y = \qquad 6.75 \text{ million samples/s}$$

$$R\text{-}Y = \qquad 6.75 \text{ million samples/s}$$

$$27.00 \text{ million samples/s}$$

$$\times\qquad 10 \text{ bits/sample}$$

$$270.00 \text{ Mbits/s}$$

(c)

Figure 4.16 (*a*) Fundamental. (*b*) If the third harmonic is added to our fundamental (3 × 135 MHz), this waveform is produced. (*c*) If some additional odd harmonics are added in the correct amplitude and phase, this waveform results. Notice that we only worried about the odd harmonics in building this square wave. Figure 4.17 shows why.

A square wave with a 50 percent duty cycle has the fundamental frequency plus the third, fifth, etc., or the odd harmonics. But notice that the second, fourth, etc., are missing (Fig. 4.17).

Thus, if only a continuous string of ones is sent, the spectrum shown in Fig. 4.17*a* would occur. Obviously, that is not what is sent. Various runs of "ones" and "zeros" are sent, which results in a spectrum that more closely resembles Fig. 4.17*b*. It takes a lot more low-frequency energy to create a signal with the duty cycle shown in Fig. 4.17*b* than the one in Fig. 4.17*a*.

Many things still happen at traditional television rates. Pairs of start of active video (SAV) and end of active video (EAV) timing reference signals occur at the horizontal line rate. The patterns encountered during the vertical interval still occur at the field rate. This ensures that considerable energy will occur at fairly low frequencies.

While a 6-MHz bandwidth might have sufficed with analog video, standard definition serial digital video requires more than 50 times the bandwidth of NTSC. Actually, in short data paths SD harmonic content can approach 1.5 GHz. 1080i's bit stream, which has a bit rate of 1.5 Gbits/s, can have spectral content over 3 GHz.

Notice that there is a narrow 270-MHz component in the spectrum (Fig. 4.18), even though it was demonstrated that no 270-MHz component would be included if a continual string of "ones" (fundamental of 135-MHz, 50 percent duty-cycle square wave) were sent. As mentioned earlier, the sharp rise in energy at this frequency is a result of crosstalk from the transmitter's clock circuitry.

4.3.3   SDI transmission

The weak link in most serial digital systems is the path from the transmitter in one box to the receiver in the next box. The physical layer used to transport the data from one "box" to the next is comprised of mostly coax. Some connectors and perhaps a jackfield might be included in an average path. However, coax provides the greatest exposure to problems for a video data stream. Coax can be thought of as an infinite network composed of inductive and resistive components in series, with distributed shunt capacitance. This works out to be a low-pass filter whose poles increase in number and move closer to zero with length. This means that the longer the cable, the greater the attenuation of all frequencies, with the roll-off increasing as a function of frequency. As an example, if a given coax had the response shown in Fig. 4.19*a,* it would show the response in Fig. 4.19*b* if its length were increased by a factor of 10.

It is obvious that another increase by a factor of 10 would not be good. Since the signal becomes attenuated as the frequency increases, the upper harmonics of the signal disappear, and the square-wave data signal starts to look more like a sine wave. Additionally, with all the low-frequency (dc) energy still available, the square wave starts to look like the one shown in Fig. 4.13. As pointed out earlier, Fig. 4.12*a* is a normal, healthy SMPTE 259M spectrum. Notice how

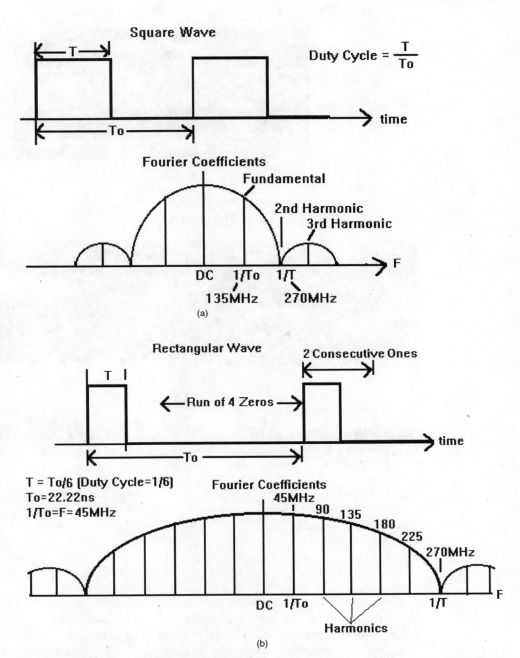

Figure 4.17 (*a*) A pulse with a duty cycle of 50 percent will have the response demonstrated here. (*b*) A pulse with a duty cycle less than 50 percent will have the response demonstrated here. Notice that many more frequency components are present.

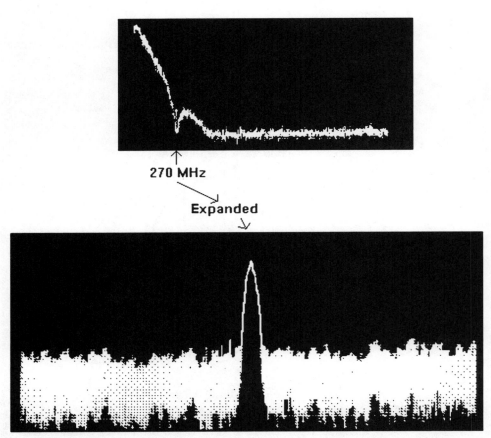

Figure 4.18 270-MHz clock component in an SMPTE 259 bit stream. This component is mainly due to clock crosstalk in the bit-stream driver or transmitter IC.

Fig. 4.10 resembles the right half of the Fourier coefficient functions in Figs. 4.17*a* and *b*. Figure 4.11 shows what the addition of 1000 ft of cable does to the same SMPTE 259M signal. Notice all the low-frequency energy left, but very little high-frequency energy. At this point, the error cliff has been reached.

4.4   Jitter

Jitter is the difference between the time when the next transition in the data stream is scheduled to occur and when it actually does occur. Jitter is another killer of sampling resolution. If jitter is bad enough, a 20-bit audio sample will have the same effective resolution as a 16-bit sample. Jitter will also add distortion. The determination as to where this crossover should occur can be determined either by the previous crossover (PLL internal closed-loop control) or by an external reference signal. Jitter is mainly caused by the SDI transmitter's crosstalk, signal saturation characteristics, and its power supply, plus

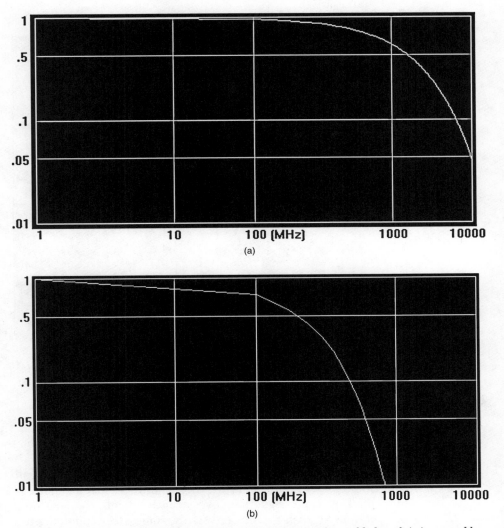

Figure 4.19 (a) Response of short length of coax. (b) Response when cable length is increased by a factor of 10.

any jitter that was present in the parallel data before it was serialized. Most digital circuits process digital video as parallel data and only serialize it right before transmission (Fig. 4.20).

The PLL clock circuit in the transmitter will also have a transient response that, hopefully, is critically damped, so that it slews to a corrected frequency quickly, without any overshoots (overdamped). Invariably the PLL will be underdamped at certain frequencies of jitter, and thus the PLL response will ring at those component frequencies.

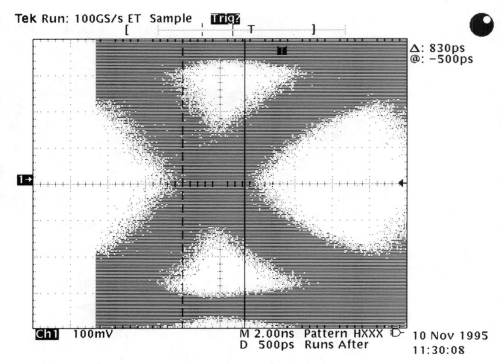

Figure 4.20 Transitions start to widen as jitter increases.

4.4.1 Measuring jitter

If an oscilloscope is used to measure the signal in the time domain by looking at zero crossings, incorrect conclusions will be gathered about the amount of jitter present. The time base (scope sweep rate) will act as a filter to cull out certain frequency components of the jitter because various jitter frequency components only occur at certain rates, and the scope is not looking at all the bit cells. If a scope triggers on every tenth bit cell, the one-tenth jitter component cannot be seen at all. Conversely, some jitter components will be seen at double their actual amplitudes.

Therefore, the time base of the scope works as a comb filter. One accurate way of measuring jitter is to extract the clock and phase-demodulate or discriminate it. This is the only time when part of the SDI signal should be considered a carrier. Once the clock and its jitter have been regenerated, the clock can be considered an FM signal with the jitter information as its payload. The baseband amplitude of this demodulated signal is the relative jitter. Some test equipment claims to take this approach, in which a bar graph, or gas gauge-type, display, which is easy to read, is used. Other test equipment offered uses the recommended approach offered by SMPTE in its standard RP-184, which is to extract a clock that is divided by some amount and then used to trigger an eye pattern–type display. The divisor is typically the same value as the word's bit rate (usually 10). This method will mask any word-related jitter, which is usually quite small.

The jitter components of higher frequencies (above 10 Hz) should have less than 0.5 ns of time-base jitter. The jitter components with low frequencies (generally called wander) could have errors as great as 6 ns. Therefore, trying to determine the jitter while looking at the total aggregate jitter is meaningless.

4.4.2 Reclocking

Most devices reclock the data as they enter the device to eliminate high-frequency jitter, but the low-frequency wander will remain. The low-frequency wander will continue to build up, which is the limiting factor to how many times a digital signal can be reclocked. The only way to stop the buildup of wander is for a digital transmitter to use an external clock. This can be done only if the serial data are run through a buffer or if they are taken back to parallel data and then converted back to serial data. Most equipment that processes serial digital data manipulates the data as parallel data. The major exception is that most routers, for complexity's sake, pass the data as a serial bit stream. Some routers reclock and some do not. Both methods have advantages and disadvantages. The disadvantage of reclocking has already been stated. The advantage is that in addition to eliminating high-frequency jitter, a form of buffering is provided, allowing the signal to be sent many places. This is needed to build large routers. A disadvantage of not reclocking is that jitter builds up and some form of bit-slicing or wideband amplification must be done to pass the bit stream.

4.5 Building, Sending, and Receiving Bit Streams

Once audio or video has been converted to parallel number sequences, they must be loaded into shift registers to march them out as a serial train of pulses representing the sampled bits. The object is naturally to have the data recoverable at the receiving end. In order for the data to be recoverable, the receiver must be able to determine the state of every bit coming in, low and high status, and what the total bit count means. As mentioned earlier, 8 bits is called a byte and 4 bits are called a nibble. Groups of bits greater than 8 are called words. Therefore, 10-bit video and 20-bit audio are called words. The receiver has to know where one byte or word stops and the next starts. It is also desirable to use as few conductors as possible to move the data, hence the predominate use of serial instead of parallel data. So one design criteria is to ensure the data are self-clocking. Video and audio are made self-clocking via channel coding, which will be examined in Sec. 4.7.

4.5.1 Convergence

Video and audio transmission have historically been a one-way proposition. The device sending video or audio—whether analog or digital—simply sent it and didn't care if the data or signal was received. The video/audio generator didn't even care if a cable was connected to its output. Computer devices evolved with a different mind-set. When communications between devices occur, an etiquette, or protocol, exists that must be followed. Any book that touches on television engineering today will have to devote part of it to the computer sciences. Convergence is combining the television and computer disciplines into a single field. As will be seen in later chapters, MPEG exists now in current television equipment and PCs. An SMPTE standard (SMPTE 305) replaces the video data in SD digital component video bit streams (SMPTE 125) with MPEG data, and it allows digital video paths, and even current routing equipment, to carry MPEG data. JPEG and later MPEG were developed for use with PCs. MPEG is often sent from device to device without ever traveling over a traditional video path. Often, it is wrapped up in other computer communications standards and shipped between two computers.

Today, many devices that look like traditional television equipment are really just computers in disguise. Most equipment today has a microcomputer at the core of its control system. The electronics industry calls these *embedded processors.* A lot of equipment today is just a PC with I/O to interface in whatever manner is necessary to handle the task, even though this equipment was built to mount in a rack and externally looks nothing like the PC. These devices are much more at home talking through the communications port than through a video BNC or audio XRL connector. As a result, a fair amount of time will now be devoted to surveying how computers talk to one another.

4.5.2   Data communications

Transfer of data involves two aspects. The first is wrapping the data with bits that tell the receiver when a new byte or word starts. In the data communications field this is generally referred to as encoding. The second aspect is setting up the rules of engagement, that is, the protocol for shipping and receiving the data. Let's look at RS-232 as an example. Although RS-232 has plenty of handshaking lines that can be used, most implementations don't use them. These lines signal to the other device that data are ready to be sent, that permission is granted to send data, or that the line is busy. Often, only the send and receive lines and a common line are used in RS-232 implementations. So, how do the two ends make sense of what the other is saying? The device doing the talking will not only send a byte of data but will also wrap that data with a start bit at the beginning and a stop bit at the end. The start bit is a space (RS-232 low state; see Sec. 4.5.3) and the stop bit is a mark (RS-232 high state). The start/stop and data bits are called a *character frame.*

Usually, whoever sets up the communications will have agreed in advance on bit rates, whether the data are 7 or 8 bits, and if one or two stop bits will be used. Two stop bits are sometimes used because it makes the receiver's job of determining when one frame has stopped and the other starts easier. Older devices sometimes sent only 7 bits because that covered the entire ASCII set of characters. If the receiver is confused as to frame start/stop location, it is said to have a *framing error*. The device that wraps the data to be sent at each end and sorts out data coming in is called a *universal asynchronous receiver transmitter* (UART). Asynchronous means that no separate clock is sent. The UART looks at the length of the start bit to determine when to sample the state of each following bit.

Now the two devices can send data to one another. Each device knows the rules for forming words, but it still needs to know the syntax for the words and how polite devices carry on a conversation. The simplest protocol for communications, which is often used in RS-232, is called *X-on* and *X-off*. X-on is sent by a device ready to receive data. X-off means the device is no longer ready, that is, the sending device can send data until the receive device's input buffer is full, at which time the receiver would send an X-off signal. When the receiver is ready again, it would send an X-on signal. X-on is the ASCII equivalent of sending a CTRL+Q, while X-off is a CTRL+S. Additionally most devices will either echo back to the sending device what they have received, or send an ACK (Acknowledged) or a NAK (Not Acknowledged).

4.5.3 Communications protocol

A more advanced streaming protocol is known as *automatic repeat request* (ARQ). The two most common implementations of ARQ are send-and-wait ARQ and continuous ARQ. In send-and-wait ARQ the transmitter sends a packet and then waits for an ACK from the receiver. This is known as half duplex where only one device talks at a time. With continuous ARQ the transmitter continuously sends packets without stopping. The receiver decodes each packet and sends an ACK with the packet number back to the sender. If an error occurs, the receiver sends an NAK; the sender then resends the packets sent after the last good one was received. XMODEM and Kermit are file transfer programs (FTPs) that use send-and-wait ARQ protocols. XMODEM is the simpler of the two; it was initially designed to transfer only one file. Kermit was designed to transfer multiple files in sequence. By the way, Kermit is named after the Muppet character. Both protocols send start bytes at the beginning of a packet. A *packet* is a group of bytes or words. A few bytes are sent that describe the packet length for error detection. XMODEM fetches 128 payload bytes per packet. Kermit can fetch up to 252 bytes per packet.

In an effort to standardize how devices communicate, SMPTE developed SMPTE RP-113, which is entitled "Supervisory Protocol for Digital Control Interface." This standard defines a "supervisor" and "tributaries." There is only one supervisor but there can be many tributaries. An example would be an editor controlling a number of VTRs, switchers, etc. Each tributary has two sets of 2-byte addresses, one for commands from the supervisor and one for polling. Periodically, the supervisor will poll all known tributaries. At that time, a tributary can send a message or make a request. The supervisor can send commands to an individual, group, or all tributaries. The supervisor can assign tributaries to groups. The tributary has five operational states:

1. *Idle.*   State until it receives a "break" character from the supervisor.
2. *Active.*   Entered after a break is received.
3. *Polled.*   Addressed by supervisor to check on status/requests from tributary.
4. *Select.*   Enters a communications mode with the bus controller.
5. *Group select.*   A number of tributaries communicate with the supervisor.

Most control systems encountered in television use RS-422 instead of RS-232 but the character frames are extremely similar. They usually consist of a start bit, 8 data bits (the LSB is usually first), a parity bit, and a stop bit. *Parity* is a bit that indicates if the data have errors or not. Parity can be odd or even. *Odd parity* occurs when the number of bits high (marks) is an odd number, so the parity bit is set high. If the number of data bits high is an even number, then the parity bit is set low. The opposite is the case for even parity. The parity to be used, if any, is decided in advance by the users. Parity can't be used to correct an error. It only indicates to the receiver that an error has occurred. If 2 bits, or any even number of bits, are incorrect, then the receiver incorrectly assumes that the received byte is OK.

The RS-422 control protocol used by various vendors sometimes follows RP-113 closely, but most of the time does not. Some protocols require a "break" to start communication but many do not. When a packet or command block is sent, a complete command with any additional arguments is included. By convention, the controlled device responds with an ACK. If no response is received, the master starts the process over again. The time limit is often only 10 ms. Conversely, if words in the command block or packet have gaps longer than a fixed time limit, and again 10 ms is common, the controlled device or slave times out and the communications sequence must be restarted. In addition to the command and any data required, the control protocol indicates the number of bytes that comprise the command, the type of command, the address of part of the protocol, and at least a 1-byte checksum in the packet. The 1-byte checksum overcomes the single-parity bits' limitation in detecting even numbers of errors. In addition to responding with an ACK, many devices return status data.

4.5.4 Computer networks

The type of communication just examined is point-to-point communication, that is, communication where two devices deal *only* with each other. Computers often work together and share information in groups, hence the computer network. Communicating with a single person in private is much simpler than talking to one or more people in a crowded room with multiple conversations being conducted. It is the same with computer communications. RS-232 was not designed for networking even though it has been used for small networks in the past. In this approach a central computer has multiple RS-232 ports or each PC has two RS-232 ports, each connected to two neighboring PCs. In the first approach the central computer acted as a resource controller. The term used today for that architecture is "server." Today's server often stores central files, allocates resources, grants network permissions, and handles connections to the outside world and other security issues. The PCs that depend on the server for those services are referred to as "clients." In the second RS-232 network approach every PC on the network was connected to another, which connected to another, etc. The last PC in the string was connected back to the first PC. This connection scheme is called a *ring*. In the computer field this is called *ring topology*. In the RS-232 client-server topology each client PC is connected directly to the server PC, and this is known as *star topology* because, when documented, the server was usually at the center with the clients around it. More on network topology a little later. Every device that sits on the network, whether a PC, printer, etc., is referred to as a *node*.

There are basically two types of networks:

Peer-to-peer.   Any device on the network can make its resources available to any other device on the network. Devices take turns being servers and clients. TCP/IP networking typically operates in this environment.

Server-based.   One or more devices on the network is the dedicated server and the rest are clients. Novell NetWare typically runs in this mode.

4.5.5   Flow control

The X-on/X-off commands for sending data over RS-234 are known as flow-control commands. When a communications session is set up between two devices, flow control is necessary. There are three types of flow control:

1. *Connectionless.*   Digital video is connectionless. The sending device ships video out not caring if it is ever received by any other device.

2. *Connection-oriented.*   In this type the sender sends a word, or even an entire packet, and then waits for an acknowledgment from the receiver. X-on/X-off and send-and-wait ARQ are examples of this method.

3. *Acknowledged connectionless.*   This type is a connectionless service, but the receiver acknowledges the received data as they are received. Continuous ARQ is an example of this.

4.5.6   Local-area networks (LANs) and wide-area networks (WANs)

The distinction between LANs and WANs can be subtle. LANs are generally networks in the same building or campus. A router or switch might be at the center. The data are usually baseband (not riding on a carrier). WANs generally involve groups of LANs at separate sites connected via modem, microwave, satellite, or the Internet.

A word here about computer routers and switches is necessary. Computer routers and switches are different than television routers and switchers. A computer switcher has the same function as a television router, with a major difference in that a television router often has the same output connected to multiple outputs and that is usually not the case in a computer switch. A computer router operates on a specific layer of the OSI stack (which will be covered shortly). While a television router has a separate control system for determining what input is connected to each output, the data stream passing through a computer router contains information on where it wants to go. The router has lookup tables that it uses to decide the output to which the incoming data stream should be routed. In simple terms a computer router is a large memory array where incoming data are stored. The router determines which output leads to the desired destination and when the output is free, it reads the data out to that port. As speeds approach the gigabit rate, many in the router industry wonder if this software approach can continue to work. Often routers talk to routers, which talk to yet additional routers. I just described the Internet. Massive router protocol has evolved to let routers work together efficiently and not have data wander endlessly from router to router.

4.5.7 Common network applications and services

Some common network applications and services are listed here:

1. *File server.*   Provides file access and management services.
2. *Directory services.*   Keep track of addresses on the network.
3. *Print server.*   Provides printing services.
4. *Electronic mail.*   Provides electronic messaging.
5. *Database server.*   Provides database services.
6. *Communications server.*   Provides access to modems, etc.
7. *Application server.*   Allows an application to be distributed among more than one PC.
8. *Network transparency.*   Allows programs not written specifically to run on networks to have access to remote drives and printers by redirecting commands issued by the program.
9. *Network administration.*Controls and sets up network servers and clients.

In order for these services to work, software must be running on each PC that interfaces the users' applications and user interface with the network port on the back of the PC. This software is called an *application programming interface* (API), and it allows applications to invoke network services. An effort is under way to standardize APIs by a group called the Open Systems Foundation (OSF). This group has developed specifications for a distributed computing environment (DCE) which would allow applications to use a standard set of calls for network service. Many computers are not manufactured with a built-in network port and thus must have a network interface card (NIC) installed. Sometimes this is called a network adapter or LAN adapter.

In the following subsections we will look at the various transmission mediums used to physically interconnect a network.

4.5.8 Network topology

Network topology is the logical shape that the network interconnections take. Common topologies include:

Star. All network devices are connected to a common central device such as a router. This topology can be expanded into what is called snowflake topology where one central router can be star connected to other routers, each of which can have a star configuration around it.

Bus. Network devices are all connected to a common physical path such as coax.

Tree. Multiple bus topologies are interconnected via what is commonly called a bridge.

Ring. Each device has a receive and a transmit port on it. The transmit port on one device is connected to the receive port on the next. This is continued until the last device's transmit port is connected back to the first device's receive port, thus forming a physical circle, or ring.

4.6 Electrical Interfaces and Transmission Lines

Interconnections between devices can be either single-ended or balanced.

4.6.1 Single-ended signals

In *single-ended signals* the information is on only one conductor. The return path is through a common, or ground, lead. Single-ended signals are said to be referenced to ground. Almost all video, both analog and digital, is made up of single-ended signals.

4.6.2 Balanced signals

Balanced signals use two conductors, one for the send path and a second for the return path. These signals are usually fed by "differential drivers." These drivers have two outputs that are out of phase with one another. When one output goes positive, the other output goes equally negative. Differential drivers are said to have good common-mode rejection, that is, if a stray signal is induced into both conductors, the receiver will ignore the unwanted interference because it only uses the difference between the two conductors.

Balanced signals also eliminate a constant problem encountered when equipment is connected together in a single-ended fashion: the ground loop. A ground loop is evidenced by a 60-Hz (power-line frequency) sine wave modulating the desired signal. It is often referred to as "hum" as it sounds like a hum when it gets into the audio. For that reason professional audio is balanced, although low-end audio equipment is often single-ended. To combat hum, or 60 cycle as it is also called, vendors offer "hum-buckler" coils. These are transformers wound to cancel the power-line frequency. The best way to fight ground loops is to prevent them. Ground loops are generally created by (1) unbalanced power loads and (2) poor grounding. Many facilities are supplied with three-phase power, with each phase or leg 120° apart. When the legs are not drawing the same amount of power, the common neutral or return line coming from each leg will be at different potentials. This situation exists because no path ever has zero resistance, and different amounts of current in each return leg means different voltage drops for each. The neutrals will, therefore, float at different amounts above ground. Current will flow in the ground connections (such as the shielding in coax) between equipment tied to different power legs, creating ground loops. The second way to minimize the problem is to make sure all equipment goes back to a good common ground via its own ground path.

4.6.3 Digital video in coax

As already mentioned, video is single-ended. Let's examine the transmission characteristics of digital video in coax. It takes 1.24 μs for a bit to propagate down 1000 ft of high-quality coax (velocity of propagation of 82 percent). An SD digital component has a 3.70-ns-long bit cell. Each cell occupies approximately 3 ft of coax at any given time. Therefore, there are 335 SD bit cells in 1000 ft of coax at any given time. Any impedance mismatches that occur along the path cause reflections in both the analog and digital domains. Only long paths with impedance mismatches have the potential to cause reflections that are bothersome to analog video signals. In analog video signals, reflections caused by improper impedance matches in cable are hard to discern. They appear as nearly imperceptible ringing during transitions, providing unintentional enhancement or cancellation in addition to the obvious incorrect level caused by nonterminated or double terminated lines.

In the digital domain a path only a few feet long with reflections could prove disastrous. In the digital domain it is not the incorrect level that usually gets the attention when incorrect terminations are applied to a path; it is the total loss of recovered video because reflections have made recovery of the embedded clock impossible. Impedance mismatches as small as 20 percent can cause errors. Even tees with short lengths of unterminated cable can cause the total loss of recovered video. Patches, connectors, or barrels can increase errors or even stop video from being recovered. Problems develop especially in older installations because most BNC connectors had a characteristic impedance of only 50 Ω until a few years ago. The amount of polyethylene or Teflon dielectric was reduced to increase the characteristic impedance from 50 to 75 Ω. By looking into a 50-Ω connector, a dielectric material can be seen along the outer contacts of the connector. In a 75-Ω connector there is no, or a reduced amount of, dielectric. Early 75-Ω connectors were manufactured by reducing the diameter of the center conductor pin, but these do not mate properly with 50-Ω connectors. However, the new 75-Ω connectors (reduced dielectric) will mate reliably with the 50-Ω connector. Most manufacturers of BNC connectors have a statement in their literature that their 75-Ω connectors will mate with the 50-Ω connector with no damage.

Coax that is robust at high frequencies is needed in a digital plant. The center conductor should be solid copper (better skin effect than stranded), and the shield should be braided (with a coverage of near 100 percent) and have a layer of foil, which has better skin effect at higher frequencies. The dielectric should produce as low a shunt capacitance value as possible, decreasing the high-end roll-off and increasing the velocity of propagation. A trade-off here is that some dielectrics accomplish this by using air pockets to lower the dielectric constant, which can lead to center conductor migration. Center conductor migration can, in turn, affect the impedance along the length of the cable, especially at sharp bends, leading to reflections.

4.6.4   RS-232

Let's now look at several common computer interfaces. RS-232 was first introduced by the Electronics Industries Association in 1964. It provided the first and most common standard for data interchange. This specification describes data interchange between two data devices. One of these devices is called the *data terminal equipment* (DTE), and the other device is called the *data communications equipment* (DCE). The DTE is generally a PC or a dumb terminal, while modems are the DCE. This standard has been amended three times (1969, 1972, and 1986). Originally, RS-232 drivers were specified to have a slew rate (rate of change) of 30 V/s, which limited the data rate initially to 20 kbits/s. The current specification is up to 64 kbits/s. The original maximum length between the DTE and DCE was 50 ft, and now the maximum length is determined by line capacitance, which should be below 2500 pF. The receiver input impedance should be between 3 and 7 kΩ. RS-232 uses up to 22 conductors, most of which are used for handshaking. Data can be sent synchronously or asynchronously. The weak link in RS-232 is that all the signals are single-ended. The advantage to being single-ended is that it takes fewer conductors to interconnect equipment. Terminology for high/low states on RS-232 came from the telegraph industry. The high state is called a mark, and the low state a space. A high state is greater than +5 V but less than +15 V. A low state is less than −5 V but greater than −15 V.

4.6.5   RS-422/449

Two other interface standards are improvements to RS-234. RS-422 defines the electrical characteristics of a balanced voltage-interface circuit. It allows for bit rates of up to 10 Mbits/s. The output voltage from an RS-422 driver is between +2 and 10 V. Current in one direction (where pin 3 is more positive than pin 8 or where pin 7 is more positive than pin 2) denotes a mark, and a space in the other direction. The pin out for RS-422, which is described in SMPTE 207, is the interface used for controlling most broadcast equipment. RS-449 was intended to replace RS-232 but it has not done so yet. It is balanced (that is, it uses RS-422 drivers/receivers) and was intended to be used at bit rates up to 2 Mbits/s. Data rates on high-quality cable can approach 10 Mbits/s.

4.6.6 Modems

The baseband data-interface schemes are often modulated by modems and sent through telephone lines. The modem, which almost always interfaces with an RS-232 line if external or the PC data bus if internal to a PC, outputs tones into the telephone line. These tones are phase-modulated using PSK or QAM modulation techniques. Early modems used FSK or QPSK techniques. Although the latest modems now have 56-kbit/s rates, their baud rate is still under 4000 Hz. This is a common point of confusion. The *baud rate* is the frequency of the carrier that is modulated. As can be seen with 64-QAM, each cycle can convey one of 64 states based on its phase and amplitude. Six binary bits would also convey 64 possible states, so each cycle carries a payload of 5 bits. A baud rate of 3200 Hz is equal to a bit rate of 19.2 kbits/s in this case.

4.6.7 Fiber optics

Fiber optics is rapidly replacing copper, microwave, and satellites as the chosen communications path. Instead of conveying information with electrons, photons are used. Fiber doesn't have the induced noise problems inherent in metallic conductors. Crosstalk problems are eliminated as well. The bandwidth is much higher than metallic paths provide. A glass fiber is about as thick as a human hair and weighs about 1 oz/km. The entire cable, including shielding as protection from the environment, can still weigh less than 10 lb/km.

Ironically, the narrower the fiber strand, the higher the bandwidth, and thus the longer the signal that can be sent. Very thin strands, 5 μm or less, are known as *monomode fiber,* which means that light that enters the transmission end of the fiber can only travel straight down the center of the fiber. Thicker fiber has rays of light that are refracted off the edges of the fiber as they travel down the strand. Multiple refraction means that the refracted ray has a longer path to travel than a ray that travels down the center of the strand. Thus, an impulse injected into one end of the fiber path arrives at the other end lower in amplitude and spread in time because not all the light rays arrive at the same time. This is known as *dispersion.*

Fiber-optic transducers are inherently nonlinear. Therefore, fiber does not lend itself well to linear analog applications. Fiber works best with FM, digital, or pulse-code modulation signals. PCM means that analog samples are converted, or coded, into numeric samples that can be transmitted digitally. Fiber-transmission transducers must have narrow-light bandwidths as the fiber caus-es light at different wavelengths to travel at different velocities, which is known as *material dispersion*. This is why lasers have been used over LEDs and the superluminescent diode (SLD). Lasers have the most collimated, or focused, out-put (LEDs the worst), which results in less refraction down the fiber path.

Fiber is expensive to install but inexpensive to maintain. Fiber can cost as little as $70,000/mi to install in rural areas and as much as $500,000/mi to install in downtown Manhattan. If a conduit is already in place, it is generally less expensive to install another fiber if the installation is under 10 km. If it is over 30 km, it is less expensive to use wavelength division multiplexing (WDM), the use of multiple transmitting transducers operating at different wavelengths, over the existing fiber.

4.6.8   Network connections

The bottom layer in the OSI model (OSI will be reviewed in Chap. 5) is the physical layer. This layer is where the actual transmission of bits occurs; it not only includes the actual transmission medium but also the transmit and receive circuitry connected to the physical network. All modulation and chan-nel coding is done in this layer.

A number of cable schemes are used:

Unshielded twisted pair (UTP).   Used for setting up Ethernet star topologies or for local telephone subscriber loops. UTP cables are usually terminated with RJ-45 connectors while telephones use RJ-11. UTP cable comes in five levels or categories. The higher the number, the better the shielding.

10 BaseT.   Operates at 10 Mbits/s over twisted pair. Base refers to baseband data as opposed to an rf-modulated or broadband signal. Baseband means the original frequencies associated with the signal are left intact. In broadband systems transmit and receive signals ride on different frequencies.

10 Base5.   Operates over coax up to 500 m/segment at 10 Mbits/s. This is usually referred to as a thicknet.

10 Base2.   Same as 10 Base5, except the length limitation is 200 m due to thinner cable. Known as thinnet.

100 BaseT4.   Operates on four pairs of category 3, 4, or 5 UTP, with each pair operating at 25 Mbits/s.

100 BaseTX.   Operates on two pairs of shielded twisted pair (STP), or category 5 or better UTP cable, with each pair operating at 50 Mbits/s.

4.7   Channel Coding

Transmission of serial data through a medium, whether over the air, down a fiber-optic cable, or just down copper, requires some modification of most signals so that the receiver can recover the transmitted signal. A true serial signal will not have a separate clock; the clock will be embedded into the serial bit stream in such a way that the receiver can extract it and use the reconstituted clock to determine the value of the received data. Also, random data will invariably have a dc component. DC is not recordable if the end receiver is magnetic tape or disk, besides the fact that low frequencies interfere with azimuth recordings. A dc component will also interfere with the phase-locked loop circuitry in the clock recovery system in the receiver. Channel coding is used to prevent these problems at the receiver.

Channel coding most frequently uses signal transition instead of levels to convey the data. The goal of channel coding schemes is to combine the serial data with a clock to produce self-clocking–modulated code. Channel coding must manage the minimum distance between successive transitions to limit the highest frequency of the signal. It must also limit the distance between transitions so that the embedded clock can still be reliably recovered.

4.7.1   Nonreturn to zero (NRZ)

Several common physical-layer signal-encoding or signaling schemes are in use. The simplest one is RS-232 where one level represents a zero and another level represents a one. This is called nonreturn to zero. However, this approach usually needs a separate clock. NRZ signals are usually encoded to enhance their performance.

4.7.2   Nonreturn to zero—invert (NRZI)

A slightly better approach is nonreturn to zero—invert on ones. In this approach the signal changes states at the start of a bit cell if that bit cell represents a value of one. This is an example of differential encoding. The cell can only be decoded by comparing it to an adjacent cell, rather then just looking at its absolute value. An advantage to this approach is that in noisy environments it is often easier to detect changes in state than absolute levels. NRZ and NRZI are the most efficient in bandwidth since most of the energy falls between zero and one-half the bit rate.

4.7.3   Biphase

A second type of signaling is called *biphase*. Manchester code is a biphase-type signal. It has a transition in the middle of each bit cell. A rising edge represents a one, and a falling edge represents a zero. The successive zeros or ones require a second edge at the beginning of each bit cell to be in the proper phase for the middle transition. A variant of this is differential Manchester code where the transition in the middle of the cell is used only for clocking. A transition at the beginning of the cell represents a zero, and the absence of a transition represents a one. Since the biphase technique can require two transitions/bit cell, the baud rate, and thus the bandwidth, is double that of the NRZ. The advantages of this approach is that the signal is self-clocking and no dc component is needed. Manchester code is used for IEEE 804.3 Ethernet, using coax cable (10 Base5 and 10 Base2), and twisted pair (10 BaseT). Differential Manchester is used for IEEE 804.5 token rings using shielded twisted pair. 10 Base5 is used for baseband networks instead of broadband networks, where 10 means 10 Mbits/s and 5 means 500 m (thick coax). 10 Base2 is 200 m in thin coax, often referred to as thinnet. 10 BaseT uses unshielded twisted-pair cable (UTP).

4.7.4 MLT-3

MLT-3 is a three-level data-signaling technique. It changes state at the beginning of every bit cell that represents a one. If negative, it returns to zero; if zero (but if negative before), it goes positive; if positive, it goes to zero; and if zero (but positive before), it goes negative. This technique, along with 4/5-bit encoding, is used on twisted-pair FDDI and 100 BaseTX paths.

4.7.5 Bit transforms

When signaling rates become high, the double baud-rate penalty incurred with biphase techniques becomes a liability. Therefore, bit transformation schemes are used. As just mentioned, a 4/5-bit encoding is used for FDDI twisted pair and 100 BaseTX. In fact, 100 BaseTX and 100 BaseFX (fiber) and fiber-distributed data interface (FDDI) fiber and twisted pair all use 4/5-bit encoding schemes, which means that every 4 bits of data are mapped to 5 transmitted bits, so that each of the 16 possible states represented by the 4 bits has two possible values that are stored in a lookup table. The transmitter decides which of the two values to use, based on the previous data values. The receiver has the same lookup table to convert every 5 bits received back to the original 4 bits. The same idea is used for 5/6- and 8/10-bit encoding. In 8/10-bit encoding the 2 extra bits added mean that four separate values for each 8 data bits are available. This ensures better dc level management as more choices are available to counter any dc drift from previous data. Fibre Channel uses the 8/10-bit encoding scheme.

NRZ or NRZI can be used in conjunction with these bit transformation schemes. FDDI optical uses NRZI, with 4/5-bit encoding, while Fibre Channel uses NRZ with 8/10-bit encoding.

4.8 Error Detection

Error detection requires a trade-off between usable data sent and the ability to detect errors. How does the data-rate/error-correction trade-off take place? Forward error-correcting codes (FECs) are sent along with the video and audio data. This extra information is used to determine if an error has occurred and how to correct that error. An in-depth discussion of FECs will not be given here, only a brief description of how they work. Common codes are Reed-Solomon and Viterbi coding, and they are often used together. When this occurs, it is called concatenated coding.

4.8.1   Reed-Solomon

Basically, Reed-Solomon coding builds arrays and adds error-correction information to the end of each row of code. In order to enhance the ability of these error-correction codes the array is not always read out the way it was written in. The array might be built a row at a time but is read out by columns, a process called interleaving.

4.8.2   Viterbi

Viterbi coding is a bit more involved. As part of the explanation, some background on the "hamming distance" is needed. If you have an initial binary number and compare it to a second number, the hamming distance would be the number of bits that must change in the first number to make it equal to the second number. Using the hamming distance allows you to take each bit of information and add additional bits in such a way as to minimize the decoding possibilities if an error does occur. A common ratio of information–to–error correcting bits is 1:3, or what is called a span of 3. At any point in time the last 3 bits at the receiving end are used to determine the value of the next 3 bits. Of four possible values, only two have the lowest possible hamming distances. One of these two possible values would signify that the next information bit is a one, and the other would signify the information bit was zero. This is known as *trellis coding* because the encoding process of selecting the set of bits that represents the actual information bit was based on decisions of what had come before. This state diagram looks like a trellis, or lattice fence.

Figure 4.21 depicts a single 2:1 conventional trellis used for decoding, which means that there are 2 bits for every original data bit. The top binary

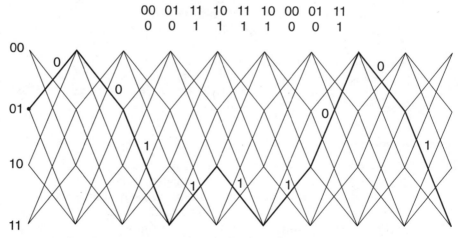

Figure 4.21   Simple 2:1 trellis. Uses shortest Hamming distance; only two possible choices (out of four). Only 1 bit may change between successive pairs of hits.

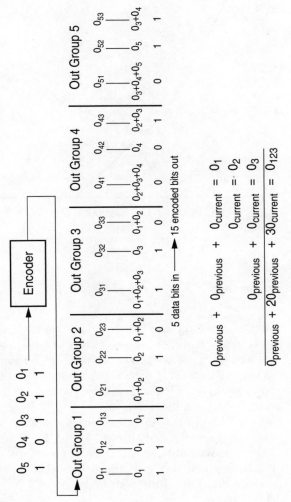

Figure 4.22 Encoding process for a convolutional encoder with a span of 3.

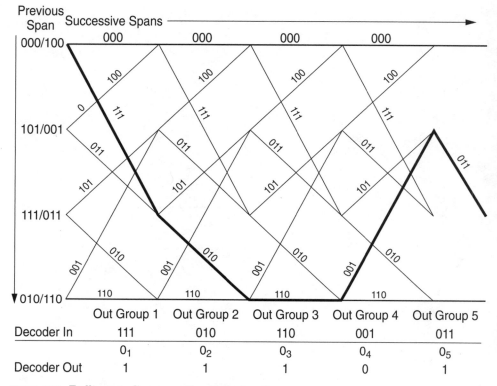

Figure 4.23 Trellis state diagram with sample decoding.

string is the bit string into the decoder. The bottom string is the decoded data stream. At every state, there are two possible new states as only 1 bit of the next two strings is allowed to change. Thus, at every state there are only two paths to the next state. If both bits change or if neither changes, an error has occurred. A positive or least-negative slope indicates a "0," while a negative or least-positive slope indicates a "1."

Figure 4.22 shows the encoding process for a convolutional encoder with a span of 3. For each data bit input, 3 bits $(0_1, 0_2, 0_3)$ are output: The first bit (0_1) is the sum (without carry) of the current and previous 2 bits, the second bit (0_2) is the current bit, and the third bit (0_3) is the sum of the current bit and previous bit. Therefore, a data string of 1 1 1 0 1 creates the string 111 010 110 001 011.

Figure 4.23 shows the trellis state diagram to decode the previous encoded string. Here there are eight possible states because of the 3:1 span ($2^3 = 8$). A span of 3 allows 2 bits to change state from one state so the next groups of 2 can be created, also making this a four-level trellis. Again, like the first decoding trellis, this one works on the shortest hamming distance. Our encoded string of 111 010 110 001 011 in produces the original 1 1 1 0 1 out.

The Viterbi algorithm takes trellis coding to a higher degree by expanding the span used to ensure correct decoding. A two-step process is used with limitations based on the hamming distance, then inserting dummy zeros, and performing the trellis lookup a second time. The result is the two shortest distances, one representing a one, and the second representing a zero, which makes for an extremely robust data stream in a channel that can teem with noise. As mentioned before, the trade-off, as in the example here, is that for every 3 bits that are sent, only 1 bit carries actual information. The other 2 bits are for error correction. Seven is a common value for the span used, with an information–to–total data ratio of 1:6. The longer the span, the longer the processing time. Viterbi coding can reduce the required S/N typically by a value of 5 dB, thus lowering the required link budget.

As is well known, the big advantage in using digital is that, once the signal is in its final domain, it stays transparent until the error cliff is reached. The error cliff is reached when the noise in the path overcomes the ability of the error-detection and -correction system in the receiver to perform its job. Generally, errors increase as the carrier-to-noise ratio decreases. Understanding and properly computing link budgets should help to avoid the dreaded cliff.

4.9 Bibliography

Boston, Jim, and Jim Kraenzel, "SDI Headroom and the Digital Cliff," *Broadcast Engineering*, February 1997, p. 80.

Kapler, Robert, "Conversion Dominates at SMPTE," *TV Technology*, November 30, 1998, p. 8.

Martin, James, Kathleen Kavanagh Chapman, and Joe Leben, *Local Area Networks*, Prentice-Hall, 1994.

Rumsey, Francis, and John Watkinson, *The Digital Interface Handbook*, Focal Press, 1995.

Watkinson, John, *The Art of Digital Audio*, Focal Press, 1989.

Whitaker, Jerry, *DTV: The Revolution in Digital Video*, 2d ed., McGraw-Hill, 1999.

Computer Network Access Control Methods

Computer network access control is the term that describes how each device on the LAN gains access to the network, and it is often referred to as the *data link*. Many common data-link access control methods are available, and they will be surveyed in this chapter (Fig. 5.1).

5.1 Asymmetrical Digital Subscriber Loop (ADSL)

ADSL operates in the 30-kHz to 1.1-MHz band, with the analog telephone system operating below it. Modulation methods for ADSL are discrete multitone (DMT), carrierless amplitude/phase (CAP), and quadrature amplitude modulation (QAM). CAP is closely related to QAM. QAM uses two orthogonal signals (I and Q) that are modulated. With CAP, the I and Q are modulated digitally. DMT is basically the same as orthogonal frequency division multiplexing (OFDM). OFDM divides its bandwidth into a large number of discrete bands, or subchannels. In ANSI DMT there are 255 subcarriers, each with a 4-kHz bandwidth, so each channel can be modulated from 0 to 15 bits/Hz. In subcarriers (or bins) with low attenuation and good S/N ratios, dense constellations are used, 10 bits/Hz being typical. In bins where the S/N ratio is poor or where interference is found, say from an AM station, less dense constellations are used. DMT symbols each last 250 μs, which means that each bin, or tone, can carry up to 60 kbits/s.

5.2 Asynchronous Transfer Mode (ATM)

The ATM standard came about as an attempt to get rid of the digital hierarchy required when the telephone companies moved voice data for the most part (see

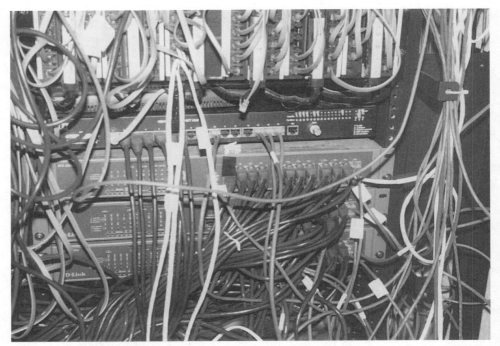

Figure 5.1   LAN hubs, RJ-45 connectors, and cat 5 cable will become more prevalent in television facilities.

Sec. 5.15). Once video and multimedia began to use the telephone networks, a standard that was continually scalable like the digital hierarchy approach was needed. The term "asynchronous" used in ATM means that packets belonging to a particular channel, or voice conversation, will not happen on a steady periodic basis. There might be successive packets devoted to the same conversation, or varying numbers of unrelated packets in between packets associated with a particular channel.

The current maximum bit rate for an ATM connection is 155 Mbits. The ATM cell consists of a 5-byte header. The first 1.5 bytes are used for the virtual circuit number or virtual path indicator, which allows two adjacent switching nodes to manage their conversation. The virtual circuit number value changes with each trip through a node. The next 2 bytes are used for the virtual channel indicator, so a routing node can handle multiple conversations with each other. Three bits of the fourth byte are used to identify whether the payload is operations, administration, or maintenance in nature. One bit in this same fourth byte signifies whether an ATM switch can discard the cell if necessary. The final fifth header byte has a CRC of the first 4 header bytes. After the header bytes, 48 payload bytes follow.

ATM data frames assembled at what is called the "convergence sublayer" can be up to 64K (1375 ATM cells) long. A TCP/IP data packet on an ATM network at the convergence sublayer would have the following data structure:

| ATM header | IP header | TCP header | User's data | TCP trailer | IP trailer | ATM trailer |

This structure is then handed to ATM's segmentation and reassembly layer. This frame is sliced up into 44 to 48 bytes, depending on the AAL type in use, and then 5-byte ATM header data are added to each ATM cell. ATM has no media access control (MAC). Therefore, unless ATM is mapped on Ethernet or Fibre Channel, each ATM node must go straight to a switch. If there is congestion at a switch, that is, too many cells arriving from different places at once, the switch starts throwing cells away. ATM has a 10 percent cell overhead. The ATM layer sits above the physical layer in the data-link layer. ATM has both circuit and path addressing. A path has many circuits running it.

Long-haul ATM providers do not generally charge based on distance. The user pays according to the amount of data dropped into the ATM cloud. A *cloud* in network terminology is the infrastructure or fabric, such as routers, switches, bridges, and copper and fiber-optic cables, that comprises the network pathways. If equipment at both ends of a path wraps the data payload in the right combination of headers and trailer bytes, the cloud will know how to deliver the data to their intended destination. This is a connectionless operation that is usually made to appear as a hard-wired or virtually connection-oriented path.

A number of these virtual type paths can be obtained from ATM vendors. The first is known as a *private virtual circuit* (PVC). The ATM vendor sets up the network so that, to the end user, a permanent path between two points is constantly connected. The end user sees it as a virtual leased line. The charges are based on the amount of bandwidth used. In terms of cost, the length of the PVC doesn't matter to the user; it is the amount of data that are sent into the ATM providers' cloud. The end user can't change the bandwidth of the circuit or the source and destinations; only the service provider can do that.

A second type of virtual circuit is the *switched virtual circuit* (SVC). This service acts like a regular voice telephone call, except for data. The source sets up a "call" with a destination when needed and data are exchanged. Both source and receive sites must be set up with SVC service of the same bandwidth and quality of service (QoS). QoS is the level of assured connectivity requested. A constant bit rate is required for video. Thirty frames of video must be delivered every second, and not just on average. This service naturally costs the most. The next level of service specifies that, on average, the desired data rate will be delivered, but it might not be a constant flow or be bursty. Lower levels of service are for users that have data they want to send only when traffic is low enough to permit it. If video files need to be transferred between servers using FTP, a lower QoS is probably acceptable, but live video requires the higher QoS. SVCs can usually be controlled and changed by the user. The cost is only for the data sent. SVCs can be used to send video and audio data at rates much lower than other means of connectivity. However, fixed monthly charges are added for access to the ATM providers' point of presence (POP) to the local telephone company and port charges at the POP to the ATM provider. An ATM multiplexer is also needed at the user's facilities.

Additionally, most ATM providers sell bandwidth in fixed amounts, and many require a minimum of DS-3 (45-Mbit/s) bandwidth.

5.3   Ethernet

This type of network has multiple devices connected to, and sharing, the same physical medium. All stations on the network receive the transmissions of all other stations. The physical medium is either coax or twisted-pair. A device attempts to transmit when data are available for transmission.

Ethernet uses a random medium control, called carrier sense multiple access with collision detection (CSMA/CD). Carrier sense means that the system listens to the physical medium before it transmits. It only transmits if it detects that the medium is quiet. Multiple access means that all devices on the physical network have simultaneous access to the transmission medium. If two devices try to transmit on the network at the same time, a collision occurs. All devices detect this collision because the message is garbled, and the signal levels on the physical medium have exceeded a predetermined level. This is the meaning of the collision detection part of the term. When a collision occurs, the sending device immediately stops transmitting and sends a "jamming" signal to ensure that all stations understand that a collision has occurred. All devices now wait a random amount of time before they try to retransmit. This waiting period is called a *back-off delay,* and it is calculated in multiples of slot time. As traffic increases, the number of collisions increase. As a device trying to transmit incurs multiple collisions, it increases the range of random numbers it generates to use as its back-off delay. Approximately 90 percent of channel utilization can be achieved using the CSMA/CD access control method. An important restriction of Ethernet is that there can only be one path between any two devices on the network, although the network can be broken up with repeaters, bridges, or hubs.

5.4   ARCnet

Physically, ARCnet uses a low-speed version of the Ethernet/token bus approach that only uses coax. It also uses a token like the token bus.

5.5   Fiber Distributed Data Interface (FDDI)

This method also uses a ring structure. Unlike token ring, though, it uses fiber-optic cable instead of twisted-pair. FDDI also uses a token, and is a token-passing protocol used over a fiber-optic cable at 100 Mbits/s. Up to 500 stations and a fiber path of 100 km is supported by FDDI. Packets of up to 4500 bytes in length are passed from one station to the next in a loop topology, where they are regenerated and passed to the next station if not intended for that station. FDDI can be set up to have two parallel rings. If one goes down, all traffic can transverse over the remaining path. FDDI-II has a hybrid mode, which has a 125-μs cycle structure to transport isochronous traffic.

5.6 Fibre Channel

Fibre Channel is a fairly new interconnection topology. Although its founders and supporters had hoped that this would become a full-fledged networking topology, it has generally become a storage solution. Fibre Channel RAIDs are being configured with up to 128 drives. Fibre Channel can be set up as either a "channel" transmission topology or as a "network" topology. Channels are generally used for storage. SCSI and high-performance parallel interface (HIPPI) are very common channel solutions for storage applications. A channel is much like the old telephone system, where a dedicated point-to-point connection is made between two devices. Channels are known for speed and reliability. The other topology use is as a network. In this topology a virtual space is generally used, where all devices are connected to each other and share the transport medium. Networks offer flexibility and reach.

The transport medium used for Fibre Channel is two interconnections between Fibre Channel nodes, either in a channel or a network configuration. Each connection, or line, handles traffic in a single direction, which is unlike SCSI, where the same bus is used for both directions. Full-speed Fibre Channel is rated at 1.0625 Gbits/s in each direction. Actual data throughput is only in the 600-Mbits/s range. This is the capability of the first generation of Fibre Channel data layer silicon, but at the physical layer the silicon was designed for a full 1-Gbit throughput. The word "Fibre" in Fibre Channel is purposely spelled the way it is to denote that it is not only intended for transport down a fiber-optic cable. There are three Fibre Channel topologies: point-to-point (only two nodes); a loop topology like token ring that can support up to 126 nodes, or ports; and a switched topology that can have up to 2^{24} items in its address space. The actual addressing in Fibre Channel consists of 64 bits, which is separate from the 24-bit addressing in virtual address space. There is a substantial addressing hierarchy used here as in Ethernet. Each Fibre Channel device has a unique address, different from that of any Fibre Channel device ever created. The top 4 bits signify the addressing authority that sanctions the rest of the addressing scheme. The lower 16 bits are locally assigned by the local controller. When used in a loop configuration, the various nodes are usually connected to a hub, which is electrically switched to actually configure a loop topology. Fibre Channel RAID systems are generally configured as loops on the front end (connection from RAID system to outside world) and often still use SCSI on the back end (connection to the actual disk drives). The RAID controller acts as a bridge between these two systems.

In the OSI level soup, or stack, Fibre Channel can end up at various places. Fibre Channel can be used to carry protocols such as ATM; conversely, ATM could be used to carry Fibre Channel protocol. Fibre Channel could be used to replace Sonet. However, Fibre Channel will probably never be used as a WAN because the telephone companies own the long-distance wires, and they use ATM. Fibre Channel is becoming fairly entrenched, though, in the storage arena because a number of drive manufacturers offer Fibre Channel interfaces that cost the same as SCSI interfaces. Actually, in many ways Fibre Channel is only a low-level protocol since Fibre Channel has no actual commands of its own. Other higher protocols, such as SCSI, which have a command set of their own must be mapped into Fibre Channel. Fibre Channel is only the transport scheme. Some RAID manufacturers actually use separate loops to divide up the drives to lower the latency. The RAID controller bridges the different loops.

To be more exact, Fibre Channel usually is mapped over three layers of the OSI model. The lowest layer is obviously the physical layer, that is, the actual physical media used, either fiber-optic cable or coax. Next up the OSI protocol stack is the data-link layer, which provides channel coding and frame management. These two layers are typically implemented with hardware. Above the data-link layer is the transport layer, which is where multiple ports in a Fibre Channel NIC card are managed and where protocols, such as SCSI or IP, are mapped into Fibre Channel frames. If there is an application program interface (API), or socket, for Fibre Channel from upper-level applications, this is the layer where it occurs. It should be noted that the Fibre Channel people tend to refer to levels instead of layers. Fibre Channel chip sets from Adaptec, Hewlett-Packard, Symbios, Q-logic, Emulex, and others support the physical and data-link layers.

The various types of ports of NIC cards are generally either N_port, F_port, or NL_port devices. N stands for node port, which means it either expects to be connected to only one other Fibre Channel node as a point-to-point topology or connected to a switched fabric. The more common NIC topology supported is NL_port, which stands for node-loop topology. Node is the same as a node-only port, while loop means that a ring topology is supported. NL, therefore, supports both types of topology. In fact, when only two NL ports are connected to each other in a channel topology situation, they actually act as if they are in a two-node loop instead of point-to-point topology. The fabric port is the topology that a switch would support, or a fabric interconnection that would convert Fibre Channel to some other protocol such as ATM.

Early physical implementations of Fibre Channel over fiber used *open fiber control*, where the NIC cards could sense when the loop from one node to another was broken and shut down optical transmit power to prevent people from looking into the fiber at the light. This is generally not done now as it was determined that the power of the light is not generally harmful to the eye. A number of connectors are used on Fibre Channel NIC cards. Nine-pin STP twin-ax connectors tend to be the most popular. SC connectors are used for fiber, single, and multimode. TNC and BNC connectors are used for coax, but these two are generally too big for most equipment profiles. There are generally two speeds for Fibre Channel at the physical layer: full speed (1064.5 Mbits/s) and quarter speed (265.625 Mbits/s). A variant is half speed (531.25 Mbits/s), but it uses the same hardware as the full-speed implementation (with the exception of an oscillator) and the costs are, therefore, nearly the same.

Physical variants of Fibre Channel include:

100-SM-LL-L [100 Mbytes/s (full speed), single-mode fiber, laser long link, long] has an operating range of 2 to 10,000 m. It uses a 1300-nm transmitter over 9-μm single-mode fiber. Note that the 2-m minimum distance is to prevent the high-power transmitter from burning out the receiver.

100-SM-LL-I has an operating range of 2 to 2000 m over 9-μm single-mode fiber.

100-M5-SL-S uses a 780-nm laser, and has an operating range of 2 to 500 m. It uses a 50-μm multimode fiber.

100-TV-EL-S has ECL transmitters and receivers and a range of 0 to 25 m. It uses 75-Ω coax.

100-M1-EL-S also uses ECL, but only has a range of up to 10 m because it uses 75-Ω miniature coax.

25-SM-LL-L is the same as 100-SM-LL-L, except it operates at a 25-Mbytes/s bit rate.

25-SM-LL-I is the same as 100-SM-LL-I, except it operates at a 25-Mbytes/s bit rate.

25-M5-SL-I is the same as 100-M5-SL-S, except it has a top range of 2000 m, and operates at a 25-Mbytes/s rate.

25-M6-LE-I has an operating range of 2 to 1500 m using a 1300-nm light-emitting diode (LED) over multimode fiber.

25-TV-EL-S is the same as 100-TV-EL-S but it has a range up to 75 m.

25-M1-EL-S is the same as 100-M1-EL-S but it has a range up to 25 m.

25-TP-EL-S has an operating range of 0 to 50 m using ECL transmitters and receivers over "type 1" 150-Ω shielded twisted-pair.

Fibre Channel uses a 8-/10-bit channel coding scheme from IBM. This scheme is also used on HIPPI and gigabit Ethernet. The object is to keep direct current offset over time at 0 V. This is done with a lookup table with two entries for every 8-bit byte value. Which of these two values is used is based on the running disparity. Two consecutive bytes with more 0s than 1s (negative disparity), or vice versa (positive disparity), are not allowed. Consecutive bytes with even numbers of 0s or 1s (neutral disparity) are allowed. Any Fibre Channel node that receives a string of bytes which break the running disparity rules is replaced by the first node that receives it with a byte that satisfies the rules. The replaced byte will probably not be the correct one, and thus the node that the data are intended for will discover critical redundancy checking (CRC) errors, and it will discard the data. If a node sends three code violations in a row, it will cause other nodes to lose sync, and it must then reacquire the link.

The Fibre Channel methodology takes each byte and scrambles it as follows. First, the bottom 5 bits are grouped together to create a 0–32 decimal value. Next, the upper 3 bits are grouped together to create a 0–8 decimal value. Then, a period is inserted. Finally, a letter is put in front to signify data (D), or control [K (Kontrol in German)]. For example,

Hex '53' =	0101 0011	
Regrouped to	010 10011	
Converted to decimal	3	19
Add a period	3.19	
Add a prefix	D3.19	

Fibre Channel data and control bytes are referred to using this convention. Although 12 special control characters are spelled out by 8B/10B encoding, Fibre Channel uses only one: K28.5 (BC hex). Active hubs used for Fibre Channel electrically maintain a ring structure and switch around nodes that are not working correctly. The way that a hub decides if a node is not working correctly is by looking for the K28.5 characters coming from each node. The start and end of Fibre Channel frames are delimited with K28.5s. The longest you can go without a K28.5 frame is 21.5 μs. In Fibre Channel, a *transmission word* is defined as a 40-bit group of four 8B/10B encoded characters. The first character is a special character, which in the case of Fibre Channel, is always K28.5. The next three would be data characters. Remember that these three characters are chosen by the 8B/10B encoding for good spectral usage and coding distance. There are also a number of primitive signals. The first of these are idles. In an *idle transmission word* the first character is a K28.5, the second is the idle character, and the last two characters are any characters that satisfy disparity. Idles are transmitted between frames. A transmitting N_port (connection to a switching fabric) inserts six idles between frames, and expects a minimum of two idles between frames. Another primitive signal is "R_RDYs," which are used to implement buffer-to-buffer flow control. Others are OFF-line (OLS), link_reset (LR), not_operational (NOS), and link_reset_response (LRR).

Three classes of Fibre Channel operation exist:

Class 1.   Circuit-switched operation, which is connection-oriented service between two N_ports.

Class 2.   Packet-switched service, which is connectionless service between N_ports with acknowledgement (receive confirmed).

Class 3.   Packet-switched service, which is connectionless service for datagrams (send and pray).

A second Fibre Channel topology is Fibre Channel–arbitrated loop (FC-AL). The purpose of this topology is to get around the limitations of point to point, that is, the number of devices and the expense of a switched fabric. FC-AL allows up to 126 node ports and one fabric port to communicate with each other without a switched fabric. Loops generally operate as class 3 devices. A problem with large loops is latency, but it is often solved by breaking large loops into smaller ones, with each loop being connected to a hub. Each loop port passes on FC frames that it receives to the next port. Whole loops or single nodes can be configured to be public or private devices. A private device might be a RAID intended to serve only one other node, say a server, on that loop. In order to communicate, all ports that want to use the loop must arbitrate. No master node controls the loop. A node that wants to arbitrate (the node with the lowest address gets access first) will get access to the loop before one that has already or recently used the loop. Like the rest of life, though, nodes don't always play by the rules. Some nodes are programmed to be "fairly unfair," which means that once the node gains control of the loop, it conducts more than one transaction. When a device is done sending its packet, it is supposed to issue a "close" packet. After that, it should start issuing "idle" packets. As soon as the idles hit the other nodes in the loop, arbitration can start. However, the "fairly unfair" node simply issues a "close" followed immediately by an "open." An "unfairly unfair" programmed node will simply give itself access to the loop as soon as it sees an idle without bothering to arbitrate. Fabric ports tend to be unfair in both ways.

After a port wins arbitration, it seizes the loop to communicate. It starts sending packets addressed to another node. The nodes along the way simply pass the packets. If the node sees that the packets are addressed to it, it reads the packets in and replaces the packets with idles. When the communications session is complete, both the sending and receiving nodes must issue a close. The loop is now ready again for arbitration.

5.7 Frame Relay

Frame relay is used mostly over WANs, but you can find frame relay encapsulated in Ethernet and ATM packets. This protocol was originally conceived by Bell Labs as part of the ISDN specifications. Frame relay supports TCP/IP, NetBIOS, SNA, and voice traffic. As with ATM, frame relay supports PVCs and SVCs. Only 2 to 5 bytes of overhead are used per frame. Bandwidth utilization is better than X.25 or IP switching. Data are organized into individual addressed units known as *frames*. Frame relay uses no OSI layer 3 (network) processing, and only a few functions in layer 2 (data link). Frame relay uses variable frame lengths. Propagation delays through a frame-relay network or cloud will vary with frame length, but frame-relay latency is generally lower than for X.25. Frame relay uses statistical multiplexing, which means that as more users or traffic appears, the frames are made shorter for each user so that all have some access to the network. Like ATM, most frame-relay pricing is distance insensitive, so the cost is based on the amount of data inserted into the cloud; it doesn't matter how far that data travel once in the cloud.

In a frame-relay header there is a virtual circuit number that corresponds to a particular destination. This number is used at the data-link layer, and identifies the destination port on the local network where the frame is currently located. The original goal of frame relay was to keep the network protocol simple to allow fast transit through a frame-relay cloud. If errors are found in frame-relay data at any point during transit or if congestion is encountered, the frame is simply dropped. The sending device will then have to resend the lost frame. Higher layers in the OSI stack at the send and receive devices will have to determine that the frame is lost and must be resent. Devices and nodes in the frame-relay cloud learn of traffic congestion from several bits in each frame-relay header. If a particular node senses high traffic via buffer queue lengths or memory usage, it will set these bits to indicate to other nodes that congestion is occurring. Upper-layer protocols like TCP also have built means of congestion detection by monitoring latency delays and lost frames.

Frame relay often carries other older protocols such as SNA. This allows diverse SNA LAN interconnectivity over high-speed frame-relay networks. However, frame relay can find itself riding on ATM networks. ATM could become the inter-LAN backbone of choice as it handles bandwidth-intensive and time-sensitive video much better.

5.8   Integrated Services Digital Network (ISDN)

ISDN is a service offered by the telephone companies. B-ISDN means broadband ISDN, and it is based on fiber-optic transport. This is usually what is being referred to when the "information superhighway" is mentioned. Standards like ATM can run over B-ISDN. Regular ISDN was designed to use the "last copper mile" or existing 24-gauge twisted-pair local or subscriber loop. The wires are kept, but the electronics are changed. Before ISDN, digital service all the way to the home did not exist. A modem, which took digital data and converted it to analog "tones," was necessary to transmit digital data. ISDN uses the bottom three levels of the ISO model.

The local loop should be able to carry 160 kbits/s over 18,000 ft, but all four wires (both pairs) must be used to reach that rate. Basic-rate ISDN has three channels—two 64-kbit/s "B" channels for the subscriber's use and one 16-kbit/s "D" channel used for out-of-channel signaling and control. The ISDN telephone instrument and the telephone companies equipment time-multiplex these three channels on the four lines of the subscriber loop. In primary-rate ISDN, which is similar to DS1, 23 basic-rate channels are combined into one primary-rate channel.

5.9   P1394

Also known as firewire, P1394 was introduced in 1989 by Apple Computer as a serial peripheral interconnect. It is now supported as part of the Mac OS. The data rate is up to 400 Mbits/s (the standard supports data rates of 100, 200, and 400 Mbits/s). It is much cheaper to implement than Fibre Channel, and is meant for use as a home LAN.

Firewire was intended to be comparable in terms of cost and performance with SCSI but without the ID switches and terminators required by SCSI. Another goal for P1394 was fair access to interconnect bandwidth between the various nodes and a deterministic latency to the interconnect system. The initial applications for P1394 were mass storage and imaging devices. When drives for P1394 actually become available, the firewire interface will allow a camera to directly transmit a digital stream to a disk drive without the help of a computer, since firewire cameras compress the video before it leaves the camera.

P1394 currently runs at 100 or 200 Mbits/s but it will soon be available at 400 Mbits/s, and speeds of 800 Mbits/s are on the horizon. Currently, there are CD-ROMs, hard disks, digital cameras, and camcorders with P1394 interfaces. Some are only at the physical layer, and some are at the data-link layer. Home networking and PC/TV systems are expected to appear soon. Windows 98 and NT5.0 fully support P1394.

Both firewire and Fibre Channel can carry SCSI commands, and this is known as the *serial bus protocol,* which was developed by the SCSI-3 Committee. However, firewire performance is much faster then SCSI. Soon firewire will have data-transfer rates 8 times that of SCSI. P1394 uses a memory-mapped approach to the I/O interconnect. The IEEE 1394 standard describes the physical and data-link layers in the OSI model. Firewire uses an arbitrated bus architecture.

The P1394 cable contains two twisted-pair wires—one for data and clocking and the other for power. A firewire node can have more than one physical connection to the bus at once. Logically, the common data-transfer space can be thought of as a bus; however, it passes through the various nodes to get to all devices attached to the bus, with the result that the firewire bus looks more like a hub and spoke system than one common bus. Since each node will supply power to the power pair in the firewire cable, a node can be turned off and still have its physical P1394 interface powered and working. The data and strobe pair are used in such a way that the receiver does not need a PLL to recover a clock. If the binary state on the data line does not change from one bit cell to the next, then the strobe line changes its state, which means that no channel coding is needed as the clock is not embedded into the data stream.

Common firewire nodes have one upstream and two downstream connectors. Cable length between any two nodes is limited to 4.5 m. Firewire's distance limitations will probably limit it to edit bays in the average facility. A maximum of 16 hops is allowed and devices cannot be interconnected to form a loop.

When the bus is first initiated, each node negotiates with other nodes with which it is connected to determine if the node is the parent or the child of the other nodes. The node that has no child relationships with other nodes (only parental relationships) is called the *root.* The root decides who gains access to the bus when arbitration occurs. The node that is the root can even be powered down, but because its Fibre Channel data-link and physical levels are still powered, it can continue to function as the root until it is physically removed from the bus. As devices are added and removed, the root recognizes this and reconfigures the bus.

There are two data-transfer methods by firewire, the first of which is asynchronous. About 20 percent of firewire bandwidth is allocated for the asynchronous method, and it is known as a transaction-based approach. One node sends data to another and the receiving node acknowledges the received data. These data might pass through many other nodes as they travel from the sender to the receiver, but all nodes ignore these data, except for the node to which the data are addressed. This type of data transfer uses a 64-bit address space, and the addressing system is much like that of Fibre Channel.

The second type of transfer method is isochronous. This method is used to move audio and video across a firewire bus. It uses only 8 address bits. The sender broadcasts the message, and all nodes are able to use the received data if desired. No nodes acknowledge receive of the data. This is a "ship and pray" approach. The data are sent only once and only at discrete time intervals, known as *isochronous cycles*. These cycles occur every 125 μs to allow for a deterministic amount of data and latency. When the isochronous cycle starts, all devices transmitting this type of data get to transmit first; whatever time is left over until the start of the next isochronous cycle can be used for asynchronous transfers. If the bandwidth is available, up to 63 virtual channels or transactions can be handled at once. There are approximately 35 firewire API commands, and approximately 12 of those are for configuration.

5.10   Small Computer System Interface (SCSI)

SCSI is the predominate method for moving data between computers and external disk drives. The SCSI specification covers not only the protocol that controls the conversation between SCSI devices, but it also specifies the electric interface or physical layer. The physical SCSI connection is slowly being replaced by other connection schemes such as Fibre Channel (FC). However, FC doesn't have a communications protocol of its own, so it continues to use SCSI protocol commands.

SCSI is primarily used for network file servers, or in an inherently multitasking system such as Unix, because SCSI can handle queued commands and it has a faster transfer rate. Macs, workstations, and VME bus-embedded systems also use SCSI. MS-DOS had no inherent support for SCSI, but Windows 95 does.

Parallel SCSI	4-Mbyte/s 6-m cable (20-m differential); eight 8-bit devices
Fast SCSI	10-Mbyte/s 6-m cable (20-m differential); eight 8-bit devices
Fast and wide SCSI	20-Mbyte/s 6-m cable (20-m differential); sixteen 16-bit devices
Fast-20	20 Mbytes with 3-m cable; sixteen 8-bit devices
Fast-40	40 Mbytes with 3-m cable; sixteen 8-bit devices

SCSI communications consist of initiators or targets. Initiators do what their name implies; they initiate communications with a target. The computer is usually the initiator and the disk (or tape) drive is the target. A typical interchange between an initiator and a target consists of the initiator checking first that the SCSI bus is not being used by checking the *busy* line. Multiple initiators and targets can share the same bus. If the bus is free, the initiator will set the *busy*. If other initiators happen to try to grab the bus at the same time, then an arbitration phase is entered. The initiator with the highest-priority number wins. The initiator that considers itself the winner will set the *selection* line. This initial bus control phase takes approximately 10 percent of the time on average.

The command phase is then entered. The initiator addresses the target and tells the target what it wants to do, such as send (write) or receive (read) data to/from the target. After the command phase, the message phase begins. This is the phase when data are actually sent between devices. In reality the communications session can bounce between message and command phases many times before the session ends. The session ends when all command lines, busy and selection, are released.

SCSI cabling has been known to cause some confusion. There are four different types of cable for use with SCSI1/2/3—A, B, P, and Q cables. The A cable is used with SCSI 1 and 2, and it is a 50-pin cable that carries nine control signals and 8 data bits plus a parity bit, all differentially. The B cable was proposed for SCSI 2 but it has not been implemented. Deemed to be used in conjunction with the A cable, the B cable was supposed to be a 68-pin cable with two additional control signals and additional bytes of data. Instead, SCSI 2 implementations use the P cable, which is an A cable with an extra byte (and parity bit) included. SCSI 3 was proposed to use the P cable and a Q cable. The Q cable is another 68-pin cable with 2 more bytes (plus parity bits) and two additional control lines added (the Q cable is to the P cable what the B cable is to the A cable). However, this will probably never be implemented since it also requires a second cable.

SCSI can be implemented as a single-ended system. When this is done, the maximum length between SCSI devices is 6 m. Since this implementation allows for cables with fewer conductors, it is used often.

5.11   Serial Storage Architecture (SSA)

SSA, along with FC, uses the SCSI 3 command layer. Although the general thinking is that parallel is faster than serial, data clocking and skewing become an issue at high speeds. SSA can only communicate with any particular device at 20 Mbytes/s, but it can carry on simultaneous conversations at that rate around a ring at up to 80 Mbytes/s. The total number of devices allowed is 129. The maximum cable length is 40 m with copper and 680 m with fiber. This loop architecture requires that each device have dual SSA interfaces. SSA uses no arbitration, A device transmits data to the next device in line with it. That device decides if the data are for it, and, if not, sends the data on to the next device. SSA is championed by IBM (Fig. 5.2).

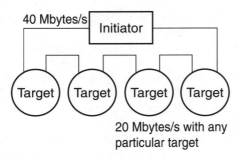

Figure 5.2 Serial storage architecture.

5.12 Slotted Ring

In this approach, which is also a random medium access control method, each device transmits fixed-length messages called *slots*. The devices are connected to each other to logically form a circle, or ring, using twisted-pair. A header at the beginning of each slot marks the slot as empty or used. If a device wants to transmit, it uses an empty slot and marks it used. The receiving device copies the data and marks the slot open. One device on the ring, usually called a *monitor,* looks for used slots that have traversed the ring at least once, and no receive device has reset the slot to unused. The monitor sets the slot to unused when necessary. This approach allows use of fairly simple NIC cards and software but is inefficient with short messages and is not commonly used.

5.13 Socket

The term "socket" refers to a group of software subroutines that provide access for applications to TCP/IP. A socket is software that allows one computer process to talk to another. These processes can be on the same computer or two computers at opposite ends of the country. A socket allows upper-level applications to communicate with other applications, either on the same or another computer. The socket software calls a function called *socket()* to open a communications path to another process. The software then binds the socket with an actual address with the function *bind()*. The created socket then either *listen()s* or tries to start communications with another process by calling *connect()*. If the process the socket attempts to connect to accepts, the communication request will call *accept()*.

Communications can now be conducted in two ways: connectionless protocol or connection-oriented protocol. User datagram protocol (UDP) is used to conduct connectionless communications. This approach is started by calling *sock_dgram*. Data will be sent in bunches or datagrams. UDP has a datagram size limit of 64 kbytes. The connection-oriented approach is called by *sock_stream*. This method is called transfer control protocol (TCP). TCP is a connection-oriented protocol. Although TCP has no size limit on the amount of data sent at a time, it does have initial setup time overhead when the connection is made. TCP is useful for implementing network services such as network login (rlogin, telnet) and for file transfer (FTP).

When the connection is made, the two applications talk to each other using *read()* and *write()* functions. When the conversation is complete, both call the function *close()* to hang up on each other.

5.14 Transmission Control Protocol/ Internet Protocol (TCP/IP)

The transmission control protocol/Internet protocol networking scheme implements peer-to-peer client-server architecture. IP is responsible for moving packet data from node to node, based on a 4-byte destination address. TCP is responsible for verifying that data have been correctly delivered from one node to another. It detects errors and lost data, and performs retransmission of the data until they are received correctly. TCP requires an approximate 10 percent return path compared to the send path for ACK packets, which means the asymmetric digital subscriber lines (ADSLs) can't be too unsymmetrical or the transmit side will be slowed down as the server waits for ACKs from the client.

IP addressing is based on the idea of hosts and networks: A *host* is anything that can send and receive IP packets on the network. Servers and clients are both IP hosts. An IP address is 32 bits wide. That provides about 4 billion unique addresses, but many of those addresses are restricted. Part of the address is a network number and part is a host number. By convention it is written in decimal form as xxx.xxx.xxx.xxx. (A period is used between each byte.) IP addresses range from 0.0.0.0 to 255.255.255.255.

There are three general classes of these addresses:

Class A. If the first byte is 0 to 126, then the last 3 bytes are for hosts and this first byte is a network address.

Class B. If the first byte is 128 to 191, then the first 2 bytes signify the network number and the last 2 bytes signify host addresses.

Class C. If the first byte is over 191, then the first 3 bytes signify the network and the last byte the host address.

If the first byte has a value of 127, then it is a loop-back address and should only be used for addressing outside a particular host. If a host number is 255, say 203.37.141.255 (in the case of a class C address), this would indicate a broadcast over the 203.37.141 network. A new version of IP is in the works that would have 128 bits of address space.

In order that each host knows which part of its address is the network address, the host is given a "net-mask." It "ands" its address with the net-mask, and what falls out is only the network address. In class C above, if a host had an IP address of 194.17.6.134 (the first 3 bytes are the network address), then its net-mask would be 255.255.255.0. This would "and" with 194.17.6.134 to be 194.17.6.0. The host, therefore, knows that it is on network 194.17.6.

To expand the number of networks an Internet service provider (ISP) might have for client use requires suballocation of a class C address, that is, not use all of the last byte for host addresses but only allow the last 6 bits for host addresses. That would still allow for 64 separate host addresses, but now four subnets could exist where only one did before. The net-mask for these four subnets would be 255.255.255.194. In the case of our previous example one net would become four with the network addresses of 194.17.6.1, 194.17.6.2, 191.17.6.3, and 191.17.6.4.

IP hands its IP address to the OSI layer (data link) below it. If Ethernet is used, the IP address must be matched to the Ethernet address, which is usually 6 bytes (and usually written in hex). This is kept in what is called the *address resolution protocol* table.

5.15   Telephone Digital Hierarchy

The vast majority of data traffic currently moves around the country via the long-distance telephone carriers. Although ATM is used in a lot of short-haul situations, generally synchronous optical network (Sonet) is the choice for long-distance operations. Sonet can be run over short distances on coax if the data rate is fairly low. The telephone hierarchy today starts with terminal equipment (TE) at customers' locations. Often the first device encountered away from the terminal equipment is a regenerator or repeater. The path between each item of equipment in the overall data path is called a *section*. The path from the terminal equipment to a central switching office (CO) is called a *line*. The CO equipment is sometimes referred to as *line termination equipment*. Today the CO is often called an add-drop multiplexer (ADM) or a digital cross-connect system (DCS). Here data streams are broken apart or demultiplexed, sorted, and remultiplexed into new data streams going toward their final destination. The overall path from one customer's TE to another's is called a *path*. This multiplexed approach means that the connection is virtual or connectionless. There is no hard connection between the two TEs communicating.

5.15.1   Sonet

Sonet today generally runs over single-mode fiber-optic cable and is approximately 8 μm thick. Since the cable is so thin, there is very little light refraction down the cable, resulting in little bunching of light pulses, which equates to a high bandwidth. Single-mode circuits can be as long as 60 or 70 mi. Today there is only a single direction of light traffic through a fiber. However, there are systems in process that will use two different-wavelength lasers so that traffic can be sent in both directions. The minimum number of fibers bundled in a fiber-optic cable today is 12. Fiber is usually manufactured in ribbons of 12 paths—

Line Rate Mbits/s	Sonet Optical Level	Sonet Electrical Level	Sonet Capacity
51.840	OC-1.	STS-1	28 DS1s or 1 DS3
155.520	OC-3	STS-3	3 DS3s
622.080	OC-12	STS-12	12 DS3s
2488.320	OC-48	STS-48	48 DS3s

Figure 5.3 Sonet line rates.

72 to 216 is a common number of fibers in a finished cable. Fiber-optic cable is expensive to install but cheap to maintain, whereas copper is just the opposite.

Sonet has standards that define the frame rate; line rates; tributary mapping; physical characteristics; and administration, maintenance, and operations capabilities (Fig. 5.3).

5.15.2 Digital service (DS)

Common digital service rates are as follows:

DS0. A single channel of voice or data is transmitted as 8000 discrete samples per second. Each sample is 125 μs in length. At 8 bits per sample, that is 64 kbits/s, which is the smallest element of service. Actually, most DS0 service only has a 56-kbit/s payload because the telephone company uses 8 kbits/s for signaling and control. Therefore, a 56-kbit/s circuit only has 7 bits of amplitude resolution, with the telephone company using one of the bits per sample.

DS1. This type is made up of 24 DS0 channels compiled into what is called a frame. A DS1 frame consists of a byte sample from each of the 24 DS0 circuits, plus 1 framing sync bit, for a total of 193 bits. DS1 has 8000 of these frames per second, which makes a bit stream of 1.544 Mbits/s. DS1 can be set up to carry 23 channels of computer data at 64 kbits/s. The twenty-fourth channel is used for signaling.

DS2. Carries four DS1 signals. A DS2 frame consists of four 193-bit DS1 frames, plus an additional framing bit, for 789 bits per frame.

DS3. Carries 28 DS1 (or seven DS2) circuits (672 DS0 channels) and has a data rate of 44.736 Mbits/s. While DS0 through DS2 are usually sent over a copper path, DS3 is almost always done with a fiber path.

Note that DS2 and above originally used plesiochronous digital hierarchy (PDH), which is nearly synchronous, with multiple levels of multiplexing (DS0 into DS1 into DS3). These are considered nearly synchronous because specific bits of information found in one frame may reside in a different location in the next frame because of bit stuffing. Bit stuffing is required if DS1 channels from far-flung locations are being multiplexed into a DS3 stream. The propagation time from the various DS1 sources to the DS3 multiplexer could vary as the weather changes or as other things slightly change along the path (for example, when sun rises and warms lines, causing them to expand and lengthen the route). If the signal were taking a little longer to arrive than before, an extra bit would be added to make up for the delay. Therefore, the size of the frame would vary. The converse would be true as the sun set or went behind a cloud. DS1 was always considered synchronous by telephone engineers because a particular bit for a particular channel is always found in the same place of a DS1 frame. Lately telephone engineers have been moving away from PDH and its bit-stuffing approach, using processes like HDLC and the use of pointers. This pointer is placed in the header of a frame and it tells the receiver how far into the frame the data starts. Because of this, synchronous digital hierarchy (SDH) is now used at all levels in new installations. Almost all SDH signals travel over fiber, except when undergoing switching.

5.15.3   Optical carrier (OC)

Common OC data rates are as follows:

OC1.   Carries a DS3 data stream. With control overhead (about 8 Mbits/s), it has a bit rate of 51.840 Mbits/s.

OC3.   Carries either an STM-1 bit stream, three DS3 streams, 84 DS1 streams, 2016 DS0 streams, or a concatenated mapping where all the bandwidth is given to a single user (no digital hierarchy at all). These users usually run ATM or FDDI over these Sonet circuits at 155 Mbits/s.

OC12.   Carries an STM-4 bit stream (622 Mbits/s).

OC-48.   Carries an STM-16 bit stream.

5.15.4   Synchronous transmission module/ synchronous transport signal (STM/STS)

STM is the standard used by most of the world. The United States uses a standard called synchronous transport signal (STS) (Fig. 5.4).

STM-0/STS1.   Carries a DS3 data stream. An STS1 frame consists of data arranged in 90 columns with nine rows, which represents 810 bytes of data. However, the first 3 bytes of each row, or 27 bytes, are used for overhead, and 8000 of these frames are sent every second, so each frame is 125 μs long. The 27 bytes of overhead are used to point to where valid data start in the frame and bytes for voice traffic between regenerators, hubs, etc. Also, control and message traffic can be sent over the path using these overhead bytes. The 783-byte payload can be used to carry ATM or other types of traffic.

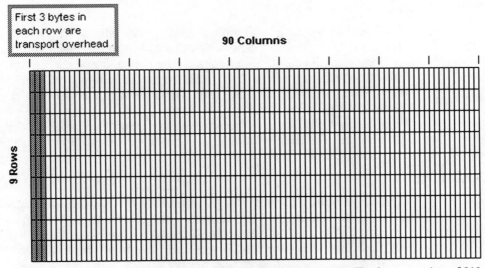

Figure 5.4 STS-1 frame structure. Each rectangle represents 1 byte. The frame consists of 810 bytes and the frame length is 125 µs. 8000 frames/s; 6.48 Mbytes/s; 51.84 Mbits/s.

STM-1/STS3. Carries three DS3 data streams. With control overhead it has a bit rate 155.52 Mbits/s. An STS3 frame also occurs every 125 µs. Instead of 90 columns, it uses 270 columns with nine rows of organized data. The overhead is 9 bytes per row, or 81 bytes per frame.

STM-4/STS12. Carries four STM-1 bit streams. It has a bit rate of 624.08 Mbits/s.

STM-16/STS48. Carries four STM-4 bit streams. It has a bit rate of 2488.320 Mbits/s.

5.15.5 T1

T1 is a transmission path that is wide enough to carry a DS1 circuit.

5.16 Token Bus

This method is physically connected like Ethernet. However, a device can only transmit when it has a special data unit called a token. This token is passed from one device to another. This and the remaining access control methods are considered distributed medium access control methods.

5.17 Token Ring

This approach uses point-to-point connections from one network device to the next. As with slotted ring, devices are connected to each other to logically form a circle, or ring, using twisted-pair. This method also uses a token.

5.18   Universal Serial Bus (USB)

USB is championed by Compaq, DEC, Intel, Microsoft, NEC, and Northern Telecom. This group of companies has published a spec that allows up to 63 devices, such as cameras, phones, your mouse, a printer, etc, to be connected on a 12-Mbit/s interface. The spec is supposed to allow for "hot" plug and play. USB is actually a variant of P1394. However, while P1394 can carry SCSI protocol, USB does not.

Once the USB device is installed, the enumeration process begins, where the host determines the features of the USB device. USB devices are plug and play.

5.19   X.25

X.25 is generally used in WAN systems. It is a connectionless, virtual circuit between two points. In voice long-distance calls, a time slot is reserved for each frame (see DS0/1/3) of that call. With X.25 a time slot is used only if there is an X.25 packet to send. X.25 was the forerunner to ISDN, frame relay, and ATM. X.25 is a switched virtual network, also known as packet switching. The X.25 data stream contains headers which tell each node along the path where it wants to go. Each node or router stores each received packet and forwards it to an available node when it finds one closer to the desired destination. This process is known as store and forward. The node owns a received packet until it can forward the packet. X.25 data put into an X.25 cloud will eventually emerge at its desired location, but it might have a great amount of latency, which won't work for voice messages. The protocol for building these packets is known as link access procedures balanced (LAPB).

5.20   Open Systems Interconnection (OSI) Layers

Let's say you have a fantastic new concept for an Internet browser. The hot idea is to put a tuner graphic in the lower corner of the user interface and let the user click on it to change, or select, new web sites. Although not too original, it may be an idea whose time has come. You write the software (code in software lingo) for the graphical user interface (GUI) to display your "tuner knob" and the other attributes the browser needs. A hypertext markup language (HTML) interpreter will have to be written because that is the language that web pages use. Code to implement a version of Java will also have to be developed. These two modules of code are necessary because the fancy moving graphics and interactivity of many web pages are written in Java. To display pictures, code will be needed that can decompress JPEG images. All this is halfway to having a new browser. Now imagine that the code had to be written to use an RS-232 port, and even more code had to be written to implement the addressing scheme used on the Internet. Now imagine the frustration at the realization that most potential customers will use Ethernet to get on the Internet. More code will still be needed (Fig. 5.5).

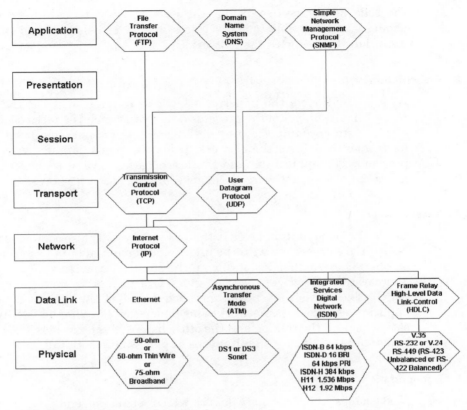

Figure 5.5 OSI software/hardware stack.

5.20.1 Application layer

In order to eliminate the scenario just described, the software running in a PC has undergone stratification. New browsers don't need to know how to communicate out all ports; they just need to be able to interface with the user and call generic services in software already running in the PC. The software that provides the services to the program is said to be on a layer *below* the application. The application literally resides on what is known as the *application* layer, and the software it calls to provide services is the next layer down and called the *presentation* layer.

The International Organization for Standardization accepted a standard developed by IEEE under guidelines from ANSI that describes this network architecture model. ANSI is the U.S. representative to ISO. This model separates the mechanics of two computer applications communicating with each other into seven layers, with the intent of making each layer independent of each other. The reason for this change was so that someone who writes a spreadsheet application to be used over a network didn't also have to write all the code necessary to talk out the port on the NIC card. The spreadsheet should simply be able to call services in a layer below it, and those services know how to deal with the network.

5.20.2 Presentation layer

As just mentioned, the uppermost layer in the OSI model is the application layer. The next layer is the presentation layer, which is a piece of software concerned with encoding/decoding data from/to the application layer or software for transmission over the network. It also performs data encryption and compression. The application and the presentation layers are often combined in the same piece of software. APIs generally reside in this layer.

5.20.3 Session layer

Below the presentation layer is the *session* layer. The session layer and layers below it are concerned with the orderly movement of data from an application on one PC to another PC, although the session layer is usually not used. Sun Microsystems, however, uses it and refers to it as remote procedures calls (RPC). The session layer was intended to define the type of dialog between applications across the network. This dialog could be simplex, where one application does all the talking and the other listens; duplex, where each application takes turns talking; or simultaneous duplex, where both talk at the same time.

5.20.4 Transportation layer

Next comes the *transport* layer, which ensures reliable communications through the network. It provides end-to-end connection services. If errors occur, it corrects and resends the necessary data. The common set of services that reside in this layer is TCP. Another common set is known as user datagram protocol (UDP), which is used by Sun Microsystems net services.

5.20.5 Network layer

The *network* layer is located under the transport layer. The data units used at this layer are often referred to as packets. This layer encapsulates the data with addressing to get it to the desired destination. The most common protocol to accomplish this is Internet protocol (IP). The packets perform analysis of the IP header, usually in software. Routers function on this layer.

Work is now being done to define a multiprotocol label switching specification to create a unified mechanism for integrating IP routers and ATM switches. In this model each router is also a switch that can rapidly forward packets that carry fixed-length labels. Labels are based on network layer routing protocols. Telecommunications network switching today entails converting from optical to electrical, switching, and then conversion back to optical. Work is currently under way to create an all optical switch, based on wavelength division multiplexing (WDM) in the fiber.

The D channel packets come into play at the ISDN network level. These packets are built using a protocol which is similar to LAPB called LAPD. When a call is placed, the packets sent from the D channel are used to route the call to its destination much like X.25 does. Unlike X.25 routing, each network node does not store and forward; it immediately tries to multiplex the packet into a stream and send it on its way. Some packets could conceivably be lost, but there is very little constant propagation delay through the path. This protocol is needed for voice traffic.

5.20.6   Data-link layer

The sixth layer down is the *data-link* layer. This layer is concerned with getting data from one device in the network to the next but not necessarily through the entire path. If data are going straight to a router, this layer is concerned only with the link to the router. A separate data-link session layer is for the sender to router link and the router to receiver link. This layer is also concerned with how data are formatted for transport and how those data are put on the network. At this point, the selected access control method is implemented. The data-level link layer can usually detect errors, but cannot do anything about them. Higher layers like the transport layer (TCP) react to bad data transmission. Bridges found in the network, either from one LAN segment to another or from one LAN access control method to another, operate on this layer. Systems that connect 10- and 100-Mbit Ethernet links tend to work on this layer.

5.20.6.1   Sublayers.   This layer is generally regarded as two sublayers—the upper half is called *logical link control* (LLC), while the lower half is *medium access control* (MAC). The upper LLC sublayer determines which medium access control is going to be used—random medium control or distributed medium access control method. Peer LLC sublayers exchange LLC protocol data units (LLC-PDUs).

The two types of LLC protocol data unit transfers are

Connectionless (or datagram).   Data are simply sent from one LLC layer to the next. No checking sequence checking, flow control, or receiver acknowledgement is provided. This is a ship-and-pray approach.

Connection-oriented.   Here a session between the source and destination is established and must be maintained while data are being transferred. All received LLC-PDUs are acknowledged back to the sender.

LLC has three types of services. The lower MAC sublayer is usually implemented in hardware in the NIC. The frame of data that emerges from the bottom of the data-link layer is referred to as the MAC sublayer protocol data unit (MAC-PDU). The header contains a MAC destination and source address. An LLC-PDU is located inside the MAC-PDU. If the source and destination are on the same bus, or ring, then this MAC-PDU remains unchanged until it arrives at its destination. If it must pass through a router or bridge, then the MAC-PDU will be disassembled at each router or bridge and reassembled into a new MAC-PDU with a new MAC destination and source address. MAC addressing is generally 48 bits long. MAC addresses can be globally unique. IEEE assigns the upper 24 bits of the MAC address to registered manufacturers of NICs. If the second bit is 0, the addressing is globally administered. If it is 1, then it is locally administered.

5.20.6.2 Ethernet. An Ethernet MAC frame generally has a 22-byte preamble consisting of

Fifty-six bits (7 bytes) of alternating 1s and 0s

A start frame delimiter (1 byte: 10101011)

A 48-bit (6-byte) MAC destination address (some implementations use only 2 bytes)

A 48-bit (6-byte) MAC source address (some implementations use only 2 bytes)

Two bytes indicating the length of the LLC data field that follows

The Ethernet MAC frame postamble is the number of bytes which pad the total length to a predetermined minimum.

The pad bytes ensure that a transmission is long enough so that devices at the far ends of the bus will detect a collision before the transmission is over if they both transmit at the same time. At a 10-Mbit/s implementation, using 48-bit addressing, a minimum of 72 bytes (46 LLC bytes) and a maximum of 1522 bytes (1500 LLC bytes) must be sent by each Ethernet MAC frame. There is also a 4-byte CRC (error checking). Packet bridging takes place at the data-link layers, multiprotocol routing takes place at the network layer, and TCP/UDP port assignments take place at the transport layer.

5.20.6.3 ATM. ATM resides on this sublayer, and is called the ATM adaptation layer (AAL). AAL protocols have been defined to support five classes of service:

Class X. Supports in-band service characteristics (control signaling).

Class A. Supports transfer of synchronous circuit services (constant bit rate).

Class B. Supports delivery of isochronous services such as variable-rate video and voice.

Class C. Supports connection-oriented data.

Class D. Supports connectionless data.

ATM has four adaptation types:

AAL1.   Provides source timing recovery. One of the 47 payload bytes in this type is used as a sequence number. This sequence number is not needed with MPEG since MPEG transport packets incorporate their own time stamps. It is also designed to transport a constant stream of bytes and provide FEC capability. AAL1 is a constant bit-rate circuit (class A).

AAL2.   Designed for real-time packet services, such as telephony with silence detection, in which multiple cells need to be transported in the same ATM virtual circuit. AAL2 is a variable-rate service (class B) that is also useful for video and audio. There are 3 bytes of overhead in the payload for this type of service. Therefore, there are only 45 bytes of user payload in each ATM cell.

AAL3.   Designed for connection-oriented data services. AAL3 has 4 bytes of overhead in the payload area.

AAL4.   Designed for connectionless service. AAL4 has 4 bytes of overhead in the payload area. AAL3 and AAL5 have a 10 percent cell overhead.

AAL5.   AAL5 is a raw cell with no overhead in the user payload area. Also, it doesn't use the CRC check byte in the header. Higher levels of the OSI model are responsible for this. AAL5 allows multiple MPEG transport packets plus an error-detection check sequence to be assembled into the AAL5 convergence sublayer protocol data unit (CS-PDU), which is then segmented into 48-byte ATM cell payloads. Padding is inserted in the CS-PDU in order to ensure that it is a multiple of 48 bytes in length. The CS-PDU trailer provides error detection but no correction. A higher protocol layer, such as TCP, has been considered as a solution. However, now there is added delay because a return response is now required. An advantage of using TCP is that it not only recovers from errors but also provides flow control, which will reduce the burstiness of large MPEG I frames.

AAL3 and AAL4 are considerably more involved than the other AAL protocols.

5.20.6.4 ISDN.   At the ISDN data-link level X.25 can be run on the B channels, but all the packets must go to the same destination. However, X.25 has the ability to talk to different logical terminations at that same destination. For nonpacket digital communications a protocol called V1.20 is used in the United States. This protocol uses HDLC (as does LAPB in X.25) or bit stuffing (HDLC is a bit-stuffing network protocol) to prevent long runs of "1" or "0" bits, and allows up to eight channels of multiplexing.

5.21 Physical Layer

The bottom layer is the *physical* layer. This is where the actual transmission of bits occurs. This layer not only includes the actual transmission medium but also the transmit and receive circuitry connected to the physical network. All modulation and channel coding are done in this layer.

5.22 Service Access Points

Higher layers request services from lower layers through service access points. Each service access point has an SAP address. Each higher layer requesting service has a unique SAP address so as to differentiate various requests. Each layer presents a set of services to the layer above it. As data are passed from higher layers to lower layers, a new envelope of pre- and postambles is added to the data kernel, which means that each chunk of data sent from one application to the next gets longer as it is handed down through the layers and shorter as it is passed back up through the layers at its destination. Data units at the same layer in two devices talking over the net are called protocol data units (PDUs).

In the next chapter we will see how digital video is wrapped. In Chap. 7 we will see how MPEG and ATSC streams resemble computer data streams much more than digital video and audio streams.

Chapter

6

Building SMPTE and AES Bit Streams

6.1   A Cacophony of Standards

Generation of complete digital video and audio bit streams is a multistep
process. It is necessary to sample the video, create parallel samples, and then
take those parallel samples and turn them into serial bit streams. The audio
must also be digitized. Often it is desirable to combine the audio and video
into a single bit stream or add time code, error detection, and even information about its origin into the same bit stream. The Society of Motion Pictures
and Television Engineers (SMPTE) and Audio Engineering Society (AES)
have developed a set of standards to accomplish this. SMPTE has standards
that describe how to sample analog SD composite and component along with
HD component video. AES has standards on how to sample analog audio and
turn it into serial bit streams. SMPTE also has the specifications to combine,
or imbed, the audio into the video. In this chapter we are going to look at the
various standards and see how they interrelate. We will also look at how
the SMPTE standards relate to ITU (formerly CCIR) standards, and how AES
and EBU standards relate to one another. We will look, in detail, at standards
that will help in understanding SD and HD component bit streams.
Standards that are important but not critical to the makeup of the bit
streams will be briefly mentioned. All the standards reviewed in this chapter
are loosely referred to as producing "baseband" bit streams, which means the
bit streams are carrying noncompressed video and audio. The exception to
this is SMPTE 305. SMPTE 310, the ATSC interface, is covered in Chap. 7. A
point of confusion here is that in the rf field baseband is any signal not modulated onto a carrier.

A note about SMPTE standards is in order here. Standards that specify how
devices interface or communicate are simply given numbers such as SMPTE
259M. Standards that recommend how a procedure should be done are labeled

"RP" for recommended process. The procedure for measuring jitter in an SMPTE 259M signal is SMPTE RP-184M. The "M" at the end of most standards simply means that dimensions are in metric units.

6.2   AES-3

The AES/EBU digital audio standard originated in 1985, and was revised in 1992. It has become an ANSI standard (S4.40-1992). The S4.40 bit stream is also known as AES-3. It uses an FM channel code, where the logic level of the AES stream changes state in the middle of a bit cell if a 1 is represented and no transition happens during a 0's transmission. As with an SDI video stream, the polarity of the signal does not matter, just when the transitions occur.

AES-3's physical layer has a 110-Ω source and receive impedance (see Fig. 6-8). The original 1985 spec recommended that receive impedance should be 250 Ω, so that one AES transmitter could drive four AES receivers. This recommendation was dropped in the 1992 revision. The source and receiver usually have transformers on the inputs and outputs. Driving and receiving signals through these transformers are usually done with RS-422 drivers and receivers.

It is possible to loop AES-3 if all receive devices have high-impedance inputs, except for the last one in the chain, and if the interconnect cables are fairly short to minimize reflections. You can pass the AES signal through coax as well. SMPTE 276 describes how this is done. An advantage to passing the AES signal this way is that since a 48-kHz sample rate has a bit rate of just over 3 MHz, an analog video distribution amplifier (DA) with a bandwidth of just 5 MHz can be used to distribute an AES stream.

The problem with this is that the original AES spec allowed signal amplitudes as high as 10 V, but the 1992 version restricts the amplitude to 2 to 7 V peak to peak. As a result, a 6- to 20-dB attenuation and matching network might need to be incorporated at the transmit side and a similar impedance-matching network at the receive end. A step-up transformer might also be needed at the receive end (Fig. 6.1).

AES samples are assembled into blocks. Each block consists of 192 frames and each frame consists of 2- to 32-bit subframes. Each subframe contains one sample of channel 1 or 2 audio. At a 48-kHz sample rate, there are 250 of these blocks each second. Each subframe uses the first 4 bits as a flag pattern to signify which half of the frame it is (channel 1 or 2) or to signify the start of a new block, which is channel 1 of frame 1. These are generally referred to as $X, Y,$ and Z preambles, respectively.

The next 4 bits are auxiliary data bits that can be used either to allow for 24 bits of audio amplitude resolution (as opposed to the 20 bits inherently provided for) to be assembled into 12-bit words at one-third the sample rate as an auxiliary intercom channel, as suggested by the ANSI spec.

The next 20 bits contain the audio sample for that channel. The last 4 bits in each subframe are flags that signify that

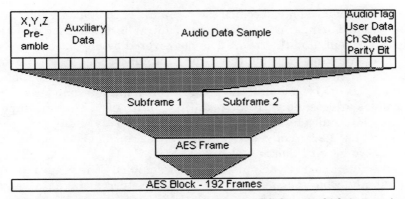

Figure 6.1 AES subframes multiplexed into an AES frame, which in turn is embedded into an AES block.

Sample data are actually audio.

One bit is for user data.

One bit is the audio channel status bit.

The last bit is the subframe parity bit.

The audio channel status bit from each subframe is combined to produce twenty-four 8-bit words for each subframe or channel per block. These status bytes contain information about synchronization status, sample rate, audio sample word length, source and destination IDs, stereo/mono channel relationship, channel reliability flags, and a CRCC check byte, among other things. This constitutes an AES audio frame, but it shouldn't be confused with the audio data stream that results when these AES audio data are inserted or embedded into a SMPTE 259 SDI stream.

6.3 Asynchronous Serial Interface (ASI)

ASI is a popular interface for digital video broadcasting (DVB) implementations. It carries MPEG-2 transport packets directly at 270 Mbits/s (no TRS sequences). If compressed payload is less than 270 Mbits/s, null packets are inserted.

6.4 SMPTE 125M

This standard's main purpose is to specify the following:

1. How component video is sampled.

2. The generation of synchronization or timing reference signals (TRS).

3. Where ancillary data (everything other than video) are added.

4. The physical interface for parallel data generated by this standard.

This standard only specifies the sampling of component video. If NTSC is to be sampled using this standard, it must first be decoded from composite NTSC to component, and that process is not covered by any spec.

The sampling rate for the Y, or luminance, signal is 13.5 million samples/s. The sampling rate for each of the color difference signals C_B (B − Y) and C_R (R − Y) is one-half the Y rate, or 6.75 million samples/s. The C_B and C_R samples are cosited with the odd (first, third, fifth, etc.) Y samples. This creates 1440 multiplexed Y and C_B/C_R samples. This sample ratio is commonly known as 4:2:2. Only active video is sampled. The sample output data sequence for each horizontal line is C_B, Y, C_R, C_B, Y, C_R, Y.... The last sample in the line is a Y. Sync pulses are not sampled. The resulting samples were originally 8 bits but are now 10 bits wide. Increasing data values for luminance data mean a brighter sample. Black is 40_H and $3AD_H$ is peak white. This allows for 877 discrete levels for luminance. Chroma is centered at 200_H because, unlike luminance, analog chroma signals have positive and negative values. The highest chroma value is $3C0_H$ and the lowest value is 40_H. Chroma has 896 discrete levels.

As with NTSC, the three primary colors—red, blue, and green—are used to produce the television picture. The complete picture is comprised of luminance and the color difference signals Cr (R − Y) and Cb (B − Y).

$$Y = 0.299R + 0.587G + 0.114B$$

$$Cr = 0.713 \ (R - Y) = 0.713 \ [(1R + 0G + 0B) - (0.299R + 0.587G + 0.114B)]$$

$$= 0.713 \ [(1R - 0.299R) + (0G - 0.587G) + (0B - 0.114B)]$$

$$= 0.713 \ [0.701R - 0.587G - 0.114B]$$

$$= 0.500R - 0.419G - 0.081B$$

$$Cb = 0.564 \ (B - Y) = 0.500B - 0.331G - 0.169R$$

The luminance and chroma matrix values are the same as ITU-R BT.601-5. Luminance and chroma difference values are both different with the SMPTE 240/260/274 specs for HD (Fig. 6.2).

In place of sync pulses groups of words called timing reference signals (TRSs) are inserted. There are two types of TRS strings, start of active video (SAV) and end of active video (EAV). As their names imply, these strings signify where the active video begins and ends. Active video and any ancillary data are not allowed to have values 000_H–003_H or $3FC_H$–$3FF_H$. Only TRS and ancillary data headers can have values in those ranges. The first three TRS words are $3FF_H$, 000_H, and 000_H. The fourth TRS word is a set of three flags. This word, referred to in the standard as the *XYZ* word, uses 3 bits to indicate whether this TRS is

Waveform Component Parade Display

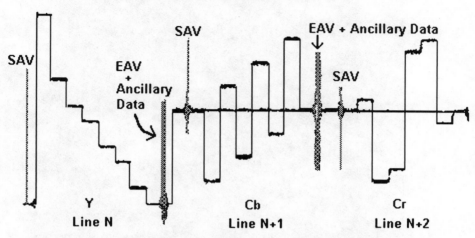

Figure 6.2   Analog representation of SMPTE 125 serial component digital video.

1. An SAV or an EAV
2. In the vertical interval or not
3. Field 1 or field 2

See Fig. 6.3.

Most of the rest of the bits in the *XYZ* word are protection bits to prevent it from having illegal values below 003_H and above $3FC_H$. The vertical interval extends from line 0 (264, field 2) to at least line 10 (273). Versions of the standard before 1995 allowed the vertical to end any time after line 10 (273) but before line 20 (283) to accommodate analog data, closed captioning, VITS, VITC, etc. This standard also calls for the vertical interval to be in full line increments unlike NTSC. Field 1 begins on line 4 and field 2 on line 266. SAV occurs immediately before the first video sample. It technically occupies the last four words in the line (1713–1716). The first video sample (C_B) is the first word in the line. EAV is immediately after the 1440th and last video sample (Y) of the line (1441–1444) (Fig. 6.4).

The space between the EAV and SAV represents traditional horizontal blanking. The number of words available during this time for use by ancillary data is 268. The header for this data is 000_H, $3FF_H$, $3FF_H$. Like the video, horizontal ancillary (HANC) data consists of 10-bit words. This is all the standard has to say about HANC. Inserting audio in HANC is covered in SMPTE 272. The horizontal lines during the vertical interval can be used for ancillary data (VANC) because early D1 VTRs, which were only 8-bit machines, used this data space. In order to stay compatible this standard specifies 8-bit VANC. As a result, VANC is not used much. However, HANC can be sent during the horizontal blanking intervals occurring in the vertical intervals. Vertical interval

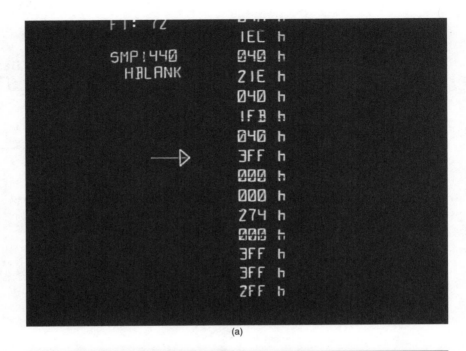

(a)

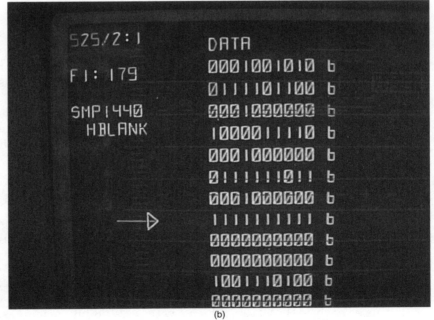

(b)

Figure 6.3 (a) Hex decimal representation of EAV sequences. (b) Same EAV sequence in binary.

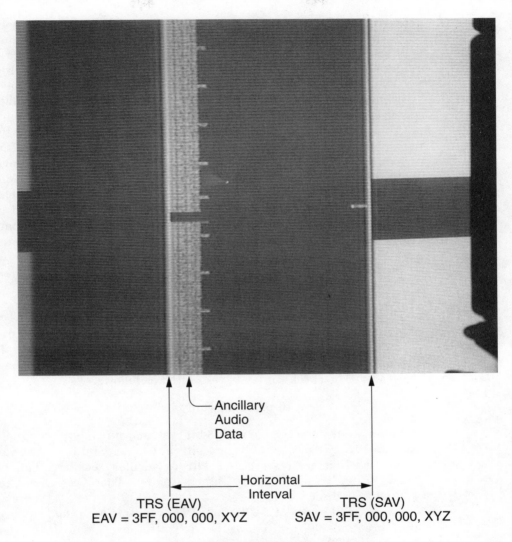

Ancillary
Audio
Data

Horizontal
Interval

TRS (EAV)   TRS (SAV)
EAV = 3FF, 000, 000, XYZ   SAV = 3FF, 000, 000, XYZ

XYZ Indicates EAV/SAV,
Field 1/2,
Active Video/Vertical Interval

Figure 6.4   EAV TRS followed by ancillary data (audio in this case) and then eventually SAV TRS.

time code (VITC) can be inserted on the Y samples only on lines 14 and 277. The chroma difference samples can be used for video index information. Each chroma sample, C_B and C_R, carries only 1 bit of a video index word. Therefore, eight chroma samples are needed to convey one index word. A chroma sample of 204_H equals a "one" and 200_H equals a "zero." These chroma values are extremely small, so if the digital bit stream is converted back to analog NTSC, color information from the video index data will not be noticeable.

The output interface calls for 11 balanced signal pairs, of which 10 represent sampled data (video/TRS/ancillary data). One pair is for a 27-MHz clock, which is the sample rate of the standard. Voltage (0.8- to 2.0-V) and impedance (110-Ω) output levels are standard ECL values. The receiver impedance is also 110 Ω. The connector specified was a 25-pin D. In addition to the balanced signals, the connector also had a system ground and cable shield connection. Rise/fall times had to be less than 2 ns and jitter had to be less than 3 ns. The clock was allowed to be as much as 11 ns late. However, because of skew between the parallel signals due to impedance and driver/receiver differences, cable lengths of more than 50 m were problematic.

6.5   SMPTE RP165

This standard describes error detection and handling (EDH). EDH uses lines 9 and 272 to embed two sets of checkwords and three sets of error flags. There is a set of checkwords for the entire field and another for just the active picture area. The checkwords are calculated as follows:

$$\text{16-bit checkword} = x^{16} + x^{12} + x^5 + 1$$

In order to stay compatible with 8-bit systems, only 8 bits are used. One bit is used for parity. Another is the opposite of the parity bit to ensure values aren't generated that are reserved for TRS or ancillary headers. This only leaves 6 bits in each word to carry the checksums. Therefore, each checkword is spread over three words.

As specified in SMPTE 272, the EDH data start with the standard ancillary data preamble 000_H, $3FF_H$, $3FF_H$. They also have a data ID of $1F4_H$, a block number of 200_H, a data count of 110_H, three words for the active video checksum, and three words for the full field checksum. Next come three words for error flags, where one word is for ancillary data errors, the second is for active picture errors, and the third is for full field errors. Each word has five error flags in it. These error flags are

Error detected here.   Error occurred between last device supporting EDH and here.

Error detected already.   The error occurred upstream from here.

Internal error detected here.   If the device has internal diagnostics, it has detected an error not related to transmission but internal to the device.

Internal error already.   Device upstream knows it has an error.

Unknown error status.   Device supporting EDH has received an error from an upstream device that does not support EDH.

Most ancillary data begin right after EAV, but the EDH ancillary data header starts just prior to SAV at the 1690th word of line 9/272. The last EDH word (1712th word) is a checksum. The next word is the start of the SAV preamble.

6.6   SMPTE RP168

This standard specifies that video switching from one source to the next should occur 25 to 35 µs after the start of line 10/276.

6.7   SMPTE RP178

This standard describes the test signals, commonly referred to as pathological signals, which are used to stress an SDI receiver's equalization and clock recovery ability.

6.8   SMPTE RP184

This standard, which describes the measurement methodology that should be used to determine jitter in a serial bit stream, is now called out in SMPTE 259. It divides jitter characteristics over time into two frequency bands. Jitter components under 10 Hz are referred to as timing jitter or wander and components over 1 kHz are called alignment jitter. The maximum jitter is a function of the period of one clock cycle, which is referred to as a unit interval (UI). SMPTE 259 calls for both timing and alignment jitter to be no more than 0.2 UI peak to peak.

6.9   SMPTE RP186

This standard describes the format of the video index data specified in SMPTE 125. As spelled out in SMPTE 125 these data are on lines 14/277. Currently, there are two classes of video index data; both use twelve 8-bit words, including three CRCC checkwords. These data are metadata about the SDI data stream (see Chap. 7).

Class 1 video index data include:

Scanning system

Signal type—RGB, YC_RC_B, etc.

Sampling structure—4:2:2, 4:2:0, etc.

Pan and scan data

Field rate heritage

Current color field—helpful if video will end up back as NTSC

Class 2 video index data include:

Film frame rate

Scene just changed

Original film material

Color encoding heritage and colorimetry

Gamma and highlight compression information

Number of quantization bits per sample

Video filtering before digital conversion

Currently, this standard is very poorly supported by manufacturers.

6.10   SMPTE 240M

This standard is the original analog HD spec. It was a 1125-line interlaced system with 1035 active lines and a 16:9 aspect ratio. It allowed either a 60- or 59.94-Hz field rate, resulting in a horizontal line rate of 33,750 Hz (at 60 Hz) or 33,716.28 Hz (at 59.94 Hz). This was an analog component system in which each of the three signals (one luminance and two color difference signals) had a bandwidth of 30 MHz. The luminance signal has a peak video level of 700 mV, while the color difference peaks are ±350 mV. All three signals had a set of sync pulses that went −300 mV negative and then +300 mV positive. The pulses had a total duration of approximately 7.9 μs. The horizontal blanking interval is approximately 6.7 μs, with active video being 25.9 μs in duration. Vertical blanking was 45 lines long with a five-line (10 vertical serrations) vertical interval.

The values that comprise luminance and chrominance are different than SMPTE 125M. These values are

$$Y = 0.212R + 0.701G + 0.087B$$
$$Cr = 0.635 (R - Y) = 0.500R - 0.445G - 0.055B$$
$$Cb = 0.548 (B - Y) = 0.500B - 0.384G - 0.116R$$

6.11   SMPTE 244M

This is the composite sampling standard, and is the composite equivalent to SMPTE 125M. We will briefly cover it as component has won out over composite. Today, this sampling standard is mainly used inside devices such as VTRs, TBCs, and frame syncs. SMPTE 244M takes NTSC (SMPTE 170 or ANSI/EIA RS-170A) composite video and samples it at 14.31818 MHz, which is 4 times the frequency of the color subcarrier in NTSC. There are 910 samples per line, with 768 for active video. The samples are timed so that they alternately fall in this with the I and Q color axes. The magnitude of each sample not only represents chroma information but also the luminance information. The digital horizontal blanking interval starts at the 769th sample and extends to the last sample per line (910), which includes NTSC front and back poaches along with the sync pulse and color burst. This area isn't actually sampled; lookup tables are used to produce representations of the blanking elements that conform to SMPTE 170 specs. Of course the device performing the A/D conversion will look at the incoming sync and color burst so as to sample the video correctly. The same thing is done for the vertical interval, which extends from line 525 sample 768 to line 9 sample 767 for field 2 to field 1 transition. For field 1 to field 2 transition the vertical interval is from line 263 sample 313 to line 272 sample 767. This conforms to one-half the horizontal line of active video at the end of field 1 and the one-half line at the start of active video in field 2. The mechanical and electric parallel interface produced is the same as found in SMPTE 125. It should be noted that there is a five-word word TRS signal string inserted at the beginning of the sync pulse (words 790–794).

6.12   SMPTE 259M

This is the standard that takes the parallel data output of SMPTE 125, SMPTE 244, or SMPTE 267 and converts the output to a serial data bit stream, commonly known as standard definition–serial digital interface (SD-SDI):

SMPTE 244M (4Fsc 525/60 144 Mbits/s) → SMPTE 259M-A

SMPTE 244M (4Fsc 625/50 177 Mbits/s) → SMPTE 259M-B

SMPTE 125M (4:2:2: 4:3 component) → SMPTE 259M-C

SMPTE 267M (4:2:2 16:9 component) → SMPTE 259M-D

SMPTE 259 specifies the following:

1. The output electrical characteristics
2. Channel coding
3. Ancillary data formatting

The output of SMPTE 259 uses a BNC connector that can pass frequencies of up to 850 MHz. It is an unbalanced signal with an impedance of 75 Ω. Both the transmitter and receiver return loss should be at least 15 dB from 5 to 270 MHz. The peak-to-peak signal amplitude is 800 mV, with no dc offset. Rise/fall times (20 to 80 percentage points) should be between 0.40 (originally 0.75) and 1.50 ns. The jitter spec was originally ±0.25 ns, but was later refined and followed RP184 specs. It is now 0.74 ns. This standard produces bit rates of 270 Mbits/s for SMPTE 125M and 144 Mbits/s for SMPTE 244.

Channel coding is scrambled NRZI:

$$\text{Polynomial for NRZ} = x^9 + x^4 + 1$$
$$\text{NRZI} = x + 1$$

This technique generates a polarity-free bit stream. The LSB is transmitted first.

Originally, SMPTE 259 specified how ancillary data would be formatted. Now it refers to SMPTE 291 for that specification. As in SMPTE 125, if called for an ancillary preamble of 000_H, $3FF_H$, $3FF_H$, only 8 of the 10 bits are used in the word to prevent values from occurring that are only reserved for TRS and ancillary data headers. It also specifies:

One word to specify that this is ancillary data = $3FC_H$.

One word to contain data ID flags which would identify this particular data.

A one-word data block number which is incremented each consecutive block if a common data ID exists. The block count would rise to 255 and reset to 0. If it isn't used, it should be set to 0.

A one-word data count, which indicates how many user data bytes follow.

User data up to 256 bytes.

A one-word checksum.

There are several progressive sampling schemes that are intended to be transported as SMPTE 259 streams: the 480p30 and 480p24 formats. However, neither has a sampling standard yet. Although they will most likely be used in the MPEG domain, they might not be used in the baseband video domain because both create objectionable flicker on displays. With MPEG (or ATSC), the display used will convert to a native display at a higher frame rate, which will be nonobjectionable.

6.13 SMPTE 260M

SMPTE 260M is the digital sampling standard for SMPTE 240 HD analog component video. It is the HD equivalent of SMPTE 125. SMPTE 260 uses the same matrix values for luminance and chrominance as SMPTE 240. The standard has an 8- or 10-bit sampling standard with 2200 luminance samples per line, 1920 of which are in active video. Like SMPTE 240, this system has 1035 active lines. It has the same 4:2:2 sampling standard as SMPTE 259, so each of the color difference signals has 1100 samples per line (960 in active video), resulting in 74.24 million words/s for luminance and 37.125 million words/s for each of the two color differences. The total sample rate is, therefore, 148.48 million samples/s. Although the standard does allow RGB sampling, each would need to be sampled at the same rate, creating 222.72 million samples/s (Fig. 6.5).

This standard has the same TRS implementation as SMPTE 259, although ancillary data preambles are the same as SMPTE 259. Three additional words

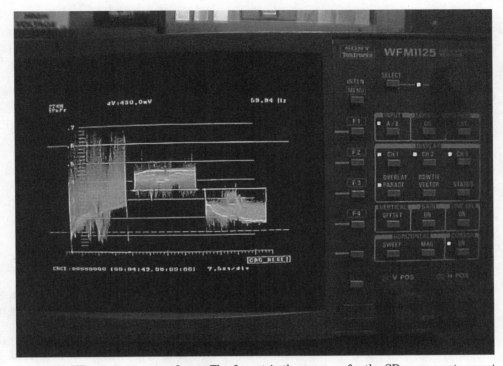

Figure 6.5 HD component waveforms. The format is the same as for the SD component, except video frequency and sampling components are much higher.

indicate data type, line number, and the number of user words that follow. Up to 272 words, including an ancillary data preamble, can be carried in the horizontal blanking interval. All the samples can be used in the vertical interval. Unlike SMPTE 125, which outputs a single multiplexed set of data words, an alternating Y and color difference sample, SMPTE 260 outputs two separate parallel outputs. One carries the luminance and the other carries the color difference samples. Both have TRS signals and both can carry ancillary data.

The physical interface consists of 31 balanced pairs of conductors, 10 pairs of which are for luminance, 10 pairs for the chroma difference samples, one pair for the clock, and 10 for a 75.24-million-sample/s auxiliary channel. The maximum cable length is 20 m. The input/output impedance is 110 Ω.

6.14 SMPTE 267M

This standard is the SD 16:9 aspect ratio spec. It handles two different sampling rates. The first is SMPTE 125M, which is the standard 4:2:2 sampled component video at 13.5 MHz. It is 2:1 interlaced video with 720 pixels/line and 483 lines/frame, which produces rectangular pixels. The CCD imager is physically constructed to capture 16:9 images, and by convention all monitors are configured to display 16:9 images. However, in between the video stream appears as a normal SMPTE 259 bit stream. Monitors not configured to display 16:9 images would display pictures with everything compressed horizontally (everyone would look thin). Since the same number of horizontal samples are used to convey a wider panoramic of picture, the horizontal resolution is decreased by one-third.

The second half of this standard allows for a different sample rate that will support the same resolution as 4:3 SMPTE 259. The sampling ratio is increased to 18 MHz. A 10-bit sampling resolution produces a bit stream at 360 Mbits/s versus the 270 Mbits/s for 4:3 video. Therefore, there are 960 horizontal samples/line. Everything else is the same (483 lines, 2:1 interlace, 4:2:2 sampling). This approach produces square pixels, so that each sample spatially represents horizontally as much as it does vertically. Most standards, such as SMPTE 125, do not have square pixels. The transport stream for this is SMPTE 259M-D.

6.15   SMPTE 272M

This standard is entitled "Formatting AES/EBU Audio and Auxiliary Data into Digital Video Ancillary Data Space," and it specifies how to make AES-3 data into the ancillary data space of SMPTE 259 (Fig. 6.6). Since horizontal ancillary data in SMPTE 259 have 10-bit words and an AES frame consists of 64 bits, some reformatting must occur. This reformatted data will start immediately after the EAV data words. They consist of three 10-bit words with the values 000, 3FF, 3FF (as opposed to 3FF, 000, 000 for EAV/SAV timing reference signals). Only the EAV/SAV and ancillary data headers are allowed to have these values (Fig. 6.7).

Next comes the data ID word. Its value indicates whether the audio data packets that follow are audio group 1 (channels 1 to 4), group 2 (channels 5 to 8), group 3 (channels 9 to 12), or group 4 (channels 13 to 16), or one of four extended audio data packets. More will be provided on the extended audio data packets shortly.

Next comes a word that is the data block number. This number was meant to be incremented with each packet. Only 8 of the 10 bits are used to count to prevent illegal values reserved for TRS and ancillary data preambles. The count value rolls back to 0 after reaching FF hex. Some equipment uses a nonsequential value here from the previous packet to indicate that a switch has occurred in the audio data to a new source. The problem here is that many products, such as some test equipment, do not increment this number for each packet, which can cause intermittent muting in some receive equipment. If this value is not incremented, it should be kept at 0 h.

The next word indicates the number of data words to follow. A maximum of 255 bytes is possible because as with the data block number, only 8 of the 10 bits are used. The number of data words will depend on what horizontal line is being embedded with this data. Here is where it gets slightly confusing.

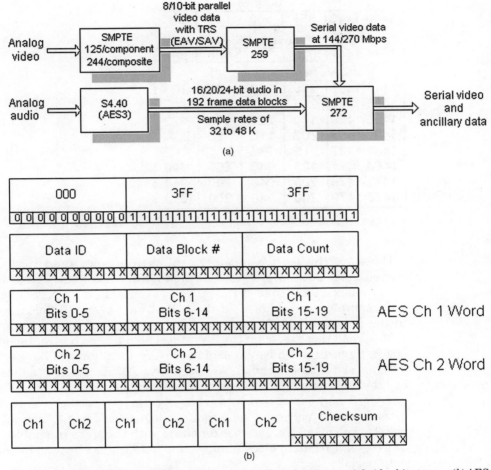

Figure 6.6 (*a*) Standards needed to embed audio ancillary data into serial video bit stream. (*b*) AES audio mapped into ancillary data space in serial video bit stream.

AES 20-bit samples are mapped into three successive 10-bit SMPTE 259 words. The extra 10 bits left over,

$$(3 \times 10 \text{ bits}) - (20\text{-bit audio sample}) = 10 \text{ bits}$$

have AES audio channel status and user bits mapped into 4 of them, with 3 bits indicating whether this is channel 1, 2, or a *Z* frame which indicates the start of 192 sample AES block, and the final 3 bits used to protect against illegal values, which again are reserved for EAV/SAV or ancillary data preambles. The result is that a single sample of channel 1 and 2 (one AES frame) requires six SDI words. One way to implement this is to have most lines carry six AES samples, which is 36 SDI data words.

	TRS (EAV)			
1440-1443	3FF	000	000	26C

	Ancillary Data Preamble			Data ID
1444-1447	000	3FF	3FF	2FF

	Data Block #	Data Count		
1448-1451	200	212	1A0	2D3
1452-1455	21E	1A2	2D3	11E
1456-1459	170	2E8	21E	172
1460-1463	2D3	11E	198	2CC
1464-1467	21E	19A	2CC	11E
1468-1471	129	040	200	040
1472-1475	200	040	200	040
1476-1479	200	040	200	040

Figure 6.7 Sample AES audio data following EAV in top row.

However, the most common sampling frequency (and the only one discussed here) used in professional digital audio for television is 48 kHz. This is not an exact multiple of the 59.94 field rate used for NTSC; it means that 8008 samples are required in the SDI stream every 10 fields, which results in the necessary 48,000 each second.

To spread these through the horizontal blanking or horizontal ancillary data space, if most lines have six samples, every 12th line will have eight.

What about 24-bit audio? Well, the first 20 bits are sent as just described, but the additional 4 bits of both channels in one AES frame are combined into one word and assembled into what is called an *extended data packet*. This extended data packet is sent immediately following the normal audio data packet. The extended data packet starts with the same preamble (0000, 3FFF, 3FFF), data ID, data block number, and data count words as the normal audio data packet.

It should be mentioned that an optional "audio control packet" is available that can be sent in the second horizontal ancillary space following the video switch point. It is used mainly to signify audio that was sampled at a rate other than 48 kHz. Each audio group (four channels) would require its own control packet. These control packets must be sent before any other audio packets are sent in this line. As stated earlier, since most professional television audio is sampled at 48 kHz, this packet is not normally sent.

6.16   SMPTE 273M

This standard is the status monitoring and diagnostics protocol, and it is intended for querying equipment for status and diagnostics via a central PC. It is modeled after SMPTE RP-116. The link between the PC, which is known as the *supervisor control,* can be via dedicated RS-232 or the commands can be wrapped in protocols like TCP/IP and shipped via a network. The supervisor can talk straight to a device that supports SMPTE 273 or to a virtual machine. This virtual machine would most likely be another PC that supported SMPTE 273 and the diagnostic set of the actual device to be monitored. This protocol defines 11 common commands and five optional commands. Manufacturers can implement commands of their own. Every command issued by the supervisor is expected to be responded to by the virtual machine or SMPTE 273 device. Strings of ASCII data can be returned by the device.

Common commands include:

- Device selection.

- Identification query by the supervisor where the device should respond with manufacturer's name, model, serial number, software version, etc. These would be strings of ASCII text.

- Command to test with the test number included as an argument. Obviously, different manufacturers' tests would be different from one another.

- Query of EDH status.

- Status of the device. The device would return an ASCII string. There are also commands to access specific device status registers.

- Request for the device to send back errors it has encountered with commands received from the supervisor.

- Large sets of binary data, such as device help files, GUIs, etc., can be uploaded from the device to the supervisor.

- A command that allows the supervisor and device to exchange strings of data that are not using SMPTE 273 message protocol.

Currently, this standard is very poorly supported by manufacturers.

6.17   SMPTE 274M

This standard amended SMPTE 260 to have 1080 active lines instead of 1035. It is a scanning and sampling standard which allows for progressive (60p) and interlaced formats (1080i). Bandwidth requirements are currently limiting any progressive implementations. SMPTE 260 specifies 90 total lines of vertical blanking, 45 per field. In SMPTE 274 that number is cut in half. Field 7 contains 562 lines while field 2 contains 566. The main reason for this change was that it provided square pixels (1920/1080 = 16/9) versus SMPTE 260 (1920/1035 = 128/69). A square pixel represents a spatial area as wide as it is tall. Pixels in SMPTE 260 represented an area that was wider than it was tall as both standards are presented 16 times as wide as they are tall. This standard allows for a 60 or 59.94 field rate.

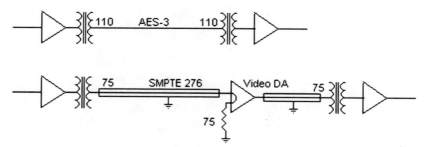

Figure 6.8 Conventional audio path versus analog video path for transporting AES audio.

6.18 SMPTE 276M

This standard specifies how AES/EBU audio signals can be sent over coaxial cable using standard BNC connectors and cable. The input/output impedance should be 75 Ω and have an amplitude of 1 V, no dc offset, and a return loss greater than 25 dB. Rise and fall times (10 and 90 percent) should be between 30 and 44 ns, which produces a bandwidth of 6 MHz. The specification calls for the receiver to work with eye heights down to 100 mV (Fig. 6.8).

6.19 SMPTE 291M

This standard is a refinement of the specification in SMPTE 125, 259, and 272 for use of ancillary data space. It provides for expanded data IDs for ancillary data. The user can, via the data ID packet (see SMPTE 259 and SMPTE 272), specify that the data block number that follows be used as an extension of the data ID, with the result that many more vendors and applications can have uniquely assigned IDs. The ID also allows for 8-bit ancillary bit streams to identify themselves. This specification also allows ancillary data to mark themselves for deletion.

6.20 SMPTE 292M

This is the HD equivalent of SMPTE 259, and it takes parallel sampled data from either SMPTE 260 (1035i), SMPTE 274 (1080i), or the SMPTE 296 (720p). It produces bit streams of 1.485 Gbits/s. There are also proposals for SMPTE 292 to transport 1080p24 and 720p24. Both would have the same pixel and line counts as 1080i30 and 720p60, respectively, but would have wider horizontal ancillary data spaces (over 3 times active video in the case of 720p24) than their counterparts. Bit rates for 1080i30 and 720p60 are made to match by having different horizontal ancillary data widths.

| | Vertical | | Horizontal | | |
Format	Active	Total	Active	Total	Frame rate
1080p24	1080	1125	1920	2750	24
1080i30	1080	1125	1920	2200	30
720p24	720	750	1280	4125	24
720p60	720	750	1280	1650	60

All four formats have the same bit rate—1.485 Gbits/s:

Total pixels per line $\times$ 2 (luminance and chroma samples) $\times$
total lines $\times$ frame rate $\times$ 10 (sample resolution) = 1.485 Gbits/s

SMPTE 292M uses the same channel coding as SMPTE 259. Since SMPTE 260 and SMPTE 292 have two separate sample sets—one for luminance and one for color difference samples (both with TRS)—SMPTE 292 TRS will be twice as long as SMPTE 259. Therefore, it has the following TRS string:

$$3FF_H, 33F_H, 000_H, 000_H, 000_H, 000_H, XYZ_H, XYZ_H$$

It has the same electrical interface as SMPTE 259, except that it has much higher rise and fall times (20 to 80 percent 270 ps). It also uses SMPTE RP-184 for jitter specs. A difference from SMPTE 259 is that it also has a fiber-optic interface spec.

6.21   SMPTE 293M

This specifies the SD progressive scanning system generically known as 480p. This is a 525-line system with 483 active lines. Each horizontal line has 1716 total samples of which 1440 (720 for luminance and 720 for color difference) are in the active picture system. Like SMPTE 260 and SMPTE 274, it provides two data samples sets out, one for luminance and one for color difference samples. If 4:2:0 sampling is used, a single 360-Mbit/s stream can be produced. It has a frame rate of 59.94 and an aspect ratio of 16:9. Like SMPTE 259, its pixels are not square. It has the same TRS structure as SMPTE 259. The standard doesn't specify but implies that the input will be SD-SDI 480i component video, which will be interpolated into 480p.

6.22   SMPTE 294M

This standard is the SMPTE 292 equivalent for SMPTE 293. It takes both sets of 480p parallel data and outputs two serial 270-Mbit/s SMPTE 259–like streams (SMPTE 294M-1) or one 360-Mbit/s bit stream (SMPTE 259M-2). The dual stream approach produces bit streams that can be handled by

SD-SDI component devices. In one bit stream the odd lines are from one frame and the even lines are from the next frame. The opposite is done for the second bit stream. Thus, on the first field bit stream 1 has only lines 7, 9, 11...525 and bit stream 2 has 8, 10, 12...524. On the next frame, bit stream 1 has 8, 10, 12...524 and bit stream 2 has 7, 9, 11...525.

A second output approach is to throw one-half the chroma values away (subsample). The chroma is sampled in a quincunx pattern, which means that on odd lines the even chroma samples are thrown away and on the even lines the odd chroma samples are thrown away, resulting in an interweaved 4:1:1 sample rate. This sampling scheme is known as 4:2:0. Some 4:2:0 schemes throw out the chroma samples of every other line while keeping the rest of the lines at a 4:2:2 sample ratio. This creates a bit stream that has a bit rate of 360 Mbits/s. The TRS, electric interface, and channel coding at the output are the same as for SMPTE 259.

6.23   SMPTE 295M

This standard describes a 1920 × 1080 scanning system (like SMPTE 292) which has a scanning rate of 50 Hz. The system can be either progressive or interlaced. In order to have the scan bit rate at SMPTE 292, 1250 total lines are used instead of 1125 as in SMPTE 292.

6.24   SMPTE 296M

This standard describes a 1280 × 720 progressive sampling and scanning system. There are 1650 total horizontal samples and 750 total lines. The frame rate is 60 Hz, and it produces a bit rate identical to that of SMPTE 292.

TRS signals are similar to SMPTE 259 and SMPTE 292. There are 362 words between EAV and SAV, all of which may be used for ancillary data. Parallel output parameters are the same as for SMPTE 274. It uses SMPTE 292 as its serial transport.

6.25   SMPTE 305

This standard allows SMPTE 259 to carry data in the place of active video. It is usually used to carry MPEG data, and is known as a *mezzanine data stream.* SMPTE 305, which was originally known as the serial digital data interface (SDDI), is also known as the serial data television interface (SDTI). The TRS structure is the same as for SMPTE 259, which means that an SDTI signal can pass through any SD-SDI component device that does not manipulate the active video region such as routers, distribution amps, etc. In fact, the SDTI stream can be viewed on an SDI monitor. However, instead of video, blocks of data will be seen. If MPEG with a group of pictures (GOP) of 2 (one I and one B frame) is used as an example, all the I frame data can be sent over the necessary lines and then the B frame data can be sent. At bit rates of 19 Mbits/s, one I and one B frame will take slightly more than one-half the active picture space. This is one way to format data. The data could be organized as columns where in the case of this example some part of both the I and B frame data would be sent each horizontal line. SDTI format data are sent as ancillary data on each line. It uses the normal ancillary header specified in SMPTE 259. With the ancillary preamble, the SDTI ancillary string is 53 words long. It immediately follows the EAV, and allows for data organization. It also contains 16 words for each destination address and 16 words for a source address. This standard is used for SDTI streams that are carrying multiple programs.

6.26 SMPTE 310

This is the standard for transporting MPEG-2 data streams. It is often used between the ATSC encoder/mux and the DTV transmitter [including digital studio-to-transmitter links (STLs)]. It is a synchronous serial interface and it uses biphase-mark coding. The peak-to-peak signal amplitude is 0.8 V. The physical connection is made with either a 50- or 75-Ω BNC connector. This standard allows data rates up to 40 Mbits/s, although the ATSC bit stream rate is 19.392656 Mbits/s.

6.27 ITU-R BT.601-5

In many ways this standard parallels SMPTE 125 with regard to sampling structure (or encoding parameters), which is essentially identical to SMPTE 125. It allows for both 13.5- and 18-MHz sampling rates for either 4:3 or 16:9 aspect ratios. This standard was developed by the International Telecommunication Union. ITU standards are followed by most of the world, whereas the United States tends to follow SMPTE's standards.

Matrix values for determining luminance and chroma are the same as for SMPTE 125.

6.28 ITU-R BT.656-3

This is the ITU standard that describes the TRS signals which are also defined by SMPTE 125. This standard also describes parallel (SMPTE 125) and serial physical standards (SMPTE 259).

7

Building MPEG Bit Streams

Broadcast television has been all analog composite until recently, but television production has evolved so that it covers all four domains, composite/component and analog/digital. Before the first DTV transmission, digital transmission to the home existed. DBS began the evolution of digital TV to the home. Now terrestrial transmission, ATSC in the United States and DVB in Europe, has been added. Eventually almost all production and transmission will be digital. Since component digital is the future, domain changes from and to analog composite should be avoided as much as possible. DTV formats are all inherently component digital. As we look at MPEG and then ATSC data streams, you will see that all video data are derived from component video.

7.1 Bandwidth Reduction

The problem with digital is the bandwidths consumed because of the bit rates required (Fig. 7.1).

7.1.1 Chroma subsampling

The first sample scheme in Fig. 7.1 (4:4:4) is used to sample RGB video. All have to be sampled at the full bit rate because the three signals are needed to create luminance. The bit rate from its video information is extremely high for SD video. The bit rate can be cut down by separating the chroma or color information from the luminance information because our eyes don't resolve as much color as luminance resolution. In component video luminance and two color difference signals are generated. These orthogonal signals represent a vector representation with the phase representing tint information and the amplitude representing saturation. The two chroma values represent a complex number which is a value that not only has a magnitude but also a direction. If the separated chroma values are sampled at one-half the rate of the luminance, the result is 4:2:2 sampled video. This is a way of reducing the raw

4:4:4 bit rate (720 × 486) = (720 + 720 + 720) × 486 × 30 × 10 = 315 Mbits/s

4:2:2 bit rate (720 × 486) = (720 + 360 + 360) × 486 × 30 × 10 = 210 Mbits/s

4:2:0 bit rate (720 × 486) = (720 × 486) + (360 × 243) + (360 × 243) × 30 × 8
= 126 Mbits/s

4:1:1 bit rate (720 × 486) = (720 + 180 + 180) × 486 × 30 × 8 = 126 Mbits/s

4:2:0 and 4:1:1 are 8 bits only (to further reduce bit rate), whereas the other two are 8 or 10 bits.

Figure 7.1 Video bit rates.

bit rate by 33 percent. Here the chroma subsampling is in the *H* direction only. Both chroma difference signals are cosited with a Y sample. The data samples are organized as follows: Cr/Y/Cb, Y, Cr/Y/Cb, Y, etc. This sample system is specified in SMPTE 125M.

The bit rate can be further reduced by subsampling the color information even more. The ratio 4:2:0 is chroma subsampled in both the *H* and the *V* directions. There are two methods for doing this. In the first method even lines are sampled 4:2:2, while odd lines are sampled 4:0:0 (no chroma samples). In the second method the chroma is sampled in a quincunx pattern. The result is that on odd lines the even chroma samples are thrown away and on even lines the odd chroma samples are thrown away. This makes an interweaved 4:1:1 sample rate. The second approach is used in SMPTE 297. The 4:2:0 chroma subsampling can reduce raw data to be processed by the encoder by 25 percent.

Other sampling schemes are as follows. The 4:1:1 has cosited chroma samples with Y: Cr/Y/Cb, Y, Y, Y, Cr/Y/Cb, Y, Y, Y, etc. Every line is sampled 4:1:1. DVC Pro uses 4:1:1. The 4:4:4 is used for image processing and high-end graphics. As we will see later, 4:2:0 is used in MPEG-2 at main level transmission. The 3:1:1 is used for HD Cam. So color subsampling is used to reduce the bit rate, which reduces the bandwidth required.

7.1.2 Interlace

Interlace was used in NTSC for bandwidth reduction. It was an early form of compression. The advantages of interlace are as follows:

More efficient to transmit in raw image content

Cheaper TV set design (at this time)

Better static resolution for a given bandwidth

Compatibility with the current system

The disadvantages of interlace are as follows:

Visual artifacts—scan lines and flicker

Incompatible with computer-generated images

More difficult to compress and upconvert

Not as good for slow motion

Although interlace is more effective in an analog transmission system for the same effective "frame rate," that is not entirely true in digital transmission. Progressive images compress more efficiently. Today's displays allow transmission of lower frame rates that are displayed at higher rates for viewing.

7.2   Compression

Even with interlace and color subsampling, though, bit rates over 126 Mbits/s still occur. Thus, in order to fit digital streams in narrow "pipes," such as disk I/O and terrestrial broadcast channels, the bit rate needs to be reduced even more. This reduction in the bit rate can be accomplished with compression. Three aspects of video can be manipulated to reduce the data rate:

- *Spatial,* the dependence or similarities between neighboring pixels.
- *Temporal,* the dependence between neighboring frames in a video sequence.
- *Coding redundancy,* the likelihood that 1 byte will be similar to another.

Later we will see that there are aspects of audio that also aid in its compression.

MPEG compression works because of limitations in the eyes and brain. Eyesight allows larger visual errors to go unnoticed at high luminance levels. When observing an MPEG compressed video stream, these errors become apparent if the display's brightness is turned down. Coding errors are also less visible around sharp edges or other high-frequency areas. In video JPEG streams errors that occur in areas with fast motion are also less detectable.

Two types of compression were mentioned in Chap. 1: lossless and "lossy" compression. Lossless compression will usually not reduce the bit rate enough. So lossy techniques were developed. Lossy compression can take place because of the limitations of the human visual and aural systems. The first technique developed was for reducing the file size of photographs digitized for viewing on computers. Photographs sampled at high rates for good resolution could easily create files of several megabytes. In order to create files that were smaller in size a group called the Joint Pictures Expert Group developed a compression scheme that came to be known as JPEG. This scheme took blocks of picture spatial information and converted them into coefficients that would describe the functions needed to create the luminance and color contours if the block were dissected in the horizontal and vertical directions. This process is similar to taking functions in the time domain into the frequency domain, where Fourier coefficients would describe the function.

7.2.1   Discrete cosine transform

JPEG accomplishes this process using a transform called the *discrete cosine transform* (DCT):

$$\text{DCT}\,(i,j)=\frac{1}{\sqrt{2N}}\;C(i)\,C(j)\sum_{x=0}^{N-1}\sum_{y=0}^{N-1}\text{pixel}\,(x,y)\cos\left[\frac{(2x+1)i\pi}{2N}\right]\cos\left[\frac{(2y+1)j\pi}{2N}\right]$$

What does this equation mean? N is the size of the block. In an 8 by 8 matrix, $N = 8$ and $1/\sqrt{2N} = 1/4$. $C(i)$ and $C(j)$ are the difference in transform values found from the previous block.

$$C(x) = \begin{cases} 1/\sqrt{2} & \text{if } x = 0 \\ 1 & \text{if } x > 0 \end{cases}$$

Pixel (x, y) is the digital value of the spatial pixel. The cosine (cos) functions multiplied together transform the pixels in their x and y spatial values in the picture to frequency components. The difference from the previous cell in the previous blocks simply means the results of the cosine operations will be added from all the pixel locations in the first column, then the second, and so forth until all column sums have been added together.

Figure 7.2 takes an 8 by 8 video (or pixel) block, starts an 8 by 8 block for frequency difference components, and then goes through all 64 (8 × 8) pixel values to generate a number (or coefficient) to go into each location of the frequency block.

Thus, Fig. 7.3 maps each spatial pixel to a frequency coefficient by looking at the values of every pixel in the block. It does this for every frequency coefficient. The cosine part of the operation assigns lower scalars to frequency coefficients in the bottom right-hand corner of the frequency coefficient matrix than to those in the upper left-hand portion. The upper left-hand corner coefficient has a scalar of 1, which is apparent if you evaluate the cosine functions with (i), and (j) equal to 0. This means the upper left-hand corner coefficient DCT(i,j) turns out to represent the dc component of the pixels in the block, scaled by the previous dc value in the previous block.

Column (x)	0	1	2	3	4	5	6	7
Row 0	Pixel (0,0)	Pixel (0,1)	Pixel (0,2)	Pixel (0,3)	Pixel (0,4)	Pixel (0,5)	Pixel (0,6)	Pixel (0,7)
(y) 1	Pixel (1,0)	Pixel (1,1)	Pixel (1,2)	Pixel (1,3)	Pixel (1,4)	Pixel (1,5)	Pixel (1,6)	Pixel (1,7)
2	Pixel (2,0)	Pixel (2,1)	Pixel (2,2)	Pixel (2,3)	Pixel (2,4)	Pixel (2,5)	Pixel (2,6)	Pixel (2,7)
3	Pixel (3,0)	Pixel (3,1)	Pixel (3,2)	Pixel (3,3)	Pixel (3,4)	Pixel (3,5)	Pixel (3,6)	Pixel (3,7)
4	Pixel (4,0)	Pixel (4,1)	Pixel (4,2)	Pixel (4,3)	Pixel (4,4)	Pixel (4,5)	Pixel (4,6)	Pixel (4,7)
5	Pixel (5,0)	Pixel (5,1)	Pixel (5,2)	Pixel (5,3)	Pixel (5,4)	Pixel (5,5)	Pixel (5,6)	Pixel (5,7)
6	Pixel (6,0)	Pixel (6,1)	Pixel (6,2)	Pixel (6,3)	Pixel (6,4)	Pixel (6,5)	Pixel (6,6)	Pixel (6,7)
7	Pixel (7,0)	Pixel (7,1)	Pixel (7,2)	Pixel (7,3)	Pixel (7,4)	Pixel (7,5)	Pixel (7,6)	Pixel (7,7)

Figure 7.2   8 by 8 pixel block.

Column (j)	0	1	2	3	4	5	6	7
Row 0	DCT(0,0)	DCT(0,1)	DCT(0,2)	DCT(0,3)	DCT(0,4)	DCT(0,5)	DCT(0,6)	DCT(0,7)
(i) 1	DCT(1,0)	DCT(1,1)	DCT(1,2)	DCT(1,3)	DCT(1,4)	DCT(1,5)	DCT(1,6)	DCT(1,7)
2	DCT(2,0)	DCT(2,1)	DCT(2,2)	DCT(2,3)	DCT(2,4)	DCT(2,5)	DCT(2,6)	DCT(2,7)
3	DCT(3,0)	DCT(3,1)	DCT(3,2)	DCT(3,3)	DCT(3,4)	DCT(3,5)	DCT(3,6)	DCT(3,7)
4	DCT(4,0)	DCT(4,1)	DCT(4,2)	DCT(4,3)	DCT(4,4)	DCT(4,5)	DCT(4,6)	DCT(4,7)
5	DCT(5,0)	DCT(5,1)	DCT(5,2)	DCT(5,3)	DCT(5,4)	DCT(5,5)	DCT(5,6)	DCT(5,7)
6	DCT(6,0)	DCT(6,1)	DCT(6,2)	DCT(6,3)	DCT(6,4)	DCT(6,5)	DCT(6,6)	DCT(6,7)
7	DCT(7,0)	DCT(7,1)	DCT(7,2)	DCT(7,3)	DCT(7,4)	DCT(7,5)	DCT(7,6)	DCT(7,7)

Figure 7.3 8 by 8 frequency coefficient block.

The cosine function was chosen because it closely matches the sinc-squared function. Continuous square waves have frequency coefficients that follow the sinc function. See Fig. 4.17a for a sinc function demonstrating the frequency coefficients for square waves with 50 percent duty cycles. This function shows that every other harmonic is missing and that the coefficients for each harmonic decrease the farther from the fundamental. The DCT transform makes the same assumption. High frequencies will have smaller coefficients.

The transform works out such that frequency components that are horizontal in nature (vertical stripes) fall in, or close to, row 0, and frequency components that are vertical in nature (horizontal stripes) fall in, or close to, column 0.

7.2.2 Quantization and lossy compression

Now that a set of blocks is no longer in the time domain, compression of the data can begin. The DCT by itself is a lossless transformation. In fact, in many cases the transformation expands the data. The 8-bit input pixel data will come out with 11 bits of frequency coefficient data. Then these frequency coefficients data undergo quantization. Each location in the frequency coefficient matrix is given a denominator (divisor) value based on its position. This can be done simply by giving the upper left-hand corner a low number (such as 3) and increasing the number by some fixed amount (say 2) as you go diagonally through the matrix down toward the lower right-hand corner, with the result that the lower right-hand corner would have a denominator of 31. The JPEG Committee actually came up with unique denominator values for each coefficient location in the matrix by trial and error. Each coefficient value is divided by its quantization denominator and rounded to its closest whole number.

As a result of this process, many of the coefficients not located in the upper left-hand corner of the frequency matrix become zero. Changing the value of these quantization values is one way to control the amount of compression that takes place. A starting quantization number over 20 in the upper left-hand corner is about the limit for any kind of recognizable picture from the process.

Next the two-dimensional matrix is converted to a serial string of coefficients (or simply a string of numbers), but it will be put into a string in such a way that most of the zeros that have been created in the matrix will come out of the string together. This is done by reading the matrix not by row, or column, but in a zigzag manner. Coefficient (0,0) is then read as (0,1), (1,0), (2,0), (1,1), (0,2), and so forth, up to (7,7). In this way the zeros that have been generated in the lower right-hand corner of the matrix follow one another out as a group of zeros. This is the first opportunity for compression.

7.2.3 Lossless compression

The next step in the process is called run length encoding, and it is basically the same process used in lossless compression schemes in current disk-doubling schemes found on many personal computers. The data stream coming from the zigzag process feeding out of the frequency coefficient matrix has a run length value tacked on it that indicates how many preceding zeros were in front of it. Only words that are not zero are kept. The words that are zero are thrown away. Next, a set of bits are inserted in the word, indicating how many bits the DCT coefficient has. The number of bits varies since the absolute value of the coefficient is not stored, only the change in its amplitude from the previous value. The relative value between successive coefficients is usually small. Hence, a type of Huffman coding is used. The changes that are most likely to occur are given short numbers of bits, while less likely values are given longer numbers of bits to describe. It has been suggested that the process resembles Morse code, where E is a dit and T is a dah, short sequences because they are sent so often. Therefore, a bit count of 1 represents a change either of ±1. Bit counts of 2 represent −2 to −3, or 2 to 3. Bit counts of 3 represent changes of either −7 to −4 or 4 to 7, and so forth up to 10 bits representing the upper half of possible change, −1023 to −512, or 512 to 1023. These large changes will seldom happen. Thus, most words will have short numbers of bits.

JPEG compression can be done up to 8 times and still maintain Betacam-type quality. A 2:1 compression can be done with essentially no loss of information (lossless compression), which means whatever is fed in would exactly match the output. At the other extreme, 22:1 compression yields VHS-type quality. Some systems use adaptive compression, that is, higher compression rates are used on fields with low information content (such as scenes with large redundant backgrounds).

7.2.4 Compression efficiency

A term that is often used to describe the average information in a message was borrowed from thermodynamics, entropy. Maximum entropy is possible only when information in a message is likely to occur equally. It is desirable to code the signal in such a manner that messages with low informational content (that is, a higher probability of occurrence) are assigned lower redundancy than messages with lower probability. What does that mean? Ever learn Morse code? What letter is most likely to be sent at any given moment in time? Well, since the letter e is found very often in the English language, "e" would be a good choice as being the next letter in a message. In Morse code "e" is signified with a single short burst of tone (carrier), called a dit. Less-likely letters have longer combinations of short (dit) and long (dah) tones. Although Morse code came first, it is considered a form of Huffman coding. Huffman coding is used in JPEG/MPEG after the DCT and quantization process to help further limit bit rates.

7.3 MPEG

JPEG was a big step forward and helped greatly. Many video server systems use JPEG. However, video bit streams needed more bit-rate reduction to fit into the paths or channels required than JPEG could accomplish without losing too much "quality." JPEG provides compression in the spatial domain, but it was soon realized that there was much redundant information from one video picture to the next, which meant that temporal compression could be done as well. This process eliminates information that doesn't change over time, in other words, from one picture to the next. So a group called the Motion Pictures Expert Group convened to develop such a system. Of course this process is now known as MPEG. It actually uses JPEG at its base but takes the process farther. It provides an improvement of about 3 times over JPEG quality.

The first iteration of MPEG was known as MPEG-1. It operates up to 1.5 Mbits/s and was intended for DVD. MPEG-2 is the standard for video compression and transport used now. MPEG-2 is a toolkit of compression techniques for generating compliant bit streams. The format and the content of the bit stream are regulated by the MPEG-2 standard. It is an asymmetrical system where most of the complexity is in the encoder. ATSC digital transmission is MPEG-2 based. MPEG-2 is now an ISO standard known as ISO/IEC 13818. This standard is actually broken up into four parts:

13818-1. Specifies system-level coding, that is, it defines the multiplexed structure for the final bit stream, including video, audio, and timing information.

13818-2. Specifies the coded representation of video data and its required decoding process, in other words, what a compressed data stream should look like and how it is decoded. MPEG doesn't specify how to create the bit stream, just like most SMPTE standards, only what the result of compression should be and how to decode (decompress) it.

13818-3. Specifies the coded representation of audio data. The same stipulations that applied to the video are applied to the audio.

13818-4. Specifies the procedures for confirming compliance of parts 1 through 3.

7.3.1   ISO 13818-2 baseband video to MPEG PES

MPEG takes baseband video (noncompressed) and moves it through a number of processing layers. First, the video is broken into blocks. The luminance and each color difference block undergo the DCT process separately.

The resulting DCT blocks are combined into macroblocks. The 4:2:2 macroblock has four blocks for luminance and two each for the Cb and Cr color difference blocks. The 4:1:1 and 4:2:0 formats are four luminance and one block each for Cb and Cr. The difference between the two is how the chroma was originally subsampled before the DCT process. Macroblocks have other information such as the macroblock address, q scalar value, motion vectors, and luminance and chroma blocks.

Next, the macroblocks are strung together to form what is called a *slice*. Usually, slices are the width of the raster (picture). Therefore, a 720×480 picture would have slices 90 macroblocks long, 60 per picture. The slice address is located at the beginning of each slice.

In the previous example these 60 slices are considered an MPEG frame. JPEG and MPEG only deal with frames, and both combine fields into frames before the DCT process. Thus, for interlaced video the blocks consist of lines for both the odd and even fields. There are three types of frames in MPEG. The process just described would generate what is called an I frame, and it is the same as a JPEG frame. The I stands for intraframe encoding. An I frame only contains its own compressed spatial information. There is no temporal information from other frames. This frame stands on its own, and is often called an anchor frame.

There are two other types of MPEG frames that don't stand on their own; they rely on information from other frames. One of these is the P (predictive) frame, which uses interframe coding. Motion compensation vectors are generated from the previous I or P frame. These vectors are encoded as part of the P frame data. Also, any errors left over after the motion vectors' predictions also undergo a DCT compression. These data are also sent as part of the P frame data, resulting in less code than an I frame.

The third type of MPEG frame is the B (bidirectional) frame, and it also relies on interframe coding. Here interpolation is used to estimate DCT coefficients from past and future I and P frames. Generally, there are more B frames than P frames and fewer I frames.

B frames are generated from the nearest preceding and succeeding I or P frame. P frames are generated only from the nearest preceding I or P frame. I frames are the starting point from which P and B frames are estimated. A measure of the accuracy of an MPEG sequence is, therefore, the ratio of total number of frames to I frames. Each slice header also indicates the type of frame (DCT type) that is being sent.

At the decoder, B frames are delayed since they cannot be interpreted until a future I or P frame is received. This implies a requirement for buffer capacity at the source and destination. The number of B frames is, therefore, usually limited and, in some cases, may be zero.

Every I frame begins a new group of pictures. GOPs can be of various lengths, with 15 a common length, but GOPs as short as 2 are used. This would be only an I and a B frame, followed by another I frame. An example of a GOP of 10 would be I B B P B B P B B P. The B frames rely on the I and P frames on either side of them; the P frames rely only on the previous I or P frame. Now in the example here the decoder will need the first P frame before it can decode either of the first two B frames, so the GOP shown previously will not be sent in the order shown. The transmit order would actually be I P B B P B B P B B.

This process proves that it is the encoder that needs lots of memory and processing power and not every decoder, which is much more efficient as decoders vastly outnumber encoders. The decoder knows the order in which to display the out-of-order frames because each comes with a time stamp embedded in the stream.

Now let's go back to the slices. Slices are assembled (60 in the example here) to create a complete picture—I, B, or P pictures. The picture sequences, like slices, have a unique start code. These picture sequences are finally assembled into a video sequence. The video sequence also has a unique start code, and contains profile and level information about the picture sequence. The number of picture sequences in the video sequence is equal to the GOP. Therefore, there is one I frame as the first picture, and the rest are either B or P frame pictures. This completes a video PES (packetized elementary stream).

7.3.1.1 Profiles and levels. The various MPEG formats are ranked by two parameters. MPEG levels define the sampling structure of the video and its bit rate. MPEG profiles define the toolsets available for compressing video. These tools are I, B, and P frames and scalable video quality. Decoders that decode a certain level and profile can decode all profiles to the left and levels below. Profiles also indicate the chroma subsampling scheme. Most are 4:2:0, but there is a 4:2:2 mode.

7.3.1.2 Scalability. Two types of MPEG scalability are available—one limits noise and the other overlays additional spatial information over a lower-resolution video PES. The noise-scalable implementation is a profile known as SNR, while the spatial overlay is known as the spatial profile.

The SNR profile is helpful for compressed video streams using heavily quantized coefficients, which results in much lost information that adds noiselike artifacts to the video data. If an additional PES stream is produced, based on the noise component of the video, a more complex decoder could use both the main PES and SNR PES to decode a higher-quality picture than a simpler decoder using only the main PES.

The same approach can be used to provide spatial scalability. Instead of a noise difference signal, high-frequency information would be sent in addition to a lower-resolution PES.

In both approaches the demodulation to reproduce the transport stream is part of the system. When used with DBS-type systems via satellite, more sophisticated decoders would demodulate all points in a QAM constellation, while simpler decoders would lump quads of QAM constellation points into lower QAM (such as 64 to 16 QAM or 16 QAM to QPSK).

7.3.1.3 Noise. Level and color adjustments are critical to ensure good MPEG encoding. Black level is critical to protect dark regions from solarization and blocking. Noise reduction can save 20 to 30 percent in bit rates for a given quality level. MPEG preprocessing should include high-frequency filtering.

7.3.1.4 Tiered compression. Naïve cascading, or concatenation or compression/ decompression, operations occur when each successive compression engine compresses the video without knowing how the original engine made its decisions on how to compress. The United Kingdom has an Atlantic project to produce a smart cascading system where an "info bus" provides coding decisions to the next coder. This becomes complicated because the info bus must be routed along with the program stream.

Snell and Wilcox have taken it a step further with "mole processing." It is called processing mole because the info bus data are actually buried in the video data. Decoders place the info bus into the bottom 2 bits of video (MPEG-2 only supports 8 of the 10 baseband video bits). A mole encoder only uses the information if the checksums are correct.

New video added to the decompressed mole video wipes out the mole data (checksum at the next encoder is wrong), and thus new decisions will be made as to when and where new video is added.

7.3.1.5 Intelligent encoding. One way to decrease the bit rate is to throw away obvious redundancy. Film is shot at 24 frames/s while television fields under NTSC are generated at 60 fields/s. In order to reconcile this disparity film projectors built for television stations used what is known as 3:2 pulldown to pull film past the illuminate gate. Every other frame of film was scanned three times versus two for the rest. The scanning sequence for these film frames is shown here:

1	1	2	2	2	3	3	4	4	4	5	5	6	6	6	
7	7	8	8	8	9	9	10	10	10	11	11	12	12	12	(30 fields)
13	13	14	14	14	15	15	16	16	16	17	17	18	18	18	
19	19	20	20	20	21	21	22	22	22	23	23	24	24	24	(30 fields)

NOTE: 60 television fields, 24 film frames.

MPEG encoders have 3:2 pulldown detection so they discard redundant images, which results in a 20 percent savings on the number of images to compress. The MPEG encoder would throw away 12 (or 20 percent) of these fields (every third field scan) and then convert the fields to frames.

Another intelligent way to limit the bit rate is by changing the compression ratio (q) to compensate for limited path bandwidth or network congestion. The accuracy of the video stream can be changed to compensate for network errors by varying the size of the GOP. More I frames limit the propagation of errors between these frames. Some encoders have motion vector processors working at 200 gigaoperations per second, using large neural network memory architecture. These systems use motion compensation algorithms which can search areas as large as 128 pixels by 64 lines. Many have 3:2 pulldown detection and use parallel SD MPEG encoding engines to construct HD pictures.

Empirical measurements have found that I frames generate 2 to 4 times as many ATM cells as B and P frames. The average number of cells per frame decreases until GOP reaches 32. After that, it increases. As the GOP increases, errors propagate among P and B frames, resulting in less efficient quantization of the P and B frames. As greater compression is used, either by increasing q or the GOP, the ratio of the standard deviation to the average number of cells per frame increases, which means that the traffic becomes more bursty, even though many MPEG codecs produce constant bit rates.

7.3.1.6 Packetization. Transport packets are the key to flexibility and extensibility; they are fixed-length packets that are relatively short and amenable to error correction and fast switching. This is best for error-prone media such as terrestrial broadcast. Each packet has a header or label that identifies the contents (packet ID, or PID). Each packet contains only one kind of data (audio, video, etc.). Packetized elementary streams (PES) come out of the video/audio coder, and this is what incoming data become. A PES has a header and a video, audio, or data payload. A transport stream consists of fixed-length transport packets, which are repackaged PES packets. A PES packet header is always preceded by a transport header when part of a transport stream.

A transport packet is 188 bytes long. It consists of a 4-byte transport header which includes an 8-bit sync word, three consecutive 1s, a 13-bit PID, 2-bit scrambling flags, an adaptation field [program clock reference (PCR), etc.], and a video, audio, or data payload up to 184 bytes (if no adaptation field).

A packet scheduler and mux permit packets into the bit stream according to need and priority. This allows dynamic allocation of the channel.

A point of common confusion occurs when transport packets are 188 bytes long, but elementary streams (PES) coming from an MPEG encoder to a multiplexer, which will build the final transport stream, can be up to 65,536 bytes long. The multiplexer will break the various long PES into transport-sized packets.

The transport stream packet header contains an indication as to whether the start of a PES packet occurs in that transport stream packet and whether the payload is scrambled. The first byte of each PES packet header is located at the first available payload location of a transport stream packet.

7.3.1.7 Statistical multiplexing. Many MPEG multiplexers look at the buffer usage of the various PES and control q values of the various encoders creating the various PES. Video PES with lots of spatial and temporal information are assigned lower q-scale values than PES with little activity. This allows PES that need a larger percentage of the bit stream (more bandwidth) to have it by scaling back the number of bits coming from more sedate streams.

7.3.2   ISO 13818-1 PES, PIDs, PMTs, PATs, CAs, and NITs all multiplexed into an MPEG transport stream

As we discussed in Chap. 1, many television stations will send more than one program in their transmitted ATSC data stream. Even a station sending a single HD video stream will still have at least two PES, one for the video and the other for the audio. When multiple PES are combined or multiplexed, it is called a *transport stream* (Fig. 7.4).

The individual PES can be identified in the MPEG transport stream by a PID. The PID is a 13-bit field in an MPEG transport packet. PID bits 0–3 define whether the PID is a video, audio, or data PID, or the PID for the program map table (PMT). The PMT is the map for all PIDs of a given program. The PMT must be sent at least every 400 ms. PID bits 4–11 of the PID are the program number, and this allows for 256 separate programs and specifies the program to which the PID is applicable. PID bit 12 is 0.

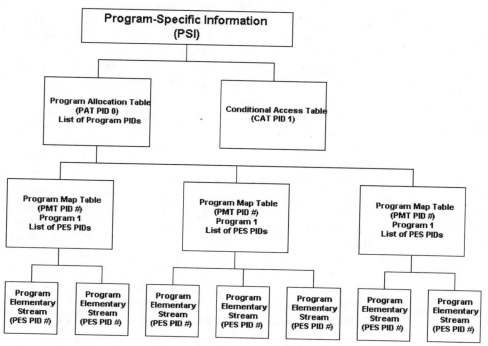

Figure 7.4 Program-specific information (PSI) hierarchy.

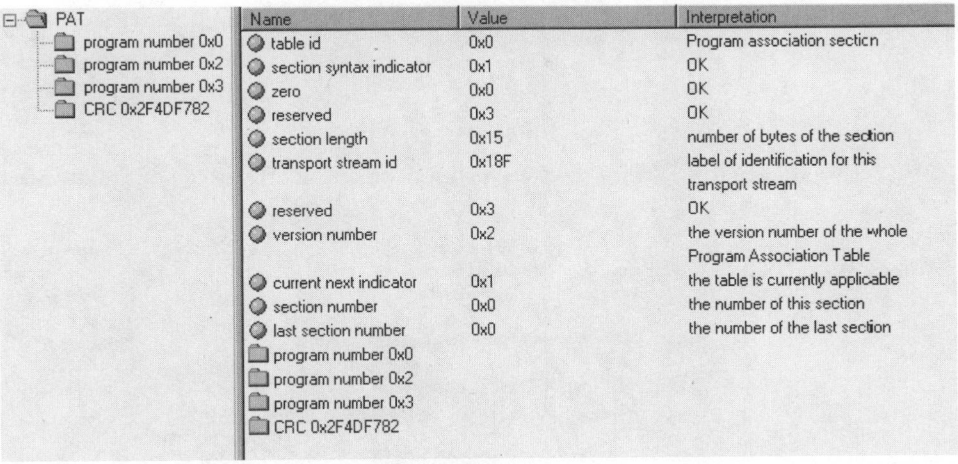

Name	Value	Interpretation
table id	0x0	Program association section
section syntax indicator	0x1	OK
zero	0x0	OK
reserved	0x3	OK
section length	0x15	number of bytes of the section
transport stream id	0x18F	label of identification for this transport stream
reserved	0x3	OK
version number	0x2	the version number of the whole Program Association Table
current next indicator	0x1	the table is currently applicable
section number	0x0	the number of this section
last section number	0x0	the number of the last section
program number 0x0		
program number 0x2		
program number 0x3		
CRC 0x2F4DF782		

PAT tree: program number 0x0, program number 0x2, program number 0x3, CRC 0x2F4DF782

Figure 7.5 Contents of the PAT as seen on an MPEG stream analyzer.

The PMTs of the various programs are listed in the program allocation table (PAT). The PAT serves as the starting point for locating the various PMTs. The maximum spacing between occurrences of sections of the PAT is 100 ms. Since it is possible for a particular PES to be used in multiple programs, some PIDs could be in multiple PMTs.

The reserved PID values are as follows:

00 = PAT

01 = conditional access table (CAT)

1FFF = null packets. If all 13 bits in the PID field are high, it signifies that this is a null packet. These are used to stuff the data stream when there are no other data to send (Fig. 7.5). In addition to PMTs in the PAT, there can be a PID that identifies the network information table (NIT). Therefore, the PAT has the NIT and various PMTs (Fig. 7.6).

The CAT identifies the PES carrying private conditional access data in the form of entitlement management messages (EMM). These two tables, along with the child tables as a group, are called program-specific information (PSI).

These tables are contained in the PSI table. PSI data can be up to 1024 bytes long. The beginning of this information is indicated by a pointer in the transport stream packet payload. PSI is as follows:

Table	PID number	Description
PAT	0	Associates program number with PMT.
PMT	Assigned	Associates PID with program(s).
NIT	Assigned	Contains physical network parameters.
CAT	1	Associates PIDs with private streams.

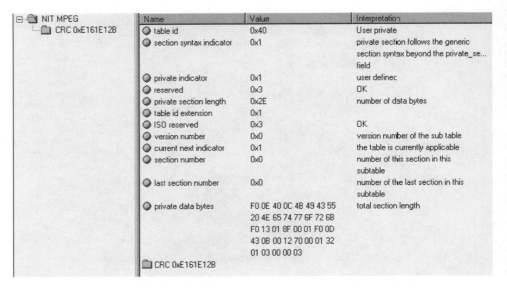

Figure 7.6   Contents of the NIT.

A PAT and a PMT are required in every MPEG-2 transport stream. They form a "mini–program guide."

An example PAT for a three-program multiplex is as follows:

Program number	PMT PID number	Meaning
2	32	PID 32 contains map for program 2.
3	48	PID 48 contains map for program 3.
4	64	PID 64 contains map for program 4.

A PAT provides correspondence between a program number and the PMT PID that carries program definition. The program number is similar to the channel number in broadcast TV. The PAT is always assigned to PID 0. ATSC programs are in the range 2 to 255 (Fig. 7.7).

An example PMT map for program 2 is as follows:

PID number	Stream type
32	PMT
33	Video and PCR
36	Audio
42	Data

Each PID number in these examples represents a PES. Together they represent one program stream. Each program has a PMT, and the PID for every program PMT is in the PAT. The PMT provides correspondence between a program number and the elementary streams that comprise it.

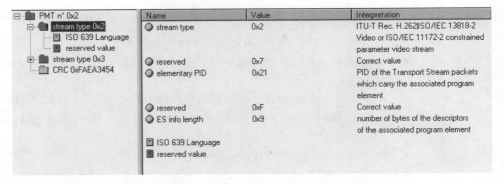

Name	Value	Interpretation
stream type	0x2	ITU-T Rec. H.262ǁISO/IEC 13818-2 Video or ISO/IEC 11172-2 constrained parameter video stream
reserved	0x7	Correct value
elementary PID	0x21	PID of the Transport Stream packets which carry the associated program element
reserved	0xF	Correct value
ES info length	0x9	number of bytes of the descriptors of the associated program element

Tree panel (left):
- PMT n° 0x2
  - stream type 0x2
    - ISO 639 Language
    - reserved value
  - stream type 0x3
  - CRC 0xFAEA3454

Below table:
- ISO 639 Language
- reserved value

Figure 7.7 Contents of a PMT. Stream type 0×2 indicates video while 0×3 indicates audio.

Packet demultiplexing is done by reading the PIDs and, based on information from the PAT and PMT, sorting the packets.

The confusing thing about PIDs is that they can be found at three levels of the table structure: the PAT, the PMT, and for individual PES. It is necessary to work down from the PAT in order to identify where a particular PID belongs in the hierarchy of the MPEG stream (with the exception of PIDs 01 and 02).

The overall system multiplexing approach can be thought of as a combination of multiplexing at two different layers. In the first layer, single program transport bit streams are formed by multiplexing transport packets from one or more PES sources. In the second layer, many single program transport streams are combined to form a system of programs. The PSI table contains information relating to the identification and components of each program.

7.3.2.1 MPEG packets. MPEG transport packets are 188 bytes long, so that they more easily fit the payload capacity of an ATM packet. The first 4 bytes are the MPEG header.

There is a sync byte with a value of 71_h ($0100\ 0111_b$). In the next 2 bytes the first 3 bits are used as flags and the last 13 bits are used for the PID. Characteristics of flag bits are (1) an uncorrectable error exists in the packet, (2) data start immediately after the header, and (3) this packet is of higher priority than other packets having the same PID. The last byte is split up as follows: 2 bits indicate the scrambling mode, 2 bits indicate whether the header is followed by an adaptation field and/or payload, and a 4-bit continuity counter increments each transport stream packet with the same PID. It wraps around to 0 after F_h.

After the header, 184 bytes of payload can be sent. This would usually be compressed video or audio data PES packets. However, the payload could be data from one of the PSI tables. Any of the 184 bytes not used would be filled with null bytes FF_h.

If the payload consists of table data, it will include information about

- Number of bytes that carry table information.
- User-generated ID to uniquely identify this transport stream in a larger network.
- Current version number of the PAT. Version count goes to 31 and resets to zero.
- Number that identifies the entry in the table that this payload contains, with 256 possible entries.
- Highest-numbered entry in the table.
- Program with which this table is associated.
- PID with which this entry is associated.

If an adaptation field is sent, it contains the following:

- The length of the field, in bytes.
- A flag to stop incrementing the continuity counter in the packet header.
- A flag to indicate that the PES payload in this packet is of greater priority than other PES data in packets with the same PID.
- Indication of a splicing point in the PES payload.

7.3.2.2 Timing and synchronization. The "master clock" is similar to "genlocking." Master clock synchronization is achieved by sampling and inserting 42-bit values of a 27-MHz program clock reference (PCR). PCR samples are inserted in selected transport packets at least 10 times/s. The decoder uses PCR samples to recreate the master clock.

Presentation time stamps (PTS) indicate when video or audio frames should be presented. *Decode time stamps* (DTS) indicate when video anchor frames should be decoded. PTS and DTS are samples of a 90-kHz clock locked to the PCR time base, and they are sent in PES headers. Lip sync is achieved by presenting video and audio frames at the proper value of the master clock (PCR time).

A video frame is 33.37 ms long. An audio frame is 32 ms (1536 samples) long. Video and audio frame boundaries are rarely (if ever) aligned.

PCR jitter can be introduced by variable delays in networks or by mux/remux operations. PCR jitter causes the decoder master clock to erroneously speed up or slow down. Depending on the decoder implementation, jitter can lead to frozen or skipped pictures, unstable color, or "wow and flutter." Additional buffers can be used to reduce the effect of PCR jitter.

Two primary factors affect jitter, that is, the variation in delivery of those packets, in isochronous MPEG packet delivery. The first is delays in starting an isochronous cycle, which are usually caused by a long asynchronous packet (they can be as long as 75 µs) intruding into the time allocated to isochronous packets. The second is changes in the order in which different channels (different MPEG packets) are transmitted. Changes in order of packet transmission are unlikely to occur unless there is a change in bus configuration. Jitter in firewire MPEG signals is eliminated by having all nodes participate in the isochronous cycle share of a common time reference. The cycle start packets that indicate the beginning of the isochronous period also contain the current value of the master clock.

MPEG audio, video, and transport encoders use a family of frequencies which are based on a 27-MHz clock. This clock is used to generate a 42-bit sample of the frequency which is partitioned into two parts as defined by the MPEG-2 specification. The first part is a 33-bit *program_clock_reference_base* and a 9-bit *program_clock_reference_extension*. The first part is equivalent to a sample of a 90-kHz clock that is locked in frequency to the 27-MHz clock and used by the audio and video source encoders when encoding the presentation and decode time stamps.

7.3.3   Multimedia MPEG

The future of television receivers will most likely fracture into various markets, based on the desires of the viewer. Dumb television receivers, such as those now in homes, will continue to be available. However, many receivers will be more than just a television receiver; they will be Web-type browsers that also display traditional television. These receivers will allow viewers to build a unique viewing experience, based on their interactivity with the options presented by the broadcaster. Not only will the broadcaster send video and audio streams to the viewer, but also data much like the HTML language sent to Web sites to build a Web page.

7.3.3.1   MPEG-4.

MPEG-4 starts this migration by breaking a complete two-dimensional picture into individual two- or three-dimensional objects and sprites. Sprites could be the floor and background, such as a wall, in the picture. Objects could be pieces of furniture or a chart that is in the scene. The "talent" in the picture would also be an object. MPEG-4 also offers conventional audio/video streaming. It shares the same timing and clocks as MPEG-2.

A group called Advanced Interactive Content (AIC), which is a spinoff of Wintel's Advanced Television Enhancement Forum (ATVEF), wants to put various multimedia platforms, such as BHTML, VRML, and BIFS (binary format for scenes), into MPEG-4 streams. BHTML (broadcast HTML), written in XML (extensible markup language), is related to Java. See Sec. 7.3.3.3.

MPEG-4 might allow a common video/audio Web-streaming format. In order to receive video or audio streams today from various sources can require multiple programs on your PC such as Apple's QuickTime, Microsoft's AVI, and Real Networks' Real Video. MPEG-4 is also geared to allow low bit rates when needed, such as over the Web, and higher bit rates when available, such as for terrestrial broadcasting. This is because breaking a picture into objects lends itself to scalable content. Instead of a PES that encompasses the whole picture, multiple PES, each carrying objects and sprites, would be used. In order to accommodate this object, elementary stream descriptors would be added to the stream.

MPEG-4 is also standardizing on something called structured audio (SA), which would allow models of various instruments to be defined and to reside in the viewer's MPEG-4 decoder. Thus, the actual sounds of a piano being played would not have to be recorded and sent, only the commands to make the virtual piano in the receiving MPEG-4 decoder replicate the real piano.

An interesting note is that MPEG-1 was introduced in 1992. MPEG-2, which is used by the ATSC, was introduced in 1995. MPEG-3 was to be the HD version of MPEG-2, but due to the similarities of tools for both SD and HD, MPEG-3 was simply included in MPEG-2. There is no MPEG-5 or MPEG-6. MPEG-7, which we will look at next, was considered so radically different from MPEG-4 that MPEG members arbitrarily skipped the MPEG-5 and MPEG-6 designations.

7.3.3.2 MPEG-7. The latest version of MPEG is known as MPEG-7. It is planned for 2001 and is the most recent standardization activity of the ISO. MPEG-7 is formally called *multimedia content description interface,* and it should be able to optimize the organization of ancillary DTV data. MPEG-7 will standardize the description of multimedia material such as pictures, sound, and moving images. These descriptions will technically be metadata. It will allow searches of MPEG-7 multimedia programs based on keywords. There will be high and low descriptors. High-level descriptors might be actors, location, scene content, etc. Low-level descriptors might be audio spectrum, color, or texture.

7.3.3.3 DASE. Many in the industry are striving to make the television display more like a Web-browser experience on a PC. ATSC's DTV Application Software Environment (DASE) group (T3/S17) is defining the software application environment for use in DTV receivers. The hope is that a set of application programming interfaces will be developed for the DTV receiver. The group is also defining a presentation engine or browser. One proposal is based on broadcast HTML, which, in turn, is based on XML. XML is likely to form the foundation of the next generation of HTML. The goals of BHTML are to improve on HTML for television use:

1. Synchronize media objects and text to each other and to frames of video.

2. Accurately position media objects on the screen.

3. Have objects persist across scenes.

4. Provide for overlays, transparent test keying, fades, dissolves, etc.

5. Work with the Java-based framework.

7.4   ATSC

ATSC video and audio are based on the following standards:

ISO/IEC IS 13818-1   MPEG-2 systems
ISO/IEC IS 13818-2   MPEG-2 video
ISO/IEC IS 13818-4   MPEG-2 compliance
Dolby AC-3   Audio

The uses of these MPEG and AC-3 standards are spelled out in ATSC's A52/53/54 documents.

7.4.1   A53/54: ATSC standard

This is the document that describes the video encoding, using MPEG as defined by ISO 13818. It provides background on the development of the ATSC DTV standard including the overall "service multiplex and transport" from encoding, multiplexing, and transport through rf transmission. It also lists video input formats and compression formats.

A53 also describes high-level C language conditional routines for processing ATSC transport streams and lists the variable names that reside in the transport stream.

A54 expounds on the information contained in A53; it also expounds on MPEG encoding methodology. The following explanation of MPEG usage in DTV is from the A54 document:

> The MPEG-2 specification is organized into a system of profiles and levels, so that applications can ensure interoperability by using equipment and processing that adhere to a common set of coding tools and parameters. The Digital Television Standard is based on the MPEG-2 Main Profile. The Main Profile includes three types of frames for prediction (I-frames, P-frames, and B-frames), and an organization of luminance and chrominance samples (designated 4:2:0) within the frame. The Main Profile does not include a scalable algorithm, where scalability implies that a subset of the compressed data can be decoded without decoding the entire data stream. The High Level includes formats with up to 1152 active lines and up to 1920 samples per active line, and for the Main Profile is limited to a compressed data rate of no more than 80 Mbps. The parameters specified by the Digital Television Standard represent specific choices within these constraints.

Here is a partial list of MPEG terms from the A54 document:

access unit   A coded representation of a presentation unit. In the case of audio, an access unit is the coded representation of an audio frame. In the case of video, an access unit includes all the coded data for a picture, and any stuffing that follows it, up to but not including the start of the next access unit. If a picture is not preceded by a group_start_code or a sequence_header_code, the access unit begins with a picture start code. If a picture is preceded by a group_start_code and/or a sequence_header_code, the access unit begins with the first byte of the first of these start codes. If it is the last picture preceding a sequence_end_code in the bit stream, all bytes between the last byte of the coded picture and the sequence_end_code (including the sequence_end_code) belong to the access unit.

anchor frame A video frame that is used for prediction. I-frames and P-frames are generally used as anchor frames, but B-frames are never anchor frames.

asynchronous transfer mode (ATM) A digital signal protocol for efficient transport of both constant-rate and bursty information in broadband digital networks. The ATM digital stream consists of fixed-length packets called "cells," each containing 53 8-bit bytes—a 5-byte header and a 48-byte information payload.

bi-directional pictures or **B-pictures** or **B-frames** Pictures that use both future and past pictures as a reference. This technique is termed *bi-directional prediction*. B-pictures provide the most compression. B-pictures do not propagate coding errors as they are never used as a reference.

block A block is an 8-by-8 array of pel values or DCT coefficients representing luminance or chrominance information.

decoding time-stamp (DTS) A field that may be present in a PES packet header that indicates the time that an access unit is decoded in the system target decoder.

elementary stream (ES) A generic term for one of the coded video, coded audio or other coded bit streams. One elementary stream is carried in a sequence of PES packets with one and only one stream_id.

elementary stream clock reference (ESCR) A time stamp in the PES stream from which decoders of PES streams may derive timing.

entitlement control message (ECM) Entitlement control messages are private conditional access information which specify control words and possibly other stream-specific, scrambling, and/or control parameters.

entitlement management message (EMM) Entitlement management messages are private conditional access information which specify the authorization level or the services of specific decoders. They may be addressed to single decoders or groups of decoders.

macroblock In the ATV system a macroblock consists of four blocks of luminance and one each Cr and Cb block.

main level A range of allowed picture parameters defined by the MPEG-2 video coding specification with maximum resolution equivalent to ITU-R Recommendation 601.

main profile A subset of the syntax of the MPEG-2 video coding specification that is expected to be supported over a large range of applications.

motion vector A pair of numbers which represent the vertical and horizontal displacement of a region of a reference picture for prediction.

packet identifier (PID) A unique integer value used to associate elementary streams of a program in a single or multi-program transport stream.

packet A packet consists of a header followed by a number of contiguous bytes from an elementary data stream. It is a layer in the system coding syntax.

PES packet header The leading fields in a PES packet up to but not including the PES_packet_data_byte fields where the stream is not a padding stream. In the case of a padding stream, the PES packet header is defined as the leading fields in a PES packet up to but not including the padding_byte fields.

PES packet The data structure used to carry elementary stream data. It consists of a packet header followed by PES packet payload.

PES stream A PES stream consists of PES packets, all of whose payloads consist of data from a single elementary stream, and all of which have the same stream_id.

predicted pictures or **P-pictures** or **P-frames** Pictures that are coded with respect to the nearest *previous* I or P-picture. This technique is termed *forward prediction*. P-pictures provide more compression than I-pictures and serve as a reference for future P-pictures or B-pictures. P-pictures can propagate coding errors when P-pictures (or B-pictures) are predicted from prior P-pictures where the prediction is flawed.

presentation time-stamp (PTS) A field that may be present in a PES packet header that indicates the time that a presentation unit is presented in the system target decoder.

presentation unit (PU) A decoded audio access unit or a decoded picture.

profile A defined subset of the syntax specified in the MPEG-2 video coding specification.

program clock reference (PCR) A time stamp in the transport stream from which decoder timing is derived.

program-specific information (PSI) PSI consists of normative data which is necessary for the demultiplexing of transport streams and the successful regeneration of programs.

slice A series of consecutive macroblocks.

splicing The concatenation performed on the system level of two different elementary streams. It is understood that the resulting stream must conform totally to the digital television standard.

start codes Thirty-two-bit codes embedded in the coded bit stream that are unique. They are used for several purposes, including identifying some of the layers in the coding syntax. Start codes consist of a 24-bit prefix (0x000001) and an 8-bit stream_id.

system clock reference (SCR) A time-stamp in the program stream from which decoder timing is derived.

time-stamp A term that indicates the time of a specific action such as the arrival of a byte or the presentation of a presentation unit.

7.4.2 A52: AC-3 audio compression standard

This document describes the decoding process for recovering AC-3 encoded audio. AC-3 is the system used for DTV audio, and it is described in Chap. 11. An AC-3 encoder takes up to six PCM audio channels, metadata control information, reference, and time code, and produces an AC-3–coded bit stream. An AC-3 decoder undoes the encoder process, but it also accepts information about the local setup.

AC-3 supports a number of sampling rates and will produce output stream rates from 32 up to the ATSC limit of 384 kbits/s. However, the main and the associated services together can be up to 512 kbits/s. AC-3 takes advantage of the limitations in human audio perception. Human hearing is less sensitive to the extremes of the hearing range, whether at the low- or high-frequency range. Sounds below a quit threshold, which varies greatly with frequency, are not encoded. Spectral masking causes louder tones to mask the presence of nearby, softer tones. Temporal masking causes louder sounds to mask the presence of softer sounds immediately before, or after, the occurrence of the louder sound. All these effects are used to reduce the amount of data traffic necessary to reproduce high-quality audio.

7.4.3   A65: program and system information protocol (PSIP)

PSIP, which is described in ATSC document A65, is intended to provide system information and program guide data. The standard describes the programs embedded in the transport stream by using a *terrestrial virtual channel table* (TVCT or simply VCT). The VCT contains the source IDs for the various PES that make up each virtual channel and a station's *transport stream ID* (TSID). Each station has been assigned a unique 16-bit-long ID by the Model HDTV Station Project after the FCC failed to do so. The VCT also points to *events information tables* (EITs). A *master guide table* (MGT) contains information regarding other PSIP tables such as packet identifiers and versions. A *ratings regional table* (RRT) will carry any content advisory descriptors. The *system timetable* (STT) defines the current time and date. PSIP also contains a service location descriptor for each digital virtual channel. In addition, A65 describes PSIP information for ATSC transport streams transmitted via cable. PSIP can potentially allow all broadcasters to switch receivers among subchannels for split or zone breaks, etc., by using the directed channel change descriptor or in the PSIP tables. Each DTV channel is paired to its NTSC parent through PSIP. The major channel ID is supposed to be the NTSC parent.

PSIP can be either static or dynamic. Static PSIP tables are generated by the encoder. There are no local or dynamic data in the PSIP table. The user sets a minimal amount of parameters such as channel numbers. With dynamic PSIP external PSIP generators are used. The information is generated externally and it is usually fed to the multiplexer via an ASI interface. The external PSIP approach consist of three components: (1) a PSIP server, which contains the database of upcoming events and controls which cable attributes are used; (2) the spooler, which is the software module that repeats each table according to the defined rates programmed by the user; and (3) the PSIP client editor where changes and new data are entered by the user. Standard PSIP data entries will consist of timing information; channel information, including RF and virtual information; program ratings; events' start time; and duration and title information. There can also be extended text tables. Such tables could include information about the category of the program, such as whether it is a movie, sports, etc., or biographical information about actors appearing in the program, and even gossip concerning those involved with the program. Additionally, demographic and geographic information could be included.

Early DTV receivers have reacted differently to lack of PSIP table data. Some models will only tune to channels with PSIP data if any channel in the market is transmitting PSIP. Without receiving PSIP, receivers might display channels incorrectly with regard to virtual channels.

7.4.4   Differences between ATSC and MPEG

The ATSC transport is based on the MPEG-2 systems spec (ISO/IEC 13818-1), which covers muxing, timing, and control. Transport packets are the key to flexibility and extensibility. They are fixed-length packets which are relatively short and are amenable to error correction and fast switching. This is best for error-prone media such as terrestrial broadcast. Each packet has a header or label which identifies the contents (PID). Each packet contains only one kind of data (audio, video, etc.). PES is what comes out of the video/audio coder, or what incoming data become. A PES has a header and a video, audio, or data payload. A transport stream consists of fixed-length transport packets, which are repackaged PES packets. A PES packet header is always preceded by a transport header.

Some constraints of ATSC PSI are as follows:

One program per PMT is allowed.

A maximum of 400 ms between PMTs and a maximum of 100 ms between PATs are allowed.

Video access units in PES packets must be aligned.

No adaptation headers are allowed in the PAT and PMT packets unless version number is discontinuous.

Some constraints of ATSC PES are as follows:

PES payload must not be scrambled.

PES header must not contain clock or rate information for elementary stream, CRC, or private data or program system target decoder fields.

Video PES packets must start with a SEQ, GOP, or picture header; must contain only one coded picture; and must carry PTS and DTS (if applicable).

For the ATSC *program guide* (PG) and ATSC document A55, PG is optional. It is carried as an MPEG-2 private section. It consists of a master program guide (if PG is sent, MPG is required), special program guides (optional), descriptive information parcel (optional), and private information parcel (optional).

7.4.5   ATSC food chain

Asynchronous compressed interconnect formats are as follows:

DVB-ASI.   270 Mbits/s, non-259M compliant

SMPTE 305M.   Also known as SDTI; 270 or 360 Mbits/s, 259M compliant

Synchronous compressed interconnect formats are as follows:

DVB-SSI.   270-Mbits/s self-clocked bit stream carrying 19.39 Mbits/s of MPEG or ATSC data

SMPTE 310M.   19.39 Mbits/s carrying ATSC bit stream

The synchronous serial interface for MPEG-2 digital transport stream is SMPTE 310. It is a self-clocking stream with biphase mark modulation. The most significant bit comes first. It is a 0.8-V p-p unbalanced signal centered at 0 V dc. It uses 75-Ω coax with BNC connectors.

7.4.5.1 Preprocessing. Most broadcasters are initially taking their NTSC programs and either feeding them directly to their SD MPEG encoders (some require a conversion to digital component first) or upconverting to their networks' HD format (if they are a network affiliate) and then sending them to HD encoders. MPEG encoders don't do well with noise, so many broadcasters are installing preprocessing equipment.

Two methods exist for filtering a video signal. The first is based on frequency, for example, low- or high-pass filters. The second is spatially or temporally (over time) based. Successive frame delays can be used to produce multiple outputs of the video over multiple frames. The video is summed and averaged to produce a transient suppressed picture, or low-pass filtering over time. This filter is known as a *transversal noise reducer.* Low-pass filtering is used to limit noise.

The problem with a transversal noise filter is that it introduces a substantial processing delay. A recursive filter is usually employed to limit that delay. This temporal filter looks for minor video changes from one frame to the next. If little change is detected, a percentage of the current frame is summed with the last frame. The amount of current frame summed with the last frame is decreased each frame to a limiting value which is determined by the desired noise reduction. When a complete scene changes or areas in the scene change, all the video comes from the current frame. The next frame uses one-half of the current frame and one-half of the previous frame. The next frame uses one-third of the current frame and three-fourths of the previous frame. In the next chapter on digital signal processing (DSP) you will see that this exponential function is employed recursively to design a low-pass DSP filter using an equivalent analog circuit response to an impulse (short-duration) signal.

A semitransversal filter is often used in conjunction with the recursive filter. A chain of frame stores, usually three, is used at the output of the recursive filter. It looks at all three frames and decides, on a pixel-by-pixel basis, which output has the least noise.

There are simpler filters, usually spatial, but some are a combination of spatial and temporal which often operate in front of the recursive filters. The first of these is the *median filter,* which has an aperture, or a fixed number of points, that is used for processing each sample. For example, a median filter could use nine points. A greater number of samples in the aperture results in additional boosts in S/N values out. The current pixel, a pixel before and after the current pixel, and the same three pixels in the previous and succeeding horizontal lines are added together, for example. The values of all nine sites are sampled and then ordered as to value from smallest to largest. The method requires use of an odd number of samples. The middle or median value would be applied as the value of the current pixel, which helps in eliminating transients. However, it also acts as a low-pass filter by reducing high-frequency picture content and any subtle texturing in the picture. In order to eliminate this problem, other circuitry must be added to look for those pitfalls and to not apply the filtering when they are found. Median filters usually operate only spatially with a single frame, but temporal applications of the process are available.

Spatial filtering is simpler still in that it totals and averages the samples in the aperture. Thus, this filter method finds the *average value*. It is generally a less accurate method than the median filter. In a median filter values at the extremes do not affect the result; in the spatial averaging filter extreme values, such as transients, have an effect on the final value.

The simplest noise-reduction filtering is called *linear* or simply the *low-pass filter* based on frequency and not spatial or temporal picture content at all. That said, these filters are not always that simple as many have extremely sharp rolloff areas. Extremely sharp rolloff filters are called *brickwall filters*. These filters require additional DSP circuitry to implement.

Digital circuitry doesn't mean that all equipment will perform equally well. The implementation and algorithms applied by various vendors will create quality differences just as was the case with analog video. How the available preprocessing is adjusted and applied will also determine how well the DTV signal appears to viewers at home.

7.4.5.2 Upconverters. Many network affiliates install upconverters to take their SD video up to HD video so that they won't need two encoders, one for SD and another for HD. Since the networks will be sending high-definition programs, most affiliates will need to have an HD encoder unless they intend to simply pass the network signal which is already ATSC encoded from satellite receiver to transmitter input. Even in the simplest implementations many affiliates will decode the network feed back to baseband video, which allows them to insert keys and switches between the network and upconverted local sources.

Just as in the preceding subsection, not all boxes will be equal. Upconversion relies heavily on interpolation. The more points used in this process, the better — in both the spatial and temporal domains. The better units use several dozen points to determine the value of each upconverted pixel. Complex processing must also be done to take two interlaced fields with action between the fields that is displaced in time. Upconverters also tend to enhance the picture. Incoming signals that already have a fair amount of enhancement might look worse coming out of the upconverter. Colorimetry is also slightly different between SD and HD.

Figure 7.8 Close-up of HD video on a monitor. Notice the block pattern on the building in the foreground.

The matrix equations to create luminance and the chroma difference signals are slightly different. Much more green is used for HD luminance than SD, and this affects the composition of the two color difference signals as they are derived from the luminance (Y) value. The upconverter should correct for this if your upconverted source is to match other HD sources (Fig. 7.8).

Noise preprocessing should be done before the upconversion is performed. The obvious should be stressed here: You never get something from nothing. Bad SD video will not look any better because it has more horizontal lines and/or more frames per second.

7.4.5.3 Encoders. An encoder takes the video and MPEG and compresses them and inserts the resultant data into a PES. An encoder can be a stand-alone box or in the same chassis as the multiplexer. As mentioned earlier, PES can be up to 65,536 bytes long; they are not restricted to 188 bytes like the final transport stream. The PES from the encoder goes on to the multiplexer.

Numerous settings must be input into the database that instructs the encoder on how to process the incoming video. Most have to do with elementary stream (called component in DVB) parameters.

Some setup parameters are as follows:

Video.   Pixels per line, if film-mode smart processing is to be used, GOP length, B-frame usage and the number between I and other P frames, and bit rate.

Audio.   Type of audio input (AC-3) for ATSC.

PES packet header.   Settings for the decoding priority of the PES, copyright information, and whether the video data contained are considered original or a copy. A name for the PES can be included.

Event scheduling.   Events like bit rate, GOPs, etc., can usually be scheduled to be changed for periods of time. The bit rate for a sporting event might be increased or lowered for simple "talking head" shows or late at night to allow increased bandwidth for data. Most systems allow renaming of the PES for that particular event.

7.4.5.4 Multiplexers. The multiplexer takes a selection of the available transport packets, including PSI and SI packets, adds stuffing packets, performs PCR resynchronization, and creates a valid transport stream that is output from the unit. The output is usually ASI, but some are now SMPTE 310 and even ATM. Many PES can arrive at a mux. It usually takes at least two PES to create a single program. More than one program can be interleaved into a single transport stream (Fig. 7.9).

The *mux* is where program (called service by DVB), transport streams, and network services are woven together. A *program* is all the PES (usually only video and audio PES) combined to create a virtual channel. With ATSC a DTV channel can be comprised of one or more virtual channels. These virtual channels can be a single HD or multiple SD channels. All the services available in a DTV channel, programs, data services, and tables about the channel end up as a single transport stream.

The network layer is located above the transport stream layer. The network layer is where multiple transport streams can be tied together as a single service. Many muxes have this layer of service because before DTV MPEG muxes were used mainly for DBS. DBS providers would have a single transport stream, consisting of up to six virtual channels on each transponder. A number of transponders were used to provide the desired amount of virtual channels. These transport streams are separate on each of the various transponders and are made to act like one continuous band of channels via the network layer. Without considering the legal ramifications, multiple DTV channels could be stitched together into one over-the-air DBS-type service using the network layer.

Multiplexers like everything else have bandwidth or data throughput rates. There can be elementary streams that are always arriving at the mux but not always being included in the transport stream. In some muxes these unused streams are still input into the mux but none of the data are selected to build

(a)

Figure 7.9 (a) Early MPEG encoders and multiplexers were quite large, (b) as were decoders. (c) Test facility set up by the Model Station Group at WRC in Washington, D.C.

the transport stream. Therefore, it is possible to overrun the available bandwidth if multiple HD, or even enough SD, encoders are always feeding the mux whether they are currently "on the air" or not. In some systems it may be necessary to throttle back bit rates from encoders not currently in use.

Some of the setup parameters for muxes are as follows:

(b)

Figure 7.9   (*Continued*)

Program.   The various PES or components that comprise the various programs are linked together via the PID for each PES. The PID or each PES that comprises a program is found in the PMT. The type of each elementary stream and whether PCRs are available are also found in the PMT.

Transport stream.   The various programs or services that comprise a transport stream are linked together, each program (or virtual channel) is given a name, and the PMT for each program is entered.

(c)

Figure 7.9   (*Continued*)

Network.   Overall bit rate for each transport stream (19.392658 in the case of ATSC); PSI table bandwidth allocation and table refresh rates (usually 0.1 s for PATs, CATs, and PMTs); and PAT, CAT, and NIT PID addressing. The NIT (network information table) is not required for ATSC.

Automation.   The process of gathering and compiling the program, commercial, and promotional information into a play list for air has historically been the task of the station's traffic department. It consisted of generating play, kill, and record lists that were turned into logs for use in on-air and acquisition activities. Now another layer of traffic-type control has to be added for DTV. These two layers will be to control the formatting of the multiplexer and the insertion of ancillary data to the ATSC transport stream. Parameters that will have to be loaded into ATSC multiplexers are bandwidth allocations, audio format data, aspect ratios, language, closed captioning, and program descriptor information. A likely additional task that will have to be handled by traffic is the amount of bandwidth, or the data rate, that various elements of transport stream consume. Many clients who buy HD or even SD programs or commercials within those programs will expect some guarantee that the required bandwidth was available.

To operators, editors, and others who will need to manipulate programming in a DTV plant, control systems, such as video servers and routers, will most likely represent material as objects. As an example, if a 720p program arrives in a plant that is mostly 1080i, the editor who needs access to the material might be presented with a graphical display that shows the material as an object at a satellite receiver. The editor might then drag and click the object from the receiver to a video server. The control system would know that the 720p material would have to be routed through a format converter to be in the plant's native format. That operation would be transparent to the user.

Each DTV station must have a transport stream ID (TSID), which resides in the MPEG PATs. In a given market the TSID must be different or unique. If broadcasters in a single market use different conditional access vendors, the broadcasters will have to carry each other's ID information. TSID can be used to map a station into an entirely different transport stream. As an example, an ABC station could map the cable channel ESPN as a subchannel. One problem with TSID is that some encoder vendors have TSID in PROM.

7.4.5.5 Gluing the pieces together. We have now talked about the pieces it takes to assemble a finished ATSC bit stream. One or more video encoders (some combination of SD and HD) are needed. They usually take digital component in, and each produces a PES out. This stream is either a parallel data signal or rides in an SDTI or ASI data stream. In order to produce AC-3 audio an AC-3 encoder is needed for each program. It produces a PES that rides on an Audio Engineering Society (AES) stream.

Next a multiplexer is needed to organize the incoming streams into one ATSC transport stream. The multiplexer might reside in the same chassis as the encoder. Also, data might arrive at the multiplexer for any number of reasons and from any number of sources. A data stream from a PSIP generator is required as well. The transport stream comes out of the multiplexer. Currently, it is either an ASI or an SMPTE 310 stream.

This stream will usually go straight to the digital studio-to-transmitter link (DSTL), but currently boxes exist that accept your transport stream and look for packets marked null. The box will take these null packets and replace them with opportunistic data. The box will also modify the PAT and create the necessary PMTs on the fly (Fig. 7.10).

7.4.5.6 Testing the bit stream (Fig. 7.11). Like any other stream in the plant, analog or digital, the ATSC has to be monitored. A number of companies offer devices that will monitor the ATSC stream. These devices should display a hierarchical presentation of the stream, and should graphically show what is in the PAT. It should be easy to burrow down to see the contents of the various PMTs in the PAT. Opening a PMT should display the elementary streams that are represented in the PMT.

It is helpful to have indications about the percentage of the overall stream used and the percentage each program uses. Additionally, the unit should quite clearly show anything that will limit a decoder's ability to successfully recover the data. Examples are stream sync loss, PATs and PMTs not being sent often

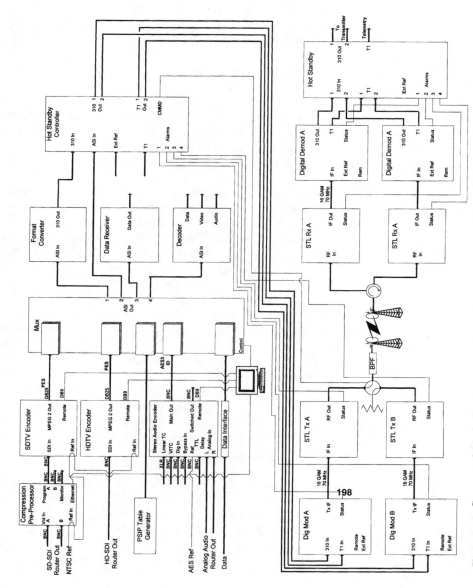

Figure 7.10 Conceptual block diagram of an ATSC path in a DTV station, covering the baseband audio and video conversion to AC-3 and MPEG data, which then multiplexes into an ATSC transport stream and microwaves from studio to transmitter.

Figure 7.11 Errors occurring in the bit stream will usually produce block patterns like the ones seen here.

enough or the scrambling control field not being set to zero, and the absence of a referenced PID in the PATs or PMTs (Fig. 7.12).

This monitoring needs to be physically near the facility's mux since that is where correction of a broken stream will need to take place. Once the stream leaves the mux, even though it might go through ASI/SMPTE 310 protocol conversion and a digital microwave link to the transmitter and then on to the transmitter's modulator, the stream should undergo absolutely no data value conversion. If any of the data change value at all, then something is broken, and very little, if any, decodable data will pass through the bad box. As a result, it is not necessary to monitor data differences between the station and the transmitter. What does need to be monitored at the transmitter is the modulated signal linearity and S/N, as we will see in Chap. 13.

7.5 Bibliography

ATSC Document A/52, "Digital Audio Compression Standard (AC-3)," December 20, 1995.
ATSC Document A/53, "ATSC Digital Television Standard," September 16, 1995.
ATSC Document A/54, "Guide to the Use of the ATSC Digital Television Standard," October 4, 1995.
ATSC Document A/55, "Program Guide for Digital Television," January 3, 1996.
ATSC Document A/56, "System Information for Digital Television," January 3, 1996.

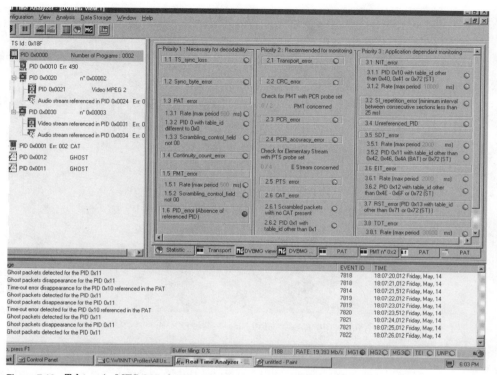

Figure 7.12   Tektronix MTS-215 showing a snapshot of an MPEG bit stream's health.

Harr, Paul, "Scientific-Atlanta's System for the Launch of DTV," *Broadcast Engineering,* December 1998, p. 76.

Koenen, Rob, "MPEG-4: Multimedia for Our Time," *IEEE Spectrum,* February 1999, p. 26.

McGoldrick, Paul, "For Every Action," *Broadcast Engineering,* December 1998, p. 118.

Nelson, Mark, *The Data Compression Book,* M&T Books, 1992.

Strachan, David, "Video Compression," *SMPTE Journal,* February 1996, p. 68.

8

Digital Signal Processing

Digital signal processing (DSP) is what digital circuitry does to mimic the function of analog circuitry. The obvious advantage to digital is that many of the analog gremlins are eliminated or at least suppressed; for example, ringing, smear, and group delay are gone. That is, until the digital signal is taken back to the analog domain. Before DTV, a totally digital television station still had to convert back to analog when the signal reached the transmission stage. Now with DTV the television data stream can remain digital right up until it is displayed on the screen at home. Eventually, digital light processing (DLP) will allow the digital chain to extend all the way through to the display. DLP is a process where micromirrors, one for each displayed pixel, up to 1.2 million, on a single integrated circuit (IC) deflect a scanned light beam onto the display screen. These mirrors can move up to 10,000 times/s. These micromechanical devices, which use the same technology that is used to etch the circuitry in ICs, hold the promise of totally digital displays.

However, digital processing is not always transparent. Functions such as filtering, decoding and encoding, and many electronic effects will still cause artifacts such as aliasing and cross-modulation. A digital video or audio stream can be thought of as a number sequence, where each sample is simply a numerical value. As long as that sequence is handed off from box to box or from tape generation to tape generation without any processing, the number sequence should stay exactly the same. This is the major benefit that digital techniques bring to television. However, when processing is applied, all algorithms or methods are not equal. The situation in the digital domain is the same as in the analog domain; some engineering is better than others. Cost still plays a role.

One-dimensional, two-dimensional, and three-dimensional filtering are terms familiar to everyone. What does that mean, though, and why does three-dimensional filtering cost more than others? In addition, how can some three-dimensional filters be better than others? In video a one-dimensional filter would look at the pixels on either side of the pixel in question in the same horizontal line, that is, in a rasterized video presentation in one dimension only, the horizontal. In a two-dimensional filter the filter acts upon the pixel of interest, based on pixels not only to the left and right but above and below, or in two dimensions. The three-dimensional filter operates horizontally, vertically, and over time, which means from one field to the next. As a result, the pixel is not only operated on based on pixels spatially around it in its field, but also based on pixels occurring in the same spot in fields preceding and succeeding the current field. The more ways a desired value can be determined, the better the result will be. However, three-dimensional operations require more hardware than two-dimensional operations. The three-dimensional filter implies frame syncs to house the various fields for use in processing. In the same way a two-dimensional filter requires storage of multiple horizontal lines, whereas a one-dimensional filter implies only storing a few pixels. Thus, two-dimensional filtering is more expensive than one-dimensional filtering, and so on with three-dimensional versus two-dimensional filtering. Cost is not the only issue though. Many filters claim to be adaptive. *Adaptive* means that the filter will switch between one-dimensional, two-dimensional, and three-dimensional filters as conditions change. Not all algorithms to determine which "dimensional" to use are equal. So engineering is still a coefficient in the equation, not just cost.

The bottom line is that not all digital and DSP will produce the same quality.

8.1 Boolean Algebra

DSP is based on the manipulation of strings of data. Each sample of data represents a value or number, which represents the magnitude of something. That "something" could be luminance or a color difference video signal. It could also just as easily be a sample from an audio channel. These samples or numerical values can be thought of as a number sequence. It is the processing of these number sequences that is the essence of DSP. However, these numbers are broken down and acted upon in even smaller elements. A number, which can correspond to 1 byte (8 bits), one word (more than 8 bytes), or even a nibble ($\frac{1}{2}$ byte or 4 bits), is ultimately operated on a bit at a time. Often the outcome of determining the value of a single bit will affect other bits, but what still matters is determining the value of each bit one by one. The value of a bit is usually determined by comparing it with another bit, and that other bit could be from another pixel. As we shall see later in this chapter, much of DSP is about adding bits and ultimately numbers together. Multiple additions are equivalent to multiplication, and that forms the heart of most filtering done in the digital domain.

In this section we will briefly refer to the rules for combining bits. This discipline is known as *boolean algebra,* and the rules describe how the basic logic gates work. At the simplest level, logic gates perform two basic operations— ANDing and ORing. An example of the "AND" operation is a business checking account where two officers of the company have to endorse a check. If only one signs the check, it is not valid. An example of an "OR" operation is the method most married couples use to set up their "joint" checking account. Either one, the husband or the wife, can endorse a check; both do not have to sign. These are the simplest boolean operations. We will look at more after we learn to count.

8.1.1   Binary counting

We have talked about number sequences. What do we really mean by that? 1, 2, 3, 4, 5,... is an obvious number sequence. However, it is well known that binary systems only deal with 0 and 1. So all numbers must be represented using only 0 and 1. The decimal system is what most people are accustomed to using. Most everyone has 10 fingers and toes so the method of counting is based on 10. When 10 is accumulated in any column, a 1 is added to the column to the left. In the binary system, which means 2, a 1 is added to the column to the left in counting from 0 to 1. Hence

00000000
00000001
00000010
00000011
00000100
00000101
00000110
00000111
00001000

This example shows how to count from 0 to 8. The column all the way to the right is considered the *least significant digit,* and the column all the way to the left is the *most significant digit.* The least significant column equals a 1 when a 1 is present in that column. The next column to the left represents a 2 when a 1 is present. The next column a 4, then an 8, etc., when 1s are present. The eight columns, or 8 bits as they would be called, can contain a number value between 0 and 255. To count higher requires more columns or bits. Digital video often has 10 bits. A 10-bit word could represent one of 1024 numbers or values, and these values represent either a luminance value or one of the two color difference signals. The number of bits used for audio is either 16, 20, or 24 bits. Sixteen-bit audio samples are able to represent one out of 65,536 values, 20-bit one out of 1,048,576 values, and 24-bit samples would range from 0 to a mere 16,777,216 values. This example illustrates how much more sensitive hearing is to level changes versus eyes to light-level changes.

In a discussion of the values used in binary systems it is not necessary to have a string of 1s and 0s to describe a binary value. However, the decimal system doesn't lend itself for use either. Since 4-bit nibbles are generally used, two of which comprise a byte, we use a counting system based on 4 bits, or 16 possible states. This is the *hexadecimal counting system.* The following example shows how to count from 0 to 15:

$$1, 2, 3, 4, 5, 6, 7, 8, 9, A, B, C, D, E, F$$

An example of 27 decimal would be 11011 binary and 1B hexadecimal. Let's review how we arrived at the binary and hexadecimal values:

Binary: 1   1   0   1   1

$$16 + 8 + 0 + 2 + 1 = 27$$

Hexadecimal: 1   B

$$16 + 11 = 27$$

How about 99 decimal?

Binary: 1   1   0   0   0   1   1

$$64 + 32 + 0 + 0 + 0 + 2 + 1 = 99$$

Hexadecimal: 63

$$(6 \times 16) + (3 \times 1) = 99$$

For 10-bit video values of 0×3FF, etc., are common. What does this mean? Well, $3FF_h$ by itself implies 12 bits or resolution:

$$0011\ 1111\ 1111$$

However, for 10-bit video 3FF is literally all bits:

$$11\ 1111\ 1111$$

With 10 bits it is not possible to have FFF; 0×## means that this is a hexadecimal value.

8.1.2 Boolean gates and functions

So now that we have learned how to count, let's see how to manipulate these values. Why do we want to manipulate the values? Let's say that we are doing a dissolve between two sources, and in the middle of the dissolve one-half the value of the two sources together is being added. If the video is going through a processing amp, or the audio through a mixer, values might be added or subtracted to the audio/video data. As we will see later in the chapter, filtering signals in the digital domain entails using a process called *multiply and accumulate,* which means multiplying a value by a coefficient, storing the result, and passing it on to another stage to multiply the resultant by another coefficient, etc. This is done to mimic the operation of an analog filter.

As mentioned, binary words, or bytes, are manipulated on a bit-by-bit basis. Now we will look at how ANDing, ORing, and other common boolean operations are implemented. First, though, let's look at the language of boolean algebra. All mathematics has a language. Most do not learn the language and,

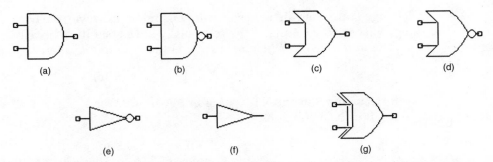

Figure 8.1 Boolean logic gates as drawn by a logic design program: (*a*) AND, (*b*) NAND, (*c*) OR, (*d*) NOR, (*e*) inverter, (*f*) buffer, (*g*) exclusive OR.

thus, are often barely literate with many branches of mathematics. Let's start with the simplest boolean sentence or equation:

$$F = X \times Y \qquad \text{(Fig. 8.1}a\text{)}$$

In algebra this equation is read "the function F equals variable X times variable Y." In boolean algebra it means "the output F is true if input X and input Y are both true." This is the AND function.

The AND function can be turned into the not AND or NAND function:

$$\overline{F} = X \times Y \qquad \text{(Fig. 8.1}b\text{)}$$

which means "the output F is not true if input X and Y are both true."

In boolean algebra "true" means a "1" or a logic high. Boolean functions are implemented in electric circuits with "logic" gates. Logic gates work with electric power, that is, they take applied voltage and current flow. These gates generally have voltage applied to the inputs which determines the voltage at the gates' output. A common type of logic gate in use for the last 25 years comprises a technology called *transistor-to-transistor logic* (TTL). These gates, which were packaged as ICs, had a handful of transistors directly coupled to each other to create various logic gates. This direct transistor-to-transistor coupling resulted in faster logic transition and, thus, higher-frequency operation than the logic gates before it. TTL logic worked with voltages of 5 V. The IC package that contained the TTL logic gates was powered by 5 V. The inputs to a TTL logic gate were expected to be above 3.3 V or below 0.6 V. If 3.3 V were applied to a TTL input, this was considered a high voltage or just a "hi." If less than 0.6 V were applied, it was considered a low voltage or just a "lo." Voltages between 0.6 and 3.3 V were not considered valid. The output of a TTL logic gate would conform to the expected TTL input levels.

TTL logic was a big step forward but consumed a lot of power, especially when a change of state at the output occurred, which could be quite often with operation at high frequencies. To limit the power draw a technology called complementary metal-oxide semiconductor (CMOS) was used. If a gate did not change state, very little power was consumed. However, early on this technology was slow. Variants of the technology have sped it up greatly.

The other basic boolean operation is the OR, and it is written as

$$F = X + Y \qquad \text{(Fig. 8.1}c)$$

which means "output F is true if either input X or Y is true." There is also a not OR or NOR function:

$$\overline{F} = X + Y \qquad \text{(Fig. 8.1}d)$$

which means "output F is not true if either input X or input Y is true."

The invert logic gate or function is simple. It takes a hi (or "1," or true) and inverts it to a lo (or "0," or false). It is written:

$$F = \overline{X} \qquad \text{(Fig. 8.1}e)$$

There is also a "buffer" function:

$$F = X \qquad \text{(Fig. 8.1}f)$$

which is usually used for three reasons. The first is due to *fanout*. A logic gate can usually only drive so many succeeding logic gates. The number of succeeding gates that the output of a gate can drive is referred to as its fanout. As a result, several cascaded buffers might be inserted to increase the number of gates that can be driven. Second, buffers are often used to change the level, or impedance, of a logic line. Inputs and outputs often run at different levels and impedance than the internal logic. Third, they are sometimes used for delay. There is always a propagation time between the time a logic level is applied to the input of a gate and the output of the gate changes level. Just as a delay is added to time various video sources to one another, the same must often be done for logic signals.

The final basic logic gate is called the exclusive OR, which is written:

$$F = X \oplus Y \qquad \text{(Fig. 8.1}g)$$

and this equation means "output F is true if input X or input Y is true but not both."

All gates have a truth table associated with them. The exclusive OR table would be

$$
\begin{array}{cc|c}
X & Y & F \\
0 & 0 & 0 \\
0 & 1 & 1 \\
1 & 0 & 1 \\
1 & 1 & 0 \\
\end{array}
$$

With a normal OR gate, the last entry would still be $F = 1$. In truth tables the inputs logic states are usually presented as a binary count, that is:

$$0\ 0$$
$$0\ 1$$
$$1\ 0$$
$$1\ 1$$

8.1.3 Math circuitry

So we have now reviewed basic boolean functions. By themselves, they don't do much. When combined, the functions perform higher functions. One such higher function, which was already mentioned, is a cornerstone to DSP, the multiply and accumulate (MAC). This high-level function is based on ADDing binary values together (Fig. 8.2). In boolean algebra the sum and carry outputs of Fig. 8.2 would be written

$$\text{Sum} = xy' + x'y$$

(where ′ is another way of writing the invert or NOT function) and

$$\text{Carry} = xy$$

which means "the sum is true or '1' if x or y is true but not both." This is the exclusive OR function. The top two AND, invert, and OR gates could be replaced with a single exclusive OR gate.

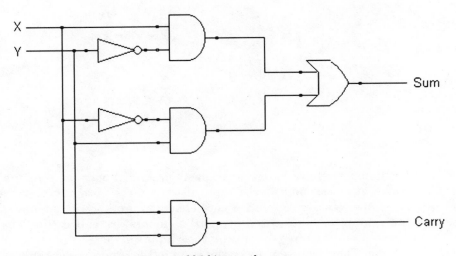

Figure 8.2 Simple logic circuit to add 2 bits together.

This combination of logic is called a half-adder because it does only half the required job. It takes 2 bits and adds them together and produces a carry, but it does not accept a carry input from a preceding stage.

The boolean expression for a full-adder is

$$\text{Sum} = z \oplus [(x \oplus y)]$$

which can also be stated as

$$\begin{aligned}\text{Sum} &= z' \, (xy' + x'y) + z(xy' + x'y)' \\ &= z' \, (xy' + x'y) + z(xy + x'y') \\ &= xy'z' + x'yz' + xyz + x'y'z\end{aligned}$$

The carry is stated as

$$\begin{aligned}\text{Carry} &= z(xy' + x'y) + xy \\ &= xy'z + x'yz + xy \\ &= xy + xz + yz\end{aligned}$$

Thus, another implementation for a full-adder.

How are these transformations made? With the following boolean algebra rules:

$$
\begin{array}{ll}
x + 0 = x & xy = yx \\
x \times 1 = x & x + (y + z) = (x + y) + z \\
x + x' = 1 & x(yz) = (xy)\,z \\
x \times x' = 0 & x(y + z) = xy + xz \\
x + x = x & x + yz = (x + y)(x + z) \\
x \times x = x & (x + y)' = x'y' \\
x + 1 = 1 & (xy)' = x' + y' \\
x \times 0 = 0 & x + xy = x \\
(x')' = x & x(x + y) = x \\
x + y = y + x & xy' + x'y = x \oplus y
\end{array}
$$

8.1.4 Sequential logic

Until now we have looked at boolean circuitry known as combinational logic. However, most digital circuitry also requires circuitry that can latch or remember a state. These memory elements are known as *sequential logic*. In their simplest state these memory elements are called *flip-flops* or *latches*. These circuits sample the state of their inputs at discrete or particular instances of time. They then remember the state of the input at the sample time, which is known as a *synchronous sequential logic circuit*. A clock is usually used to

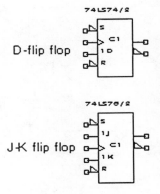

Figure 8.3 Two common flip-flops (latches) as displayed by a simple logic design program. The D flip-flop has set (S) and reset (R) inputs that force the top output either high or low. The bottom output always assumes the opposite state. The output will also assume the value of the D input whenever a positive transition occurs at the clock (C) input. The flip-flop always remembers the last set, reset, or D-input event. The bottom flip-flop has two inputs (J and K) that provide additional flexibility in controlling the output.

D-flip flop   74LS74/2

J-K flip flop   74LS76/2

determine the sample point. Sequential logic can also be asynchronous, which means that the device latches the new input state dependent on a new latch command and not a periodic clock (Fig. 8.3).

In practice combinational logic and sequential logic are both used to implement a function. Sequential logic is used to latch digital video and audio so that all the digital video and audio words can be marched through processing circuitry. Large groups of sequential logic are integrated to create shift registers. Large shift registers can be combined to create large timing buffers. These buffers are expanded to form the essence of time-base correctors and frame synchronizers. In simple terms digital video is clocked into the buffers synchronized to the incoming video. However, the video is clocked out based on local video rate requirements. The sequential gates are organized as a circle. Video or audio samples are loaded into registers at the incoming data rate and read back out at the local rate.

8.1.5 State machines

In the most basic sense a sequential logic system is usually implemented to create a "state machine." A state machine can be divided into two parts, data-processing and control logic functions. The data-processing function inputs data, acts upon them, and outputs them based on commands issued from the control logic. The control logic issues commands based on the status it receives from the data-processor side of the state machine.

A state machine's operation is often described by a flow diagram. In Sec. 6.4 we learned that digital component video has timing reference signals that are generated based on some unique data values, namely, 0×3FF and 0×000. The string 0×3FF, 0×000, 0×000 indicates that a timing reference signal (TRS) has occurred. A fourth word indicates whether the TRS is occurring during active video or the vertical interval, whether the TRS indicates the start or end of an active video line, and whether the current field is odd or even. A processing circuit acting upon active video would have to know the state of the data stream arriving, namely, whether the current data were active video or not. Figure 8.4 shows the flowchart of a state machine.

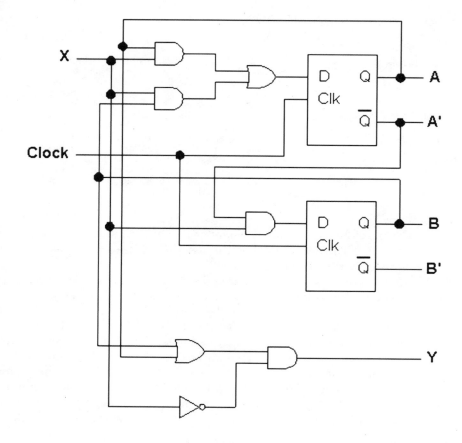

Next State		Input	Present State		Output
A	B	X	A	B	Y
0	0	0	0	0	0
0	0	1	0	1	0
0	1	0	0	0	1
0	1	1	1	1	0
1	0	0	0	0	1
1	0	1	0	1	0
1	1	0	0	0	1
1	1	1	1	0	0

Figure 8.4 Flowchart of a state machine. The next output state after the next clock can be predicted by the input and the present state.

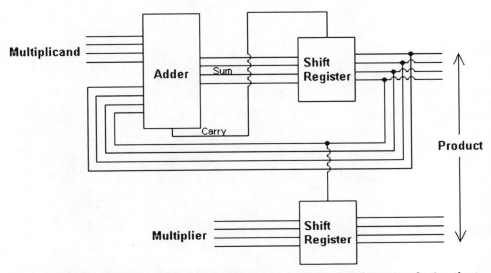

Figure 8.5 Binary multiplier. Successive additions are performed with carries, forcing the top shift register to shift all outputs down by one. The bottom or most significant bit from the top shift register forces the same downward shifting by the bottom shift register.

8.2 Building Blocks

Implementation of DSP applications requires combinational logic and adders. However, as mentioned, the MAC circuit is an important DSP building block. We accumulate by adding; we multiply with the circuit shown in Fig. 8.5. A few other blocks are needed to process and manipulate video and audio data. The most common is the parallel to serial and serial to parallel circuitry. Almost all data are shipped between boxes as serial data. However, processing and manipulating serial data are extremely difficult. So if a box is to act upon the serial video or audio data it receives, it first converts it from serial to parallel data, processes that data, and then converts it back to serial data for shipment out the box. The devices that perform these transformations are made up of sequential logic devices strung together as "shift registers" (Fig. 8.6).

Another important function is multiplex and demultiplex circuitry. Many strings of data are usually interwoven to create a complete serial data stream. Let's take the case of an SMPTE 259 serial stream. Not only does it have video data, but TRS strings are included, and often audio ancillary data are added between the EAV and SAV TRS strings. Sometimes error detection and handling (EDH), video index, closed captioning, and vertical interval test (VITs) signals will be added during the vertical interval. All these various data streams will be combined into one data stream by a process know as *multiplexing*. A multiplexer is sort of a railroad yard switching system that lets each train of data move from its particular side track onto the main line at the appropriate time. These separate trains combine to form a continuous train. Of course once this continuous train reaches its destination, it must be separated back onto the appropriate side tracks again, a process known as *demultiplexing* (Fig. 8.7).

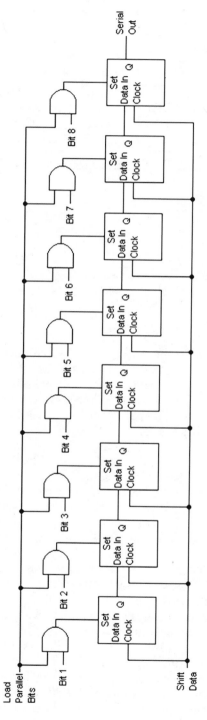

Figure 8.6   Parallel in/serial out shift register.

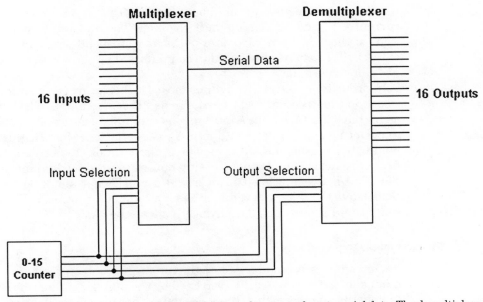

Figure 8.7 The multiplexer takes parallel data and converts them to serial data. The demultiplexer reverses the process.

8.2.1 DSP engines

Many generic DSP devices are available off the shelf, although many systems are based on proprietary chip sets. Makers of these chip sets often refer to them as DSP engines. The generic DSP chips are offered by a number of manufacturers. These ICs are specialized microprocessors geared to handle large amounts of data throughput. While most dedicated DSP chips rely heavily on hardware (with lots of microcode) for speed, generic DSP chips rely on generic hardware and specialized software written for the application at hand. This specialized software used to be written in assembly language, but today it is written in high-level languages such as C, using compilers supplied by the DSP chip manufacturer.

The dedicated DSP IC usually has a hardware organization of a fixed size. The hardware usually consists of shift registers. The data are clocked from register to register as they pass through the DSP. At each register stage some operation, often the multiply and accumulate function, is performed on the data. The number of registers, which are referred to as taps, can range from a few to well over 100. The more taps, the more complex the operation that can be performed. At 50 taps fairly complex operations can be handled. Creation of filters with multiple poles takes many more taps then a simple single-pole filter.

Today's generic DSP IC uses an architecture commonly called Harvard architecture, which has separate internal buses for the processing of data and DSP control operations. This is an enormous advantage over the common microprocessor, which often uses the same bus for both operations. Some DSP ICs use as many as eight internal buses. Bus widths can be as wide as 64 bits, but currently 16 and 32 are the most common. The multiply and accumulate sections of these ICs can have data paths as wide as 96 bits. Today, over one-half of the DSP ICs in use are found in computers. The generic DSP business is over $5 billion/year today.

In order to increase throughput most DSP ICs can perform multiple operations, and instruction caches as large as 2K can be found. Clock speeds can be as high as 200 MHz, but simultaneous command execution allows a single IC to perform on the order of 1.6 billion instructions/s. Most DSP ICs allow commands to be issued once and repeated many times. Today, many of the bottlenecks are not due to data processing but are caused by data movement around the IC or between other DSP ICs that comprise a system. Some implementations leave the data in a common memory and allow other DSP components to look at and change but not physically move the data.

8.2.2 Programmable logic devices

ICs are often referred to based on the number of boolean gates they contain. The number of gates is generally restricted by the number of pins on the IC package. In general, the more pins, the more gates that are integrated into the IC. This integration is often referred to as *small-scale integration* (SSI), *medium-scale integration* (MSI), *large-scale integration* (LSI), and *very large-scale integration* (VLSI). SSI packages often have two to six boolean gates—two if they contain flip-flops or registers and six if they contain buffers or inverters. MSIs contain things like small multiplexers or demultiplexers. LSIs are often called application-specific ICs (ASICs) and these devices perform a specialized function such as an IC that demodulates an NTSC RF signal into video and audio. VLSI is typified by microprocessors and DSPs.

MSI and LSI components can be comprised of large arrays of elementary boolean gates found in SSI packages. The boolean gates would be arranged as an array with gate inputs and outputs connected through electronic fuses. Programming these devices would be done by "blowing out" the fuses from paths not connected to leave only desired connections. In this way the elementary gates in the IC can be combined to create more advanced functions such as the various mathematical circuits required for DSP operations. These ICs are known as *programmable logic devices* (PLDs).

Three types of PLDs are available. They consist of an AND array and an OR connected together to provide an AND-OR sum-of-product implementation. The connections between the AND and OR arrays are determined by "programming" the device. They differ by the placement of fuses in the AND-OR array.

The most common is the read-only memory (ROM). The programmable version of the ROM is the programmable ROM (PROM). The PROM/ROM consists of two parts, a decoder and output OR gates. If the PROM is an 8-bit output device, then there are eight output OR gates. The decoder takes the binary value of the input bits and enables one output. As an example, a 4 by 16 decoder has four inputs and 16 outputs. A 5 by 32 decoder has five inputs and 32 outputs, a 6 by 64 decoder has six inputs and 64 outputs, and so on. Only one decoder output can be true at a time. Each decoder output has a link to each output OR gate. In the case of our 4 by 16 decoder with an 8-bit output that's 64 links. In the case of our 6 by 64 decoder that's 512 links. Output OR gates that are not true or high for a particular decoder output are blown open during programming. These devices are used as permanent memory for computer operations. The inputs are an address while the output is data. The PROM/ROM is an IC that contains a decoder linked to OR gates. The decoder can be thought of as a system of AND gates that have many outputs, with only one true at any time, based on the combination of inputs.

The next device is the programmable array logic (PAL) device, where the input to the AND gates is programmed and the connection between the AND gates and the OR gates is fixed. The most versatile PLD is the programmable logic array (PLA), which has fusible links on the input of both the AND arrays and the OR arrays.

There is another type of logic array called the complex PLD (CPLD). It is closely related to a device known as a field-programmable gate array (FPGA). These arrays usually consist of sequential logic devices. The basic building block is often an 8 to 1 multiplexer driving a flip-flop. These blocks can be made to perform all the basic boolean functions. Unlike regular PLDs, these devices start out with all open connections and programming creates the desired connections. These devices can literally have large arrays (many rows and columns) of logic blocks. Between the columns of blocks is a crosshatch of channels or paths. Horizontal paths go to the inputs and outputs of each block, but no horizontal path makes it completely from one column to the next. Instead, both input and output paths straddle vertical paths. Connections between blocks are made by programming connections to and off the vertical paths.

8.3   DSP Design

Next we will get an overview of how DSP is accomplished. We will start by looking at how a camera processes light information output from a CCD block into analog composite, or digital component video using DSP. The explanation provided here is not meant to be rigorous mathematically, and it is not intended to make one able to design or even analyze digital signal processors. What you should get from this is a method to think about what a DSP system is doing in more familiar analog terms.

8.3.1   The digital camera

Figure 8.8 is a function block diagram of a typical camera. This block should make it obvious that a DSP camera has to perform most of the same functions as a non-DSP camera. Although the functions are the same, the approach is

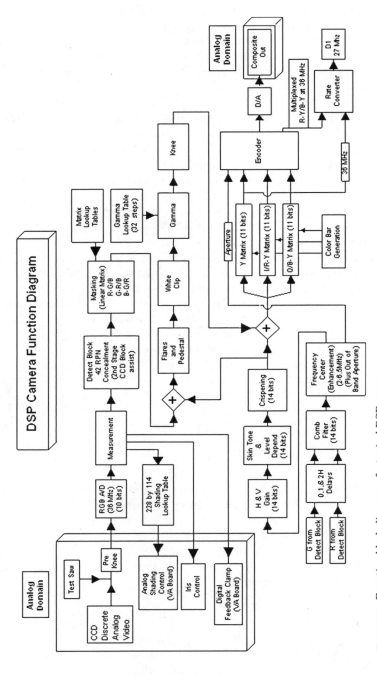

Figure 8.8 Function block diagram of a typical DSP camera.

216

different, which is what we will discuss now. In the next chapter we will look at camera functions in detail; the following DSP techniques can help:

Shading.   Values are added to pixel values based on the pixel's location.

Masking.   Correction values are added to each channel's (R, G, B) pixel value to correct for imperfections in the prism's ability to separate the primary colors.

Flares.   Correction values are added based on pixel intensity due to optic imperfections.

Gamma.   Values are added based on pixel intensity to achieve the desired intensity transfer function. Multiple gamma lookup tables are generally available.

Image capture smoothing.   Correction values are used for pixel imperfections in the CCD block.

Knee.   Values are added to achieve the desired intensity nonlinearity above the desired pixel intensity.

It is helpful to think in digital camera terms. Analog tube cameras were all continuous analog domain devices. The pickup tubes developed continuous analog signals from incoming light, and the electronics behind them processed and combined the signals into a composite signal out. With the introduction of CCD imagers, the camera moved from the continuous analog domain to the discrete analog domain. Discrete means that the signal is no longer one continuous entity but a series of samples.

8.3.2   CCDs

Looking at a signal leaving a CCD imager (light through the cameras optics is converted to an array of discrete voltages and then marched out in order from the picture's top left to bottom right) reveals a series of analog values that can change only at a predetermined sample rate. The sample rate in the case of a CCD camera is dependent on the number of cells in the CCD imager. A typical high-quality SD "professional" camera would have 523,152 pixels. HD camera pixel counts are near 2,000,000. The CCD block can actually be thought of as a sample-and-hold circuit set up in an array 1038 wide by 504 tall (actually 980 by 494 of it is used). Photon energy is captured in those cells and converted into a volume of electrons that represents the amount of light that hit that cell. These 523,152 pixels appear 30 times/s. That's approximately 15.7 million pixels, or discrete samples, every second. Black is placed between them every so often to create horizontal and vertical blanking spaces. Therefore, these pixels are clocked out at a rate of 18 MHz.

8.3.3   Nyquist rate

A horizontal line from these cameras can change its analog value every 55.555 ns, or 980 times per active line. This is a discrete signal. Along with discrete

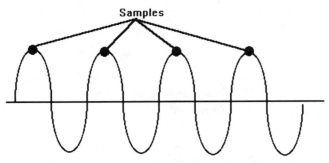

Figure 8.9 A signal sampled once per cycle appears as a dc signal.

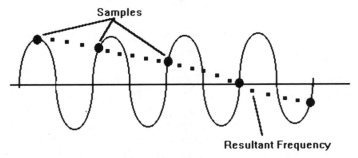

Figure 8.10 Samples taken slightly more than once per cycle appear as low-frequency signals.

signals comes a potential problem. If the Nyquist sample theorem is applied to this situation, it means that a scene can't be shot with more than 490 vertical lines (remember the picket fence from Chap. 1) in it ($1/2 \times 980$) and not have the dreaded aliasing problem.

As seen in Fig. 8.9, if it is assumed that the waveform is the result of four scanned vertical lines on a resolution chart and if it is sampled once per cycle, the resultant samples will all be equal. The lines will appear to be a straight dc line. Figure 8.10 shows what happens if the sampling is done at a slightly faster rate. The resultant discrete signal appears as a much lower frequency than what is actually there. This is called *frequency foldback* or *aliasing*. Figure 8.11 shows sampling at the Nyquist rate (twice per cycle). Although this would produce a series of samples resembling a square wave, as shown in Fig. 8.12, this is OK because the final output is low-pass filtered so that only the fundamental of the square wave is left at the output. The result is that the sine wave is back.

To reduce the problem of frequencies above one-half the sampling rate, filtering is done optically (and electrically) at the very front of the camera to minimize the amplitude of high-frequency picture patterns (lots of lines) in the scene that causes this problem. This is accomplished with the creative use of multiple reflections within pieces of glass. This process works as an optical low-

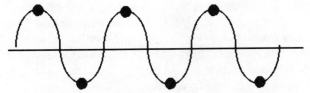

Figure 8.11 Sampling at Nyquist rate.

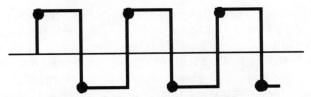

Figure 8.12 Discrete output at Nyquist rate. Low-pass filtering at receive end would convert this back to original sine wave.

pass filter. The 18-MHz clock rate is no accident; this contrived rate will become important later.

8.3.4 Preprocessing

We have moved from the discrete analog domain to the discrete digital domain by taking the analog values coming out of our CCD array and converting them into 10-bit binary numbers. This discrete analog sequence is now a discrete number sequence. Most cameras do this after some initial processing while still in the analog domain, as Fig. 8.8 demonstrates. Things that must be processed at high rates (shading and clamping controls), be analog-driven (iris), or have the number of bits required to do the job cut down (pre-knee) are still done in the analog domain.

By taking video highlights as high as 600 percent normal and compressing them down to 226 percent of normal while still in the analog domain, the number needed to handle that amount of overexposure is lowered by 2 bits. The 10 bits then handle the range from 0 IRE (20 hex/32 DEC) to 226 IRE (3FF hex/1023 DEC). Incidentally, 100 IRE equates to 1D6 hex, or 470 decimal.

8.3.5 Sampling

We double-sample each CCD analog signal (R, G, B) at 36 MHz because the green sensor is spatially offset by 1/2 pixel from red and blue. The green is delayed by 1/2 pixel when it is converted to a digital signal; therefore, a 36-MHz (2 × 18 MHz) clock is needed. Additionally, the green channel usually contains information not available from the red and blue channels. This is how manufacturers are able to quote resolutions of over 800 TV lines for these cameras. This 36-MHz sample rate eliminates aliasing from the higher information rate produced by the green channel spatial offset.

We are now in the digital domain with three 10-bit parallel number sequences marching through the camera (R, G, B). Now we will explore how a digital system processes these number sequences as compared to an analog camera. The snap answer to the question just posed lies with digital adders, and subtraction by twos complements addition, right? And the answer is, well, right, but with some additional explanation required.

8.3.6   Digital equivalent to analog

Just as earlier IBM-compatible PCs could perform very complex mathematical functions with no math coprocessor by multiple additions to accomplish multiplication, or many summations to mimic integration, many DSP systems work the same way. So if that is all that had to be done in video signal processing, the explanation could end here. But even in these cases the speed of these adders has to be very fast to accomplish this action because each individual parallel byte of video is in one location of the video DSP pipeline for approximately 28 ns. Nonetheless, most of the blocks shown in Fig. 8.8 perform nothing more than addition (or subtraction).

The main exceptions are areas that must act as filters. Looking at the detail blocks in the lower left of Fig. 8.8, instances of bandpass filtering can be seen. Also, the encoder block must low-pass the two chroma channels to meet output requirements. Looking at the lower right corner of the block diagram, you can see we must get from a 72-MHz parallel byte rate (36 MHz Y plus 18 MHz R-Y and 18-MHz B-Y rates) to the equivalent of SMPTE 125 4:2:2 at 27 MHz.

As mentioned earlier, the way to think of DSP is to start with its analog equivalent. In fact, many DSP designs start with an analog perspective and then transform, or equate, the analog design to a digital design. The same will be done here.

In the analog domain it is usual to think of functions in terms of frequency. Fourier in the nineteenth century found that overall thermodynamic cycles could be explained by breaking out the individual temperature cycles that combined to make a complex thermal function. The same can be done with electric functions (that is, video waveforms). What he found is that if you add enough harmonic sine waves of the proper amplitude and phase with a sine wave at the fundamental rate, it is possible to build any periodic (repeating) function, as we saw in Sec. 4.3.2.

8.3.7   Crafting the digital transfer function

A designer of wideband systems (like video cameras) might first decide on the highest frequency function that his system needs to pass, and then create a system that will pass the required harmonics to realize a reasonable representation at the output. The designer will also want to limit the passband in the system, mainly to minimize noise.

In the simplistic explanation, a transfer function for the system which would yield the desired passband would be designed. This is usually done in the frequency domain because the math is much tougher if it is done in the time domain. It is possible to design a transfer function in the time domain, but the creation of such a signal, or just about any signal, will take both capacitance and inductance. Just about any circuit path has both.

The big problem is that capacitors tend to integrate the signal over time, while inductors differentiate the signal over time. The tools to cope mathematically with such circumstances are called *differential equations*. Not even engineers enjoy such endeavors. To wade through such a task, manipulations are used to reduce such problems to algebraic exercises. Hence the entrance of Laplace transforms. This branch of mathematics lets the designer take standard wave shapes and transform them to an algebraic function, which is referred to as a *transfer function*. It does this by moving functions that are dependent on time to functions that are dependent on frequency.

Simply put, a transfer function is just a big fraction representing the output versus the input:

$$\text{Transfer function} = \frac{\text{output}}{\text{input}}$$

This transfer function is reduced to simplest form, and the roots of the numerator and the denominator are found. Roots in the numerator are called zeros, and roots in the denominator are called poles. Poles make the response curve in a Bode plot break upward; zeros make the curve break downward. In Fig. 8.13 there are two poles (X) and two zeros (open bullet). Generally, the more zeros, the faster the roll-off. A transfer function cannot have more poles than zeros.

Once poles and zeros are found, actual values for inductance, capacitance, and resistance and the amount of gain required for the system can be determined. This is obviously a gross oversimplification of what actually happens. It should be understood that, in reality, each block on a camera block diagram would have its own individual transfer function done first (actually many blocks are broken down even further into subblocks with their own transfer functions). Once every block's and subblock's function is determined, they can all be multiplied together (as in any cascaded system) to receive a system transfer function. But the result would probably be extremely cumbersome to use or understand.

8.3.8   Design help

8.3.8.1   Spice.
Today, such analysis is not done by humans very often. Other tools have replaced the manual analysis to check designs. The most well known is a computer program called Spice. It takes all the components in a circuit, and using a process known as nodal analysis (based on Kirchhoff's current law), it will produce both time-domain and frequency-domain analysis of the proposed circuit design.

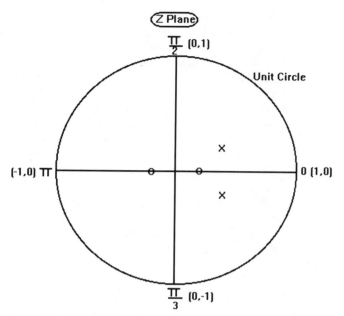

Figure 8.13 Z plane (from Z transform). Open bullets are zeros, Xs are poles.

This program has been in widespread use for 20 years, but it is being succeeded by more advanced analysis packages. While Spice was mainly used at the subblock level of design to simulate the result of a proposed design, today engineers can describe the external specifications of a proposed design and have software propose the actual circuitry needed to realize the desired result in many cases. One of the best known of these programs is called Verilog, which is known as a hardware descriptive language. Another is VDHL.

8.3.8.2 VDHL. This program is the original top-down system methodology. It is a software standard for modeling electronic systems. VDHL was developed in the early 1980s, and uses a language to describe inputs and the associated required outputs. A compiler is then run to create a simulation of the system. The goals for VDHL were to provide a description of digital hardware systems at every behavioral and physical level from gates to entire systems and to be a design and documentation tool. The compiled version imitates the actual system. The basic building block in VDHL is the "design entity." Design entities can be simple logic gates or complete systems. This design language is a lot like the C software language where more complex functions are created from simpler functions. With VDHL more complex design entities are created from simpler ones. Also, like most high-level languages VDHL allows branching and conditional control. Another type of VHDL design unit is the architecture design unit, which describes what an entity does or shows what it is composed of.

8.3.8.3 Verilog. This is the other major modeling program. It was developed during the mid-1980s and is used much more often than VHDL. VHDL is now mainly used by defense contractors as the U.S. Department of Defense (DOD) requires documentation by VHDL. Verilog has a much larger device library than VHDL. Verilog describes a digital system as a set of modules, where each module has an interface to other modules as well as a description of the modules' contents. Like VHDL, once a module is defined, it can be used in larger modules. Two fundamental data types are used in Verilog: nets and registers. *Registers* are logic devices at the lowest levels and modules at higher levels. *Nets* are the interconnection between registers. However, Verilog can also be used to describe the external specs (outputs versus inputs) for behavioral modeling. Here no effort is used to describe how the module is implemented in terms of logic gates. This is done early on in the design process to understand how a system will react. From the purely behavioral model the design will migrate to a purely structural model when the design is done. In between the two types of models will be mixed.

8.3.9 Digital design—mimicking its analog equivalent

Circuitry that does its processing in the digital domain uses a different set of manipulations. Instead of the Laplace transform, something called a Z transform is used. This topic was covered initially in Chap. 4 but will be reviewed here. As with Laplace transforms, it is possible to move from the time domain to the frequency domain and vice versa. Instead of a horizontal frequency display (such as might be seen on a spectrum analyzer), the Z transform rolls the spectrum response into a circle, as shown in Fig. 8.13.

Thus, just like the analog designer, the digital designer calculates the desired response, designs a transfer function in the frequency domain of the Z transform, and then performs an inverse Z transform to get back to the time domain. At this point, the designer could use the function, which is now described in terms of time, instead of frequency, to find its impulse response. This is done by passing a very narrow pulse through the analog equivalent circuit. In the case of a low-pass filter the time response might look like the one in Fig. 8.14.

The response the circuit had to an impulse can be related to the response it will have to a square wave via the use of convolution. The output response of a square wave through the circuit will result if the square wave is convoluted through the response, and this is graphically demonstrated here. First, the time response is reversed in time, as shown in Fig. 8.15. Next, the square wave is moved through the response in time steps equal to the sampling rate. At each point, the area of overlap between the square wave and the response is integrated. The area found by integration is used as the magnitude at that point in time. This is the response characteristic that the digital circuit must mimic (Fig. 8.16).

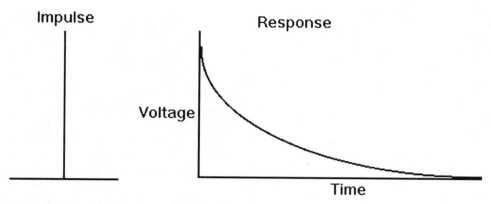

Figure 8.14 Impulse in the time domain (left), sample response (right).

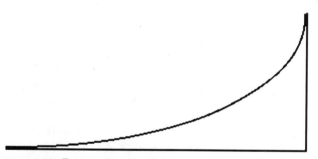

Figure 8.15 Response reversed in time.

8.3.10   Finite impulse response (FIR) and infinite impulse response (IIR) circuits

The filter derived from moving into the Z domain could be realized with the circuit shown in Fig. 8.17. Notice that this circuit is only 8 bits wide. Most cameras use a minimum of 10 bits. In some sub-subsystems, where more precision is desired, such as the detail circuitry, 14 bits are often used to minimize the rounding errors. The circuit shown in Fig. 8.17 is actually quite simple as digital filters go. With only the last four samples summed at any one time, this circuit would only be able to mimic a single-pole low-pass filter. Additional samples would have to be added concurrently to mimic filters with more complex functions, which means that the array of registers would get wider. To handle more bits, the array would get taller.

The multipliers are what allow this circuit to act like a low-pass filter. Going back to Fig. 8.16 and examining the resultant response obtained from convolution, it is apparent that the first sample is the highest. If that is called unity (or 1), each succeeding sample is something less than unity. In fact, the second is two-thirds of the first, the third is two-thirds of the second, and so on. Therefore, to mimic

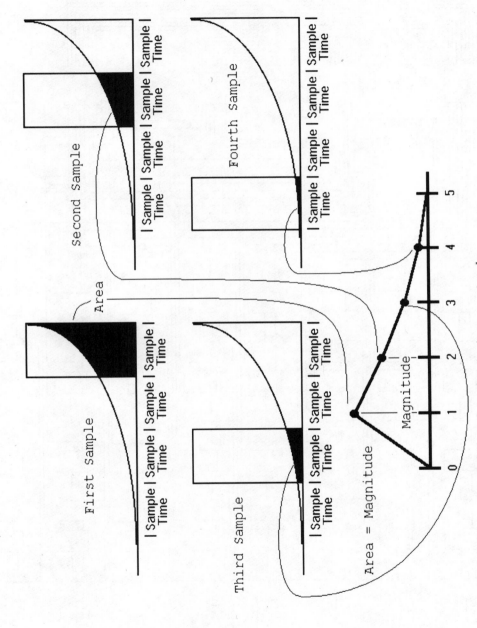

Figure 8.16 Process of convolution to determine circuit response to pulse.

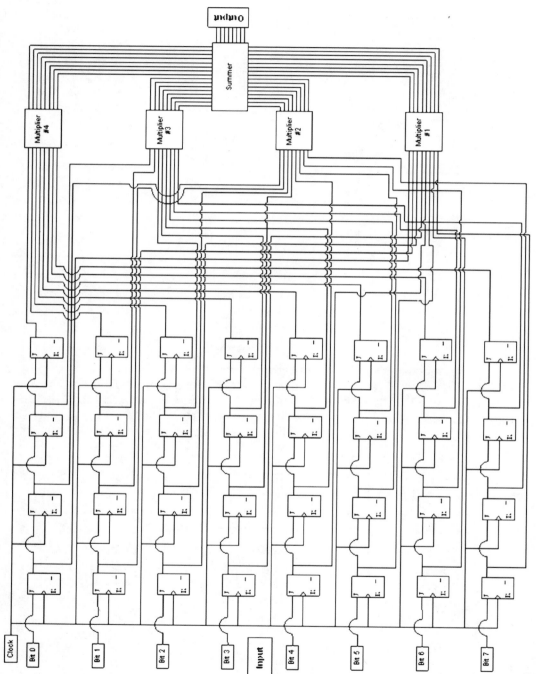

Figure 8.17 Digital low-pass filter.

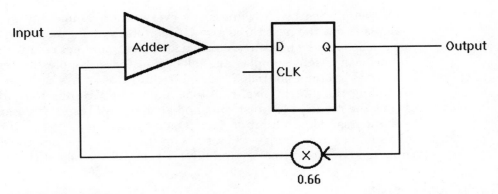

Figure 8.18 Simple IIR filter.

the required response, multiplier 1 is set to 1, multiplier 2 is set to 0.66, multiplier 3 is set to 0.44, and multiplier 4 is set to 0.15. Now if an impulse arrives at the filter (one sample long), it will produce an output that acts like a simple one-pole low-pass filter. This result can be extrapolated to mean that this digital circuit will have a transfer function just like its analog equivalent.

This simple circuit is actually much more complex than it looks because the boolean algebra and the associated gates to implement it are not shown for the multipliers and the summer. This filter is known as an *FIR filter*, which means that it will respond to a given input for a finite length of time. In the example filter the response is exactly four clock (or sample) periods. The function in Fig. 8.17 could have been built much simpler using an IIR filter (Fig. 8.18). Obviously, the IIR is a much simpler approach. Remember that the transfer function had a 0.66 multiplication factor from one sample to the next. Therefore, the multiplication shown in Fig. 8.18 would work for this simple low-pass filter. Note that the input, output, multiplier, adder, and register in reality are all 8 (or 10, or even 14) bits wide. The IIR approach is good for simple functions, whereas the FIR approach is needed for more complex functions.

8.3.11 Rate conversion

Now let's touch on rate conversion. In a typical digital camera a parallel data stream of 72 Mbytes/s must be converted down to the 27-MHz rate needed by the digital VTR permanently docked to the back of the camera. To get from 72 to 27 MHz, it is possible to divide by 2 and then multiply by 0.75 to arrive at 27 MHz. You might remember at the beginning of this chapter that 18 MHz as the CCD clock rate was no accident: 18 MHz is four-thirds of 13.5 MHz, which is the original sample rate of the luminance channel in 4:2:2. It would appear that all that is necessary is to subsample (or throw away some current samples—a process known as *decimation*) the incoming stream at a three-eighths rate and the result is a parallel bit stream at the right rate. This can be accomplished by throwing away every other sample and at every sixth sample throw the next three away. The process would then start over. Interpolation would be needed so strange artifacts would not result. With interpolation every eight samples are taken and three samples are created with weighted information from all samples. But wait—that's the same as initially sampling the 9-MHz pixel rate (one-half of 18 MHz) at 13.5 MHz! That is below the required Nyquist rate, and aliasing will result. It doesn't matter if we undersample up front or sample adequately initially, and throw some of the samples away later—the results are the same. The rate conversion must undergo low-pass filtering (digitally) to prevent aliasing at this point.

8.3.12 DSP trade-offs

What has been demonstrated is that a digital camera can still be thought of in terms of analog performance. When processing takes place in the digital domain, the complexity of the circuitry increases greatly. In fact, it wasn't until IC technology had gotten to the point where large blocks of circuitry could be reduced to reasonable sizes that we could even think of taking the digital approach. The main advantage obtained in using the digital camera is long-term stability, ease of setup, ability to store and recall different setups, and much greater ability to ensure that all cameras of similar designs match one another.

8.4 Bibliography

Boston, Jim, "Using DSP Technology," *Broadcast Engineering*, August 1997, p. 54.
Proakis, John G., and Dimitris G. Manolakis, *Digital Signal Processing*, Macmillan, 1992.

Processing Digital Video

In this chapter we're going to look at how most digital video is generated and processed. From the time a digital video signal is generated to its transmission out of a facility, most of the processing that is done on it has an ancestor in the analog domain. The path through a television plant today has generally been broken up into four subsections. Here we are going to look at the first two of these sections—acquisition and production (also known as manipulation). In Chap. 10 we will look at storage and in Chap. 13 we will look at the fourth subsection—transmission.

Acquisition literally means acquiring video images and sound. In a television facility acquisition means the arrival of these signals into the facility, either via microwave or satellite, or it could mean a person carrying film or videotape into the building. It also applies to the program production in the facility's studio. We are going to look at acquisition beginning with light entering a camera lens. We will follow it as it traverses through the camera and exits the back of the camera. From there it hits either the production or storage section or phase. These two phases can come in either order, and often the storage phase is encountered more than once. A camcorder goes straight from acquisition to storage as the camera's output goes straight to the docked VTR. The output from a camera in a studio usually goes through processing equipment in a control room before it goes on to transmission, in the case of a live broadcast, or storage. *Storage* means recording onto videotape or into a disk-based video server.

Once in storage the material is either considered "finished" and the storage is simply used for time shifting or delaying the program to its scheduled "air" playback time, or the material could still be considered "raw." *Raw material* means that additional production processes are needed before the material is considered finished. These additional production processes usually take the form of editing and layering of additional graphics. Editing occurs when excessive or unnecessary material is removed. Editing can also mean that special effects are added, either during a scene or the transition from scene to scene. This process is known as postproduction. Postproduction can take place in the same control room used to capture the raw studio production or it can take place in a room or facility geared for postproduction editing.

Editing for television is done in one of two stages. The first stage is offline editing. High-quality television equipment for any type of production is expensive, and the people skilled in the use of that equipment are expensive as well. As a result, the amount of time spent using those resources is often limited. The first phase of editing uses equipment of lesser quality to view the raw material and to decide roughly what needs to be done to create the finished program. Almost all television editing equipment produces what is known as an edit decision list (EDL). Both high- and low-cost editors will generate this type of list. This list is transportable, which means an EDL generated on one editor can be saved and taken to another editor. This system was pioneered by a company named CMX.

CMX was a joint venture of CBS and Memorex. Memorex made early disk storage devices. CBS wanted a better way to edit television material. The joint venture's first product, 20 years ahead of its time, was a nonlinear editor developed in the early 1970s. The product was ahead of the technology, and CMX soon developed computerized systems (minicomputers) that controlled VTRs and allowed accurate editing and, more important, accurate previews of edits before they were actually made. It was soon realized that all the previews could be made ahead of time and when all the artistic decisions were made, the computerized editor could be given the command to assemble the finished piece. This decision process could be done using lower-quality copies of the high-quality original, resulting in lower-priced VTRs and editors with just enough capability for generating an EDL where needed to edit offline. Today, many lower-quality nonlinear editors are used as replacements for the VTR and separate editor.

Once the EDL generated by the offline process was competed, it could be taken to a facility where online editing was performed using the offline-generated EDL. Online editing is where the expensive equipment and people take the raw material and transform it into a finished project. Offline editing does not have to take place, but many higher-end productions use the offline process first.

9.1   Equating NTSC to a Digital Video Stream

Most people who make the transition from one technology to another like to equate the new with the old. The move to digital television from its analog ancestor is no different. The digital television signal has the same requirements placed on the analog signal. Obviously, the digital bit stream must convey video information just like the analog signal. The digital video signal has the same number of active video lines as its analog equivalent. The only difference is that the digital signal always starts and ends with a full line, whereas in analog NTSC field 1 ends with one-half line and field 2 starts with one-half line. In addition, digital video doesn't have all the historical baggage to carry. Moreover, NTSC has equalizing pulses and wide vertical serration pulses that were needed to trigger early receivers, while digital signals dispense with those.

Composite digital television, which is rapidly fading into obsolescence, resembled its analog equivalent. It consisted of samples that literally represented all parts of the analog signal. These samples at 4 times the color subcarrier rate included not only active video but also blanking, sync pulses, and even color subcarrier burst. Sync rise and fall times were actually generated by obtaining values from lookup tables instead of samples from the actual video. In addition, ancillary data, such as audio signals, could be placed in the sync pulses.

Unlike composite, component digital video signals have vertical and horizontal intervals that do not much resemble their analog NTSC ancestor. First off, the digital video, composite or component, is actually nothing but a sequence of numbers. When the numbers occur and what their value is determine whether a particular number in the sequence represents a piece of video information or something else. What can the something else be? Once again, as in composite, component can carry ancillary data during horizontal blanking time. Since there are no sync pulses in digital component television, timing reference signals (TRSs) indicate when the blanking intervals start and stop. These TRSs consist of four words. The first three are illegal values for video and other data. These are $3FF_H$, 000_H, and 000_H, which signify to receiving equipment that a TRS has arrived. The fourth word indicates whether this TRS is occurring at the end of a horizontal line's active video or at the beginning of the active video for the line. The signals are known, appropriately, as end of active video (EAV) and start of active video (SAV), respectively. This fourth word also indicates whether the current horizontal line is occurring during the vertical interval and whether it is field 1 or field 2.

9.2 Component versus Composite Digital Video

Converting an analog signal to a digital signal does not eliminate the ills that can befall any signal; the same is true for digital video. Digital video shines brightest where the signal is going to pass through paths that do not process the video. Digital video DAs, router, frame syncs, and switching that involves no effects (particularly mixes and dissolves) should provide absolutely no degradation. If they do impact the digital signal, something is broken. VTRs, such as D1, D2, and SD-D5, that involve no compression should be transparent, no matter how many generations the video undergoes. However, effects, any type of video processing, and compression all take their toll, albeit less than on their analog counterparts, on the digital signal.

The creative and QA processes of professional television ensure that a fair amount of processing is done on almost all television signals. As a result of this, component video turns out to be much more robust than composite. Composite, although much cheaper to implement, has many of the same problems that analog composite has. The composite curse that component video was supposed to fix still haunts digital composite—color/luminance cross-modulation. Anytime the color information is combined with the luminance information, everything downstream that has to reextract the two will cause a degradation in video quality. Under the best scenario, the only device would be a color monitor or an NTSC encoder as the video makes its way to the home receiver. In that case the classic "cross-color" artifacts will not be that bad, namely, "tweed" suits that ring with a rainbow of color and a slight loss in luminance resolution. However, if that signal has to be taken apart (composite to component, also known as decoding) and reassembled (component to composite, also known as encoding) a number of times, the signal slowly degrades into uselessness. Many devices process internally in the component domain, even if the signal in and out is composite. However, suppose the signal is digital. It doesn't matter; all the transfer function rules (signal out versus signal in) apply to digital just as they do in analog. As we saw in Chap. 8, many digital designs are the result of mimicking the analog equivalent.

9.3   Acquisition

We will start this section by looking at the television camera. Digital cameras need to do almost all the same things that their analog parents did. The major difference in today's camera as opposed to cameras from 10 or 50 years ago is the replacement of imaging tubes with solid-state charged coupled devices (CCDs). The need for high voltages to drive the deflection circuitry to scan the tubes is gone, but much terminology still lingers from their use.

9.3.1   Analog versus digital processing

The all-digital camera is still a myth. The CCD, as we will see shortly, is still an analog device. It is digital-like in one aspect, though; it samples discretely in time, which leads to digital sampling issues surrounding the CCD. The CCD puts out analog but discrete samples. Some processes are easier in the analog domain. The CCD can output a signal representing video information that can be more than 6 times greater than normal signal levels. These high-level signals often contain useful scene information. Instead of simply clipping the signals at some value, it would be better to gently lower them to be closer to the proper level. To do that digitally would require that 3 bits of data be added to each sample to increase the sample resolution up to the "bright" values. Most digital signals have 10-bit words. It would be wasteful to use 3 of those bits for scenes with bright lights when most scenes don't have them. So most cameras use analog circuitry to lower the highlights. As a result, some analog processing is done up front. Many cameras now have 12-bit processing, while many others have been doing internal 14-bit processing to minimize rounding errors for a number of years. The more bits a sample has, the better the signal-to-noise ratio. Digital cameras today can claim signal-to-noise ratios of 65 dB, whereas analog cameras topped out in the mid-50-dB range. Most compression systems don't like the unpredictable aspect of noise.

Up to now digital cameras have tended to cost more than the analog equivalents. In professional television today one of the largest blocks of acquisition equipment has been in the broadcast and cable news engines. Digital has found a niche in postproduction [electronic field production (EFP)] because it allows unlimited layering and multigenerations and because individual camera budgets have made digital acquisition possible. A large production house might have a half dozen cameras, whereas a large news department in a local station could literally have dozens of cameras. News cameras [electronic news gathering (ENG)] also tend to be "stressed" by their users and environments more than their EFP counterparts. Moderately priced cameras have been the order for ENG. As a result, almost all ENG acquisition has been analog but that is changing, especially with the compression tape formats coming into use.

9.3.2 Film versus video

This argument has gone on for over 20 years. Some things are out and out more efficient and faster with video than film. Anything live obviously must use a video camera; the same is true for anything "nearly" live. Before the development of videotape, west coast network feeds were done using kinescope, a process where a CRT was used to expose film with live video from the east coast feed, and then quickly processed and loaded onto film projectors for playback to the west coast 3 h later. Videotape was the obvious kinescope replacement when it arrived. Many horse racing tracks around the country have internal television systems that provide patrons with live shots of each race along with instant replays. However, the main reason most of these system exist is so the race judges who use television replays can scrutinize any perceived infractions during the race. Before VTRs became affordable for this application, film was shot and, in many cases, loaded into processors that feed the wet, negative image through a conveyer system straight to the judge's stand. Film stock is expensive enough to have limited how many photographers a television news department could put on the street. The amount of film processors in a market also limited how much film could be processed each day. Not all stations had their own film-processing operations. Camcorders lightened the cameraperson's load and allowed much more raw footage to be acquired since videotape could be reused.

So video quickly took over in areas where it made operational and economic sense. Many early videophiles predicted that the end would soon arrive for film. However, film still made sense and was simply better than video in a number of areas. Most commercials and even a majority of prime-time television are still shot on film. First, the resolution of 35-mm film is still considered the best HD acquisition format. At this time, there is no clear dominant HD tape format, so 35 mm is as safe a bet as any. Film tends to crush black or dark areas of the scene less than video cameras, plus film handles a greater contrast ratio than video cameras. Video cameras tended to be noisier and harsher than film. Film also lends itself to progressive scan which many think is the future. Film also seems to have one other human aspect going for it, the people who insist on using it. The film photographer or cinematographer tends to use different production techniques than video photographers, partially due to the technology limitations that, until recently, were placed on video cameras. Television cameras had required large amounts of light and a large number of technicians who set up and rode herd on the pictures coming out of the video camera. The problem was that sometimes the creative side didn't need the technical corrections that many technicians would insist be made. Picture highlights didn't always have to be at 100 percent, low light areas at 7.5 percent, and gamma at 0.45 to produce the desired creative effect. In the television technician's defense, video pictures often ended up being broadcast on television stations, and those stations had a set of FCC transmission standards that were supposed to be met. Film people felt they were totally removed from any government technical restraints.

Many from the video side are finally learning what film people have known for decades about production. Just because you have a lens that can zoom doesn't mean you need to use that feature. Often, trucking a camera provides a better sense of depth and place than zooming. Low amounts of light with the accompanying narrow depth of field often help a scene. The best picture from a high-resolution camera isn't always the sharpest. Now that many film people are discovering the video camera, many are bringing these techniques to the video side. The video camera has reached a point where it makes sense to use video to acquire material that will eventually end up on television. Much of this is due to better marketing to the film industry by video manufacturers. Another reason is that video cameras are now made to operate like their film cousins. The video camera or camcorder today doesn't need a technician's continual intervention to ensure correct operation. Additionally, different setups can be loaded into the camera by a creative type with minimal technical knowledge. Those setups can easily be recalled in the field.

However, a major reason that video is expanding into the film realm is cost. To rent a high-quality camcorder and buy the needed videotape can be one-fifth the cost or less than the cost to rent a 35-mm camera and buy the associated film stock that can only be exposed once. The video process can still be less than one-half as much as using the 16-mm film.

9.3.3   Coping with aspect ratios

We are now entering an era of the dual aspect ratio. Most will probably continue to view television in customary 4:3 ratio for many years to come but from now on many shows will be shot in the new 16:9 aspect ratio. This poses a number of problems for the production people. In many ways this is similar to the same problems people are experiencing doing HD productions. HD is very impressive during wide shots of events; it displays more of the spectacle. Close-ups are not as impressive because the close-up often displays far more than viewers care to see. SD needs to use more close-ups to capture enough detail to convey the action. Many cameras used for sporting events today have 70:1 zoom lenses to get in as close as possible. So HD and SD often seem to have two diametrically opposed camera coverage requirements. Many wonder if HD/SD sporting events will ever be able to be done with only HD equipment, which is also downconverted for simultaneous SD use. It might be that these events are what is known as a side by side in the remote business, an HD production truck sitting side by side with an SD production truck.

Now throw in the aspect ratio and we have just too many scenarios to cover uniquely. All HD is 16:9, but SD can be either 16:9 or 4:3—two trucks at a venue and maybe three but not likely. By discounting the HD versus SD production, crews will have to cope with satisfying two different aspect ratios simultaneously. The rules will most likely be to compose for 4:3 but protect for 16:9, which means that all pertinent action will be in the 4:3 scene but, at the same time, not shot in the 16:9 frame. So a cameraperson is framing a 4:3 main action shot while still checking that lights, mic booms, and distracting shadows are not in the 16:9 shot.

9.3.4 Studio versus the field

Before we look at the difference, let's ask what is a high-end camera? First, such cameras support the infrastructure required to act as a studio camera, including functions such as intercom, teleprompter power and video, return video, and microphone feed. They also have a fully integrated viewfinder. Additional attributes include full operator control of the video transfer function such as gammas, flares, knees, etc. It should be possible to set up the camera for a particular shoot and store those settings. This function is normally referred to as *scene files*. Many high-end cameras have multiple scene files, along with setup files to compensate for particular lenses. Setup files can usually be moved between different cameras.

The high-end cameras also have very good spatial resolution and processing sample resolution, and their sensitivity is usually greater. This is an important consideration for anyone using a video camera in a film-style shoot, where light levels have dropped more dramatically than television. The higher-end cameras also support better lenses.

Currently, there are two types of studio cameras: hard (studio configuration) and soft (field configuration). Hard cameras can be the customary large camera head case or a field camera in what is called a studio buildup kit. This buildup kit provides a field camera with the same functionality as a hard camera. Sometimes the buildup kit actually has a larger camera head case to house the smaller field camera. The second big delineation between cameras today is standard definition versus high definition. Many question the need for

HD cameras in the studio as most sets do not have the spectral content to warrant high-definition acquisition. However, high-definition cameras do ensure the highest possible resolution and quality. HD acquisition also means 16:9 format material.

9.3.5 Lenses

The very first step in acquiring a video picture is to get light into the camera via a lens. Early television cameras only had fixed lenses. There were usually four lenses (three if one position was used for capping the camera). Standard lens sizes were 35, 50, 90, and 135 mm. When the first zoom lens was introduced, it covered the same focal lengths as these first fixed lenses. These 5:1 zoom lenses soon gave way to 10:1. The second-generation zoom lens provided wider wide shots as the focal length at the low end extended below 20 mm. Soon the 15:1 lens became common. By the mid 1980s, the 30:1 was in use for sporting events. As already mentioned, the 70:1 zoom lens is now in use. High-quality lenses, especially high-quality lenses with large zoom ratios, have become extremely popular. Often the lens is larger than the camera on which it is mounted, even if the camera is a studio-type camera. The physics used in optics has not kept pace with the physics of converting light into television pictures.

Lenses have a number of attributes that must be matched to the imagers in use. Lenses produce different image sizes, and the lens used should be matched to the size of the imager in use. Many lenses contain a prefix letter(s) in the model name to indicate the image area size. The image the lens makes which falls upon the imager(s) is round. Any light outside the image circle has light aberration. Smaller-image-size lenses are generally cheaper than larger-image-size lenses. Using a smaller image lens on a larger imager will result in aberrations near the image's edge.

In television the brightness of an image is related to the f-stop number. Lenses generally have f-stop values of 1.4, 2, 2.8, 4, 5.6, 8, 11, 16, and 22. These numbers all have a ratio relationship with the square root of 2 because the amount of light entering the lens is proportional to the square of the diameter of the lens. An f-stop change of 1, say 4 to 5.6, reduces light through the lens by one-half. Conversely, an f-stop change from 11 to 8 doubles the amount of light. Thus

$$\text{f-stop} = \frac{\text{focal length}}{\text{effective aperture}}$$

The brightness is inversely proportional to the square of the f-stop number. Lenses with the same f-stop setting might not have the same brightness because different lenses transmit different amounts of light. As a result, the film industry tends to use a value called t-stop which takes this into account. Two different lenses with the same t-stop values have the same brightness.

Zoom lenses have an attribute known as f-drop, which is the light change that occurs when a lens is zoomed from all the way out to all the way in. As the lens is zoomed in, the entrance pupil changes size to allow more light in. When the pupil is as wide as it can go, the light level will start to drop. To reduce the size and weight of the lens, it is common to allow some f-drop. The auto-iris circuitry in the camera will usually open the iris to prevent this effect from affecting the video levels. Many studio lenses have focus groups that are wide enough to minimize f-drop.

Many early HD cameras used 1-in CCD imagers, while recent HD cameras use $^{2}/_{3}$-in imagers. The physics required to produce smaller high-quality images becomes more difficult as the size of the image decreases.

9.3.6 Tubes

Early three-tube color cameras used 4-in image orthicon pickup tubes to produce a noisy picture with unstable colors (Fig. 9.1). Next came four-tube color cameras which used a single $4^{1}/_{2}$-in image orthicon to provide resolution equal to the best of the monochrome cameras, plus three smaller 1-in vidicons to process scene color information (Fig. 9.2).

By 1966 cameras no larger than RCA's original TK-10 "Field Portable" (Fig. 2.1) were delivering sharp, clean color pictures with light levels well under 200 fc. The first camera which did this was Norelco's PC-60 (soon to become the

Figure 9.1 Image orthicon tube on the left, vidicon in the middle, and plumbicon on the right.

(a)

Figure 9.2 Cameras used to capture images from film and slides.
(*a*) This specialized camera was called a "film chain." An RCA
TK-27, it used an orthicon for luminance and three vidicons for
red, green, and blue. (*b*) Norelco PC-70 (*courtesy of Tim Stoffel*).
(*c*) PC-70 camera head with cover open to expose assemblies
housing the three plumbicon tubes. A prism between the lens
and the tubes splits incoming light into red, green, and blue
(*courtesy of Tim Stoffel*).

PC-70). (PC stood for plumbicon color). These cameras were three-tube cameras. RCA followed within 3 years with a comparable camera (TK-44). (See
Fig. 9.2*c*.)

By 1976 cameras that could be considered portable appeared (RCA's TK-76),
and they could match the PC-70/TK-44 in picture quality. In 1980 RCA introduced the first computerized camera (TK-47). This camera made auto setup
and operation possible. Ikegami, Hitachi, and Philips (previously Norelco)
soon followed with similar products.

9.3.6.1 Preamps (Fig. 9.3).   Cameras generally consist of three subsystems (not
including the encoder). Subsystem 1 consists of three video preamplifiers. In
the highest-performance cameras the preamp is a specially designed, specially
packaged high-performance, high-gain transistor and associated circuit components, mounted in or on the deflection yoke at the edge of the pickup tube

(b)

Figure 9.2 (*Continued*)

faceplate. The input to this subsystem is directly connected to the light-sensitive layer on the tube faceplate through a platinum or other rare metal feed-through electrode.

This physical arrangement counters the tendency of the tube's preamp coupling circuit to act like an antenna and pick up stray noise. The short length involved here lowers the signal shunting capacitance of the transistor input circuit and improves the resolution (increases the frequency response) of the pickup tube/preamp subsystem.

In recent circuit design evolutions of the pickup tube, its power supply and preamp function as an integrated feedback subsystem to improve the handling of hot highlights in the picture. This function is usually known as automatic beam optimization (ABO) or comet tail suppression (CTS). The purpose of these circuits was to temporarily increase the electron beam current in the pickup tube when "hot" video was encountered.

9.3.6.2 Vidicons. The original widely used tube was called the image orthicon. It has not been installed in new cameras since the mid-1960s. These tubes were generally large, mechanically complex, and prone to disaster, and they were sticky (high image retention) and tended to put halos around highlighted areas (hence the field mesh tube).

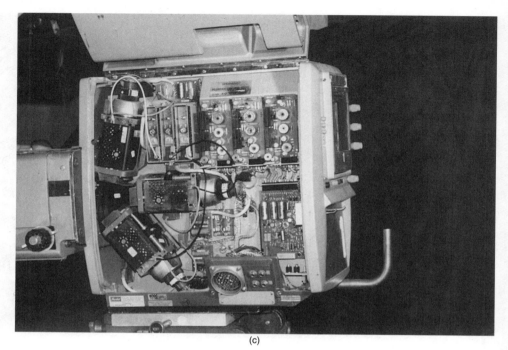

(c)

Figure 9.2 (*Continued*)

The vidicon was the ancestor of all later tubes. In all designs a hot cathode emits electrons which are sent to a photoconductive target by the combined potentials of the control grid and the accelerating grids shown in Fig. 9.4. The quantity of electrons landing on the target and then conducted through the glass wall to an external preamplifier is determined by the magnitude of the positive charge created by the light in the image focused on that target as the beam sweeps over it.

The vidicon was much smaller, lighter, less expensive, and simpler then the IO. It also had lower resolution and sensitivity. The photoresistive layer had a resistance of more than 50 MΩ with no light shining on it. However, when light (which has been focused by a lens in front of the tube's transparent conductive film) strikes the photoresistive layer (through the transparent conductive film), its resistance drops considerably. The photoconductive mosaic was a layer that was charged up by the electron beam from the cathode.

Charges in one area of the photoconductive mosaic would not bleed off into surrounding areas of the mosaic. Thus, wherever light from the lens struck the photoresistive layer, the charge induced onto the mosaic layer would bleed through the conductive film and out the tube (this was the video information). The electron beam had to make many horizontal sweeps (horizontal lines) across the mosaic. Each sweep was just below the previous sweep. After 262

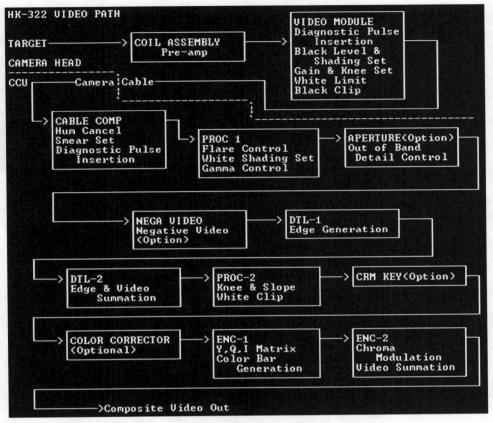

Figure 9.3   Block diagram of the Ikegami HK-322 (early 1980s).

sweeps, the beam was returned to the top of the mosaic to start over. This was done 60 times/s.

The electron beam was made to sweep horizontally and vertically by means (usually) of a magnetic yoke. The tube was placed in the yoke, which resembles a cylinder. Varying electric currents were passed through the yoke which varied the magnetic field out of the yoke, thus varying the sweep or position of the electron beam. The electron beam was generated by heating the cathode to a high temperature using a filament by passing current through it. The beam went to the mosaic, which was more positive than the cathode.

Grid G1 was used to control the amount of electrons allowed to travel from the cathode to the target (mosaic layer). If there were too few, not enough were available to send through the photoresistive layer in highlight areas. Too many electrons resulted in the beam becoming too wide, limiting the resolution and the tube's life. Grid G2 was used to accelerate the beam toward the target. (It has a high positive voltage in relation to the cathode.)

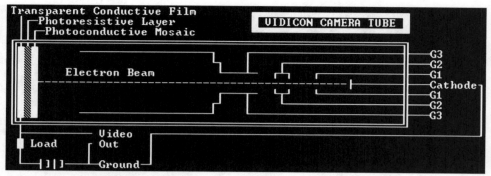

Figure 9.4 Elements of vidicon tube.

Grid G3 was used to focus the electron beam and keep it as narrow and defined as possible.

9.3.6.3 Plumbicon.   The plumbicon tube was the next major tube development, and it was probably the most widely used tube ever. It became available around 1969. This tube was an advancement over the vidicon. It was basically the same as the vidicon, with the exception of the target or photoconductive area of the tube.

The advantages of the plumbicon over the vidicon are as follows:

1. Very low dark current (current flowing into the external preamplifier in the absence of light on the faceplate), which maintains black as black.
2. Low image persistence (reduces lag).
3. Gamma (light on faceplate-to-signal out) is very predictable and close to unity.
4. Has a much longer life.

With no light on the faceplate, the target (or photoconductive layer) takes the form of a *P-I-N* junction diode that was reverse-biased by a target-to-cathode voltage of approximately 40 V dc. When light strikes the photoconductor, electron-hole pairs are generated within the intrinsic region within the diode and they are swept toward their respective *N* and *P* pole surfaces.

When the electron-hole pairs are swept toward their respective pairs, the result is the partial discharge of the photoconductor. As a consequence, a pattern of positive charges was produced in the *P* material which is directly proportional to the light intensity falling on the tube's faceplate. The scanning beam then deposits enough electrons to replenish the charge lost due to the photocurrent on the next sweep. The current flowing within this series circuit, consisting of the *P-I-N* junction layer, the scanning beam, and the target current supply, was called *signal current*. Thus, the photoconductive layer was discharged by the photocurrent and recharged by the beam.

A disadvantage of the plumbicon tube was its lack of sensitivity to red light. Some plumbicons are more sensitive to red light; these tubes are called extended red tubes. In these tubes the target layer was thickened, along with the addition of sulfur in the target. The disadvantage of an extended red tube was increased lag and "stickiness."

Plumbicon's dark current will not vary with temperature where vidicon's will. A plumbicon's sensitivity was independent of the cathode-to-target potential (target voltage). As a result, the signal out was usually controlled by the amount of light allowed through the lens by the lens iris.

9.3.6.4   Saticon.   In 1975 another major derivative of the plumbicon appeared. This tube was called the saticon. The name was derived from its target layer constituents of selenium, arsenic, and tellurium.

The major advantages of the saticon are as follows:

1. The resolution of the 18-mm ($^2/_3$-in) tube approaches that of the 30-mm (1-in) plumbicon.

2. It has reduced flare and is stable under highlight conditions.

3. It has little dark current or noise in the blacks.

4. It has good lag characteristics.

5. It has a long shelf life.

Other than for special target materials and construction, it was basically a vidicon. Just as the plumbicon was an enhanced vidicon.

9.3.6.5   Diode gun.   A more recent improvement in the plumbicon was the diode gun plumbicon. This tube lowers beam resistance, resulting in fewer beam landing errors. The diode gun also has a thinner target which lowers target capacitance, resulting in higher frequency output (higher resolution). This tube was also the first to use a small button as the target connection instead of a target ring, which also helps lower output capacitance.

As illustrated in Fig. 9.5, the grids of the diode gun tube are arranged in such a way as to make the electron beam as narrow as possible, which helps to improve tube resolution.

9.3.6.6   Pickup tube characteristics

LAG (decay lag).   *Lag* is the time it takes for an image to disappear from the output of the tube, after it is removed from the tube's faceplate. The quantitative specification for lag was a percentage that compared the signal current from the tube 50 ms (three frames) after illumination cutoff to the signal current generated by a highlight illumination before cutoff. Lag has a strange phenomenon associated with it. The higher the highlight signal current, the faster the decay. Thus, the higher the overall scene illumination, the less visible lag in fast pans and zooms.

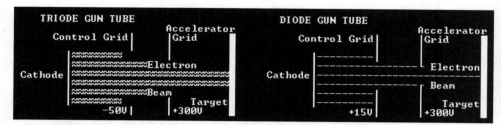

Figure 9.5 Main difference between triode and diode gun tubes is the potential of the control grid.

Comet-tailing. This is a special kind of lag. "Hot spots" in a scene cause this severe lag problem, which occurs because the target's excessive positive charge from a highlight takes many frames of the scanning electron beam to resupply the discharge due to peak current limitations in the tube power supply. Some tubes had special internal devices to try and handle this problem, and they were called anticomet-tail (ACT) tubes. Today, this problem is usually handled by the camera itself by increasing beam current when needed. Tubes $^2/_3$ in in size have fewer problems with this because the available light is concentrated into a smaller faceplate area.

Bias light. Most cameras have a light aimed at the faceplate of the tube that keeps the tube's target uniformly illuminated even when no image comes to it through the camera's lens. Some tubes have a bias light built into the tube itself. The bias light keeps a minimum, controllable amount of signal current flowing at all times. This current can be adjusted with external camera circuitry to be at reference black. The end result of bias light is less noise in the blacks. This light is usually in the camera's prism assembly.

Shading. Camera tube shading is the difference in signal current from the center to the corner of the tube, when faceplate illumination is constant. This problem is usually caused by electron beam landing errors on the target.

Alignment. Tube alignment is necessary to correct slight mechanical and electric misalignments encountered in yoke and tube manufacture. Usually an additional coil is built into the deflection coils to adjust for this parameter. The align coil aligns the electron beam through the center of the tube grids before the beam is deflected by the deflection coils. If alignment is off, corner focus, the geometry, beam size, and registration can be off. Increased lag is also possible.

Flare. The target in a tube has varying degrees of sensitivity relating to light of different wavelengths (or colors). Thus, a tube's target absorbs, scatters, or reflects light in relation to its frequency (or color). The reflected or scattered light can be refracted throughout the tube faceplate, contaminating the light acting upon the tube's target. Red is affected the most. Many tubes are supplied with a antihalation button to minimize the problem. This "button" is a glass cap over the tube's faceplate which minimizes light refraction.

Most cameras have external adjustments which pull black levels down as overall levels increase to counter the effects of flare.

Resolution. This is the amount of information that a tube can put out per scene. It is usually measured information in a horizontal line. Vertical resolution is normally limited by the 525-line scan rate of the NTSC television system. Resolution is directly related to frequency response. A rough conversion between the two is that for every 80 vertical lines in a picture the frequency bandwidth required is 1 MHz. Resolution is checked with all image enhancement off.

9.3.7   CCDs

The CCD is today's method of converting light into electrons, and is analogous to its tube ancestors. CCDs were introduced in the early 1980s, with RCA displaying the first professional ENG camera. Much initial amazement was generated as the camera was demonstrated looking directly into bright lights or even at the sun—something that most tubes could not do. The bright images would end up permanently displayed because the image was burned into the photoconductive faceplate of the tube. The CCD didn't have this problem.

9.3.7.1   Frame transfer CCD. CCD stands for charged coupled device, and it is an array of MOS capacitors which build up a charge based on the amount of light that falls on the sensor. That charge is allowed to build up over a field's time. Then during the vertical interval, the charge (electrons) is moved out of each cell via a read-out gate (ROG). What happens next with the charge is what delineates different types of CCDs. The earliest CCDs were known as frame transfer (FT) devices. During the vertical interval, the collected charges in each sensor were shifted down (vertical) through the cells below it until all the charges had been moved down and out of the light acquisition array. Below this first array was a second complete array which was masked so that no light could fall on it. Thus, during the vertical interval, cells with light information from rows (horizontal lines) were shifted rapidly from one cell to the next in a shift register–type action. The charges from each cell ended up in place in a second array. This second array then shifted the charges out cell by cell (pixel by pixel), with each row representing a horizontal line. The string or sequence of charges marched out of the second array represented the video information. The CCD charge transfer has replaced scanning.

The problem with this first architecture was that even though the stored charges were shifted down into the second array fairly quickly, light would continue to fall on the cells. Bright lights caused a visible smear to appear in the vertical direction as the light would continue to add to each charge as it was shifted through rows where the bright light was incident. Early CCD cameras had mechanical shutters which covered the CCDs during the vertical interval to correct this problem.

9.3.7.2. Interline transfer CCD. Instead of this mechanical approach, an electric fix was soon implemented. A set of shift registers was added in between each cell in a row. These additional shift registers were covered with aluminum so direct light would not contaminate them. This architecture is known as *interline transfer* (IT). During the vertical interval, the ROG would dump its charge into the shift register cell next to it. Then, instead of a complete additional array, the video charges could be shifted down vertically line by line during the following field. However, this approach came with some baggage. The first problem was that it still produced some vertical smear because: (1) some light was reflected around the cells in the array and found its way around the aluminum shield and (2) the cells leaked charges to a slight degree and cells with large charges would contaminate nearby cells. As a result, as charges were shifted down line by line, the charges that were shifted through these areas would have an unwanted charge added. However, vertical smear was still much less than the FT without the mechanical shutter. The second problem was that a smaller area of the array was now devoted to collecting light, which obviously made the array less sensitive, but produced a larger problem.

9.3.7.3 CCD aliasing. Unlike the image tube, the CCD is a discrete device. While the voltages out of the tube's target were continuous in nature, the CCD has discrete samples. An individual pixel is sent for a finite time, after which the next pixel is sent. Instead of a continuous function or curve, the video from the CCD appears as discontinuous steps. This takes us back to the chapter on digital sampling. Nyquist rules start to appear to CCD imagers where they didn't to tube imagers. However, IT architecture makes this problem worse. CCDs with no interstitial space between the cells bent the Nyquist rules somewhat. Instead of the maximum allowable frequency (remember the picket fence from Chap. 1) being one-half the samples, or in this case cells per row or horizontal line, the frequency would be three-fourths. However, when the cells became separated with IT, the one-half Nyquist rule took over. To eliminate this problem, an optical low-pass filter is placed in front of the prism. This filter is based on double refraction. When a quartz crystal with an optical axis that is not parallel to the normal to its surface is the same thickness as the spacing between CCD cells (pixels), optical low-pass filtering results because as rays of light pass through the optical filter, double refraction occurs. In essence a thin object is spread out to cover at least two pixels, preventing the breaking of Nyquist's rule. However, the roll-off slope of this filter is now steep and bright objects with high-frequency content can get through, so aliasing is seen on all CCDs in some scenes.

9.3.7.4 Green sensor offset. IT did provide another benefit. CCDs that have gaps between the sensor elements, which IT inherently has, allow the use of spatial offset. What this means is that when the red and blue sensors are mounted onto the prism, these sensors are sited so that their sensors are sampling a scene in between where the green sensors are sampling. Unlike tubes, the CCD imagers are soldered onto the prism assembly. The scanning techniques used on tubes drift over time. Also, rough handling of a camera can physically move the tube within its deflection yoke, which means that the images produced by the three imagers don't exactly overlay each other. This is known as misregistration. In other words, the geometries of the three images aren't the same. Now that scanning isn't needed for CCDs and the CCDs are mounted so that they can't move in relationship to the prism or the other two images, geometry and misregistration are no longer issues with CCD cameras. As a result, red and blue imagers are permanently mounted so that their samples fall between green samples. Many SD CCDs today have between 500 and 1000 pixels of cells per row or horizontal line. By taking Nyquist's rules into consideration, horizontal resolution for these CCDs could not be above 500 lines. Yet many cameras claim specs of over 900 lines. How can this be? Because of the spatial offset. The blue and red sensors add horizontal information that the green imager misses.

9.3.7.5 Frame interline transfer CCD. In order to improve upon the IT the first two approaches were combined, that is, the FT and IT became the frame interline transfer (FIT). The interline attribute and the second array of the FT were included in one device. At the start of the vertical interval the ROG does exactly what it did in the IT, that is, transfer the charges from the cells to the interline transfer registers. However, the difference is that those transfer registers rapidly move the charges down into a second array just like the FT had. This eliminates any contamination from bright lights. The vertical smear is eliminated, but the cost of these devices is high, often 3 times as much as an IT device. High-end camcorders have FIT CCDs, while most low-end ones do not. There are many studio cameras that can use IT because the studio environment can be made to eliminate the situations that cause the vertical smear. However, many high-end cameras used for remotes must be FIT to keep the clients of such equipment happy.

9.3.7.6 Hole-accumulated diode CCD. The next improvement in CCD was the hole-accumulated diode (HAD) architecture. Earlier CCDs had overflow gates that were used to eliminate excess charges. These gates took up real estate. The HAD CCD allowed the substrate under the sensors to handle this function so that the actual area of the CCD surface used for sensors increased from under 25 percent to over 30 percent. As a result, the number of sensors per line increased from approximately 500 to over 1000. The HAD architecture also added another feature, the electronic shutter. The substrate could be commanded to drain all charges. The electronic shutter function keeps all sensor wells drained a certain percentage of the total time. The larger the percentage of time that the cells are drained, the higher the shutter speed. This can help capture rapid movement for use with instant replay.

9.3.7.7 Microlens CCD. To increase CCD sensitivity, microlenses are placed on top of each sensor to gather the light in the area of each sensor. Light of 2000 lux requires a lens f-stop setting of f8. Pixel counts in professional CCD arrays have been in the 500,000 to 700,000 range for SD cameras. HD CCD pixel counts of 2 million or more will be the norm. CCDs have another benefit for the future. Most believe progressive scan will eventually become the trend in television; it is only a question of when. CCDs now have dual readouts that deliver both interlaced and progressive scans.

9.3.7.8 Cell compensation.   CCDs are never perfect and some are less perfect than others. Not all the cells in any CCD array are equal. Invariably, some will be more sensitive than others. A few might even be dead on initial power-up, and some will most likely die over the CCD's life span. Few CCDs even begin life perfect. In addition to some being dead on power-up, many cells vary in sensitivity. CCD blocks usually have blocks of nonvolatile memory where offsets to compensate for the varying sensitivities of individual cells can be stored. However, as already mentioned, some cells will not function at all, and over time additional cells will fail. Early CCDs used to be extremely sensitive to gamma rays. The higher the altitude of the camera's operation, the more likely a ray will take out a CCD cell. Airplane travel used to cause the most cell failures. However, almost all CCD blocks have the automatic ability to hide a cell that goes bad. Some simply use the information from a cell next to the failed one. Some blocks use more elaborate systems where all surrounding cells are used to interpolate information for the missing cell. When does the camera check to see if cells have gone bad? Usually when the camera is black-balanced. The camera caps itself to perform this adjustment so the camera knows what the value of each cell should be.

9.3.8 Camera video processing

The second subsystem consists of three channels of signal amplifying and processing electronics that condition the CCD or image tube output signals for encoding. Gamma, black, stretch, level suppression, contrast expansion, etc., processing is done now. From the names of these processes it is obvious that they adjust the amount and type of nonlinear response of the electronics to linear changes in light intensities on the pickup tube faceplates.

Vertical and horizontal enhancements take place in this subsystem. Horizontal image enhancement is the simpler of the two. A peaking circuit brings up the high-frequency response of the channels to compensate for the slower buildup and collapse of the output voltage caused by the infinite dimensions and slightly ellipsoidal shape of the electronic scanning beam. This circuit also brings up the ringing (multiple ghosting) inherent in the electronics as well as high-frequency noise.

Vertical image enhancement requires video bandwidth signal delay lines and electronics that are costly in terms of materials, power, and space. Since many of today's cameras process the video digitally, RAM or shift registers can be used to delay the video. Vertical image enhancement may be included only as an option.

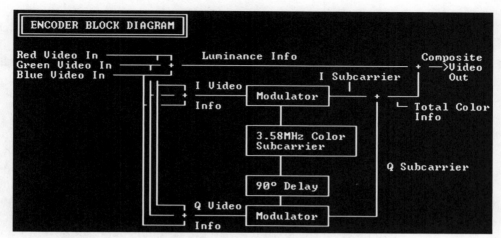

Figure 9.6 The three video signals are combined by the encoder.

High-performance cameras use a design called "contours out of green." High-frequency details from the green channel only are enhanced and then mixed equally into all channels before encoding. Today many cameras derive contours from both the green and the red channels (Fig. 9.6). The third subsystem consist of the sync, blanking, subcarrier generating circuitry, and the three pickup tube power supplies.

9.3.9 Camera system control

The camera system control accepts input from the operator or from automatic subsystems such as the auto-iris circuitry. The system control circuitry processes these requests and outputs control commands. Historically, cameras have been analog devices. Control over analog circuits since the implementation of microprocessors in cameras has been through a circuit called the sample and hold. We will look at that shortly. As we saw in the previous chapter, analog circuits are giving way to DSP techniques.

How does a microprocessor control the adjustments and operational modes of a camera? First, like most microprocessor systems, all the circuitry that controls the operations or operating parameters is memory mapped. What does that mean? It means that each circuit that the microprocessor wants to control has a specific address. When the microprocessor wants to change something in a particular circuit, it sends out the address for that circuit, along with the new information (on the data bus). Addressing of different circuits is done on the address bus, where new operating values or instructions are sent on the data bus.

Many circuits only have to be turned ON or OFF to be controlled. This is done by sending a high or low on the data bus. A high (or 1) usually means that the circuit is to be turned ON; a low (or 0) usually means that the circuit is to be turned OFF. The data bus that controls the ON/OFF signals consists of eight parallel lines, that is, for every address eight ON/OFF lines are available. Thus, in reality eight ON/OFF circuits are controlled per address.

What if the circuit is not an ON/OFF type but is what is usually called an analog circuit? An analog circuit is one where a voltage (which is changed as required) is used to control the circuit. Analog-type circuits control things such as blanking, registration, iris, etc.

How are varying voltages sent by the microprocessor to circuits that require such voltages? Well, if each such circuit is given its own unique address, the entire eight-line data bus can be used to specify a value between 0 to 255 with a binary count. A binary count works by having a high (or 1) on a data line signify a numerical value weighted by its particular line. For example, if the first line is high, it equals 1, if the second line is high, it equals 2, if the third line is high, it equals 4, if the fourth line is high, it equals 8, if the fifth line is high, it equals 16, if the sixth line is high, it equals 32, if the seventh line is high, it equals 64, and if the eighth line is high, it equals 128. Add the weighted values of any line that is high and the result is its binary value. All lines that are high will produce a value of 255.

The digital count on the data bus is converted to a voltage by a digital to analog (D/A) converter. This converter can put out any of 256 different voltage levels that can occur between any two voltage values. As an example, if the low-voltage value was 0 and the high-voltage value was 5 V, the converter would put out a value of 0 to 5 V, with 254 possible values in between. That would make each step in value equal to approximately 0.02 V. If the low- and high-voltage range is −6 and +12 V, the 256 possible values would increase by 0.07 V. Once a voltage is obtained from the data on the data bus by the D/A converter, it must be stored as this voltage is only available for a short time and it is not updated often. This voltage is stored in what is called a sample-and-hold (S/H) circuit (Fig. 9.7).

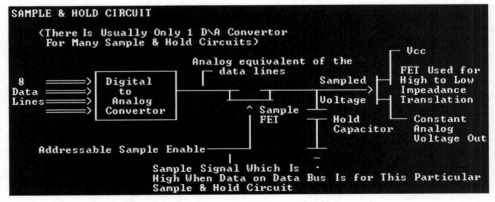

Figure 9.7 Digital data are converted to analog voltage and held by capacitor.

A sample-and-hold circuit consists of a gate (usually an FET) which is addressed by the microprocessor. This gate lets the voltage through when it is addressed by the microprocessor. This voltage charges a low-leakage capacitor so that it is the same voltage as the D/A output. After the gate is no longer enabled (or addressed), the capacitor will continue to hold its charge because the capacitor feeds a high-input impedance device which provides a constant control voltage out. There is usually only one D/A converter for many sample-and-hold circuits.

Overall camera architecture has evolved greatly over the last 20 years. Before the development of the ENG camera, high-end television cameras usually consisted of three boxes, the camera head that everyone saw, a camera control unit (CCU), and a power supply chassis. Initial video processing took place in the head and ended in the CCU after a trip through the camera cable. Control of parameters in the head was with control lines through the camera cable, one line for each item to be controlled. Early cameras had fat camera cables that handled all the control lines. In order to minimize the number of conductors in the cable time multiplexing was introduced. All the control signal values went down a single conductor, one after another. Each value was stored in a sample-and-hold circuit until a new sample arrived.

Frequency multiplexing was introduced to reduce camera cable size even further. In this process all signals go down the same conductor because each signal is modulated onto a carrier at various frequencies. This camera cable is known as *triax*. Triax is coax with two shields instead of one to minimize interference. Power for the camera head is also sent down that same conductor. Some cameras run on fairly high voltages (low currents are involved to limit IR losses) so elaborate safety circuits are used to check for the proper load before power is applied.

Digital video is generally not sent between the camera head and the CCU because of the bandwidth requirements of digital signals. One digital SD component video signal consumes the entire bandwidth of the triax, allowing no room for any control signals, let alone return video to the head or any intercom traffic. Some digital cameras that are on triax must convert the digital video back to analog for the trip down the triax. HD digital video requires so much bandwidth that a single bit stream can only travel a few hundred feet down coax or triax. As a result, most HD cameras use one or more fiber-optic cables in the camera cable. The cable will also have several additional conductors for power and control. The thickness of this cable is comparable to triax. Depending on the length of triax to be used, diameters range from 8 to 14 mm.

In most studio cameras today all the video processing is done in the camera head. Now the CCU is mostly a remote control device. Almost all cameras available today will work as stand-alone units just as all EFP or ENG cameras or any camcorder can. Studio cameras in outdoor remote applications are referred to as "hard" cameras, while handheld ENG/EFP cameras are referred to as "soft" cameras. A hard or studio camera usually has infrastructure built into it that soft or ENG/EFP cameras do not. For example, intercom, return viewfinder and teleprompter video, and externally available ac power are found in a hard camera but not in its soft counterpart. While it used to be easy to tell a soft from a hard camera, mainly by size, it is no longer that simple. Many hard cameras are small enough and ergonomically built to be used as soft cameras. Conversely, many soft cameras will dock with accessories that provide the hard camera infrastructure. Many studio cameras today are soft cameras pressed into hard or studio camera service. One remaining use for the hard camera in all places is for remotes because of the large lenses hung on them for sports events. Many camera operators prefer a large camera on a pan tilt head to a smaller one because the extra mass lends itself to better camera moves. In the studio many camera operator positions are being lost to robotic camera pedestals which are generally cheaper to purchase if the camera weight is reduced, and as a result, soft/EFP cameras are being used as hard/studio cameras.

9.3.10 Effective use of the tools

Camera shader used to be a technical position that required some ascetic ability. The term "shader" comes from the fact that early imaging camera tubes tended to change sensitivity from left to right and top to bottom. Controls to balance these levels were called *shading controls*. Early cameras tended to drift a lot electrically, and had to be set up often. The process of making sure all color cameras looking at the same scene agreed as to color, known as *color matching,* was an art onto itself. However, cameras matured and evolved into highly stable devices that did not drift, and the setup became automatic. Even when shooting different scenes with varying color temperatures, the cameras could be adjusted for these differences, which were then loaded into memory for easy recall. The age of the camera shader has faded, much more so than this author thinks is good for image quality's sake. Many television productions have no adjustments made to the cameras at all during the shoot, depending on the automatic correction circuitry of the camera to adjust to varying light levels and optical flares. The camera used to be a team operational effort. One person would point, zoom, and focus the camera to compose shots that the director wanted. A second person made the camera look its best from the camera control position. Actually, the camera shader would control multiple cameras. At times, that operator might even be called upon to degrade the performance of the camera to achieve a particular artistic effect desired by the director. The use of the camera in producing electronic effects has been moved downstream to other production, or postproduction, gear.

Actually, a form of the camera shader lives on. Today, they are known as colorists. These are people who make sure that the video images being captured from film and recorded on tape are faithful to what was originally intended when the film was shot. Or, as is sometimes the case, to restore what time and storage have caused to fade. The devices that do this were originally specialized television cameras which were simply repackaged for this particular purpose. Television stations used to have a number of them. The camera was installed as part of a system that usually also comprised two film projectors, a slide projector, and mirrors to direct the light from a particular film/slide projector. This system was known as a *film island*. Most television stations no longer have these as all program material, commercials, and graphics are either played back from videotape or video servers.

Production houses that specialize in film-to-tape transfers evolved from a system that used modified cameras to one either using CCD imaging strips or lasers to scan each frame of film. Unlike the camera, which required a projector that pulled each film frame into the gate and stopped it there for an instant before pulling in the next film frame, the new devices kept the film moving at a constant speed. This process is much easier on the film and the projector. It is also much easier to keep the film in an exact place, and have each frame horizontally and vertically the same as its preceding frame.

The colorist now has an operational cousin, the compressionist. These people review the decisions made by MPEG encoders, which sometimes make poor automatic choices. To ensure that HD is released from a network or that a 35-mm film is transferred correctly, colorists manipulate many of the MPEG attributes that we have looked at in this book. MPEG compressed video can be considered just another data stream and, therefore, can be recorded on a lossless data recorder and not a video recorder. This means that the storage system will not be making any compression decisions. All current HD video tape recorders conduct some type of compression.

9.4 Plant and Facilities Processing

Once video is generated, it usually must be transported around a facility, either for storage or routed to the transmission system out of the facility. Even in a situation where a camera is being transmitted live, it will usually pass through some processing equipment on its way to air. The shot of your favorite news anchor during a live newscast will usually travel from the camera through a video distribution amplifier (VDA), and then through a production switcher in the studio control room (Fig. 9.8). The path continues through another VDA to the master control switcher, and then through another VDA to the television station's studio-to-transmitter microwave link (STL) to the transmitter. That's five intervening boxes between the camera and the station's transmit antenna. Today, a trip through a video router after the studio control room and master control would replace those two VDAs. Also, a bug generator to key the station's logo (lest you forget) in the lower right corner would be inserted after master control (Fig. 9.9).

The studio control room switcher would be used to key the anchor's name over the video of the anchor or to generate a split screen of the anchor in the studio talking to a reporter in the field. In addition to the control room production switch, the camera video might have to go through a digital effects device (DVE) to accomplish this. The video from the camera in the field shooting the

Figure 9.8   1980s vintage production switcher control panel (top); 1990s equivalent (bottom).

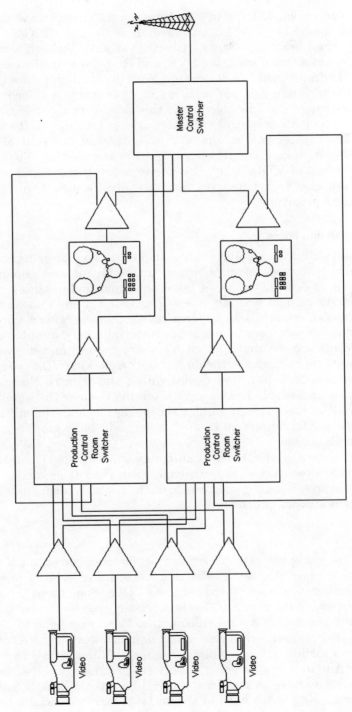

Figure 9.9 Block diagram of a simple television station. As we can see, a number of different devices are necessary to tie a television facility into a complete system.

reporter would have an even longer path. That video would go from the camera to possibly a VDA and switcher in the ENG/SNG (electronic news gathering/satellite news gathering) vehicle, through a microwave or satellite hop to a frame sync, the DVE, and then the control room production switcher. The frame sync is a device that locks the incoming video to the video present in the facility (or "house" as it is called). If this is not done, then most DVEs and switchers won't be able to mix this video with other sources.

The master control switcher is the gatekeeper to the station's transmitter. The master control operator would select the control room for news segments during the newscast and the commercial playback device, either a system of VTRs or a video server, during the commercial break. After the newscast is finished, the master control would then select the source of the next program.

9.4.1   Distribution amplifiers

A distribution amplifier is the simplest video processing box encountered. There are two types of distribution amps—fanout and equalizing/reclocking. The fanout does just that—it takes one input and produces multiple outputs. The studio camera has only a few outputs but it often has to feed many sources. A camera might need to appear on monitors in one or more control rooms, production switchers in one or more control rooms, a router input, etc. Equalizing VDAs are for analog video. They are used to adjust passband problems that were introduced. On the digital side, reclocking VDAs convert the incoming bit stream back to actual digital values and then re–channel code the video bit stream. This eliminates some jitter and restores the signal to its original bandwidth, which results in nice sharp edges again. The other approach is to pass the digital bit stream through a very wideband amplifier, but this approach is seldom used.

Some digital VDAs offer automatic input timing. Digital systems tend to have more propagation delay through them than their analog equivalents. Format converters, such analog composite to digital component, etc., can add a significant amount of delay.

9.4.2   Processing

Processing is a catchall for modification and correction of the video. As we saw in the last chapter, a television camera does a lot of processing on the video images generated by the camera's CCDs. Sometimes processing must be done outside of the camera. This is especially true for analog video paths but generally not true of digital video paths. Most processing of digital video occurs in switchers and digital video effect units. However, with ATSC many facilities are now installing processing units ahead of their MPEG encoders (see Chap. 7).

Any unit that is going to perform processing on the digital video stream is going to have to convert the serial bit stream back to parallel words. Digital processing is too hard in the serial domain, especially with the clock speeds encountered. As a result, the video is passed through most units that act on the video values as a parallel number sequence. As we will see in Sec. 9.4.2.5 on switchers, this posed some problems for early digital switchers. Most television engineers have been trained to time video signals as close to each other as possible. Signal timing has been an important activity in television facilities since the beginning. The techniques have changed over time as the video signal has evolved.

Originally, video was not only just black and white, but noncomposite also. Noncomposite doesn't mean component; it means video without sync pulses. Today there are ICs in most devices that act as sync generators for the device. These sync generators on a chip genlock (lock to an outside signal) to the house reference sync generator. Early on, sync generators were big and expensive, and it made little economic sense to try and include a sync generator in each device. As a result, the facility's master sync generator provided the necessary drive and sync pulses to the equipment. Because of this most equipment didn't bother to tack sync pulses to the video coming out of each individual device. Only video was output; nothing could be found in the picture blanking areas. Sync pulses were added at a central point, usually in a production switcher, or even at the master control switcher. Video timing was fairly simple—the video was overlaid so that blanking start/stop coincided.

When sync generation became less expensive, sync pulses were included in the video from each device. Some switchers would strip off the sync pulses and the initial timing procedure continued; some other switchers did not strip off sync. Now system timing for each source was set so that the leading edge of the sync pulses coincided. Video phase would also have to be set so that blanking start/stop also coincided. Timing within a microsecond was usually sufficient.

Then color came along and timing got much more stringent. A display determines what color is to be displayed by the color subcarrier's phase relationship with the reference color burst information sent on each line. In the case of NTSC, $1°$ is only 0.75 ns. Most critical observers can detect an error of only a few degrees. As a result, video, sync, and color phase had to be adjusted. However, NTSC (along with PAL) color brought other subtleties not noticed until videotape editing arrived, especially with the advent of the TBC.

9.4.2.1 Time-base corrections (TBCs).

As we will see in the next chapter, VTRs are as much mechanical as electric in nature. Since they are mechanical, they cannot be made to play back video with the stability that video requires. The video scanner, which recovers video from the tape, still has jitter in the range of at least a few microseconds even with the advanced servo technology available today. The servo system controlling the scanner continually speeds up or slows down the scanner, oscillating around the correct velocity based on tach and reference signals. Unless techniques such as those in consumer VHS and industrial Umatic are used, namely, the color heterodyne process, color will not be viewable. Also, the overall signal will not be usable in any effect with another source. Two-inch quad VTRs solved this with analog circuitry. Modules with large networks of inductors and varactor diodes were employed to vary the delay of the video off-tape encounter. The varactors acted as variable capacitors in the LC (inductance/capacitance) network. A control voltage applied to the varactors varied the capacitance, which in turn varied the amount of delay the video experienced throughout the network. The delay was inversely proportional to whether the video arriving from tape was early or late. The result was video that was in time with the other video sources in the house.

In the early 1970s solid-state memory became fast enough and was priced low enough that it was used instead of the LC network. Analog video coming off a tape was converted to digital and stored into memory. The clocking for writing into memory was created from the incoming video. Video was then read out of memory using clocking derived from the master reference in the facility. Thus, video was fed in at a rate that varied with the mechanics of the VTR and came out at a rate locked to the house. This process was known as *time-base correction* (TBC).

The memory was organized in what is known as a ring buffer system, which is analogous to a merry-go-round where you only ride a maximum distance of three-fourths of the way around. People walk up to the ride at varying rates: groups of two, three, or four; as singles; etc. Sometimes no one arrives for a short while. When a particular person walks up to ride, he or she has to get on a horse such that there are no unoccupied horses in front. If no one has arrived in a while, a rider might have to walk a ways around the ride to get the next unoccupied horse, which means that rider gets a shorter ride. If a group of people have arrived right in front of that rider, the walk to the first unoccupied horse will be shorter, resulting in a longer ride. At three-fourths of the way around, riders are forced off the ride regardless of the length of the ride. Bunches of people get on, but a steady predictable stream gets off.

Now, obviously, the memory does not physically rotate. To duplicate this on our merry-go-round would be a strange and unsatisfying ride. Imagine that the gate where you get on and off moves but not the ride itself. The entrance gate moves to the next unoccupied horse and you get on. You sit on the unmoving horse while the gate where you get off steadily moves from one horse to the next, letting people off. Electronically, the moving gates are replicated by memory addressing and READ/WRITE commands.

Video was originally sampled at 3 times the subcarrier rate, with 8-bit resolution, but soon 4 times the subcarrier rate and 10-bit resolution became the norm. However, memory was still fairly expensive so 1.5 to 3 lines of horizontal video were all the buffering offered, which meant that small disturbances to the VTR's scanner would be outside the correction ability of the TBC. In very rapid order TBCs with dozens of lines appeared.

However, the TBC exposed a common error in color television system timing, that is, the timing relationship between color subcarrier and horizontal sync (SC/H). In NTSC the color subcarrier phase and a particular horizontal line have the same relationship every other time. Let's look at line 80 as an example. In the first field, the start of the horizontal sync for that line and the color subcarrier have a particular relationship. Line 79 preceding it and line 81 succeeding it have timing relationships 180° out of phase. In the next field lines 263 through 525 are scanned. In the third field the SC/H phase relationship is exactly 180° from what it was in field 1. Then a fourth field occurs again at the upper lines. Finally, in the fifth field the SC/H phase relationship is the same. While two fields are necessary for an interlaced black-and-white picture, four fields are required for an NTSC color frame. PAL is even worse—it requires eight fields for each PAL color frame.

Most TBCs insist on producing correct color frames. Some accomplish this by decoding the color information and reencoding it so that it is always correct, but this usually degrades the quality. Most accomplish this by sliding the luminance video picture half a color subcarrier when the SC/H phase seems incorrect. This produces a noticeable shift in the picture, and the problem shows up during editing. If video being inserted or assembled onto a tape has a different SC/H phase (more than ±90°), then the TBC will shift the video already on the tape when the new video starts and/or stops during playback.

9.4.2.2 Frame syncs.   It was not long before TBCs had enough memory to store entire fields or even a frame of video. This created an opportunity to eliminate a practice that had plagued television stations since the beginning. All the video sources in the house could be locked together but how could signals arriving from outside the station, such as the network, be locked. Most sync generators are able to lock to an external video signal. As a result, some television stations actually locked the station's sync generator to the incoming network signal.

This process worked fine until something happened to the network signal. Then all sources in the house would take a hit. If a recording being performed, it would be affected. It got worse. When the ENG revolution allowed microwave shots to become a part of the newscast, news and production people wanted to be able to use the shot in effects. This meant that the entire station would be genlocked to the ENG camera providing the video being microwaved back to the station. Often something would happen to the camera, something as simple as a battery going dead, that would play havoc with the entire station.

Enter the frame sync. The frame sync worked just like the TBC but had enough memory to take a video signal that was not vertically or horizontally in phase with the house and lock it to the house. External sources could now be treated like sources generated in house.

9.4.2.3 DVEs.   The frame sync continued to evolve. DSP techniques emerged that allowed data stored in memory to be manipulated. The simplest operations were instituted first. These operations involved simply reordering the readout of memory or not reading certain locations. This process made it possible to slide a picture on and off the screen or reduce and enlarge it in size. From this point processing techniques exploded so that effects were limited only by the imagination of the person driving the box.

9.4.2.4 Format conversion.

In today's digital plant many analog devices and paths continue to exist. As a result, the bridges between the digital and analog domains must remain. Most of these bridges are now decoders which take NTSC composite analog video to digital component. Where analog is required, encoders are used after digital processing. Decoders are much more complex devices than encoders. Separation of chroma from luminance when decoding frame syncs is often used to provide the increased quality of three-dimensional filtering. As a result, digital component signals are sometimes timed ahead of the NTSC signals because encoding from digital component to NTSC generally uses simpler devices with no built-in frame sync, just some delay.

9.4.2.5 Switching.

Switching can be the most involved, yet common, process performed on the video. In its simplest form *switching* is simply the selection of the video source that will be fed downstream. However, some switchers perform many more functions than simple source selection. There are four types of switcher operation: plain old switching or "cuts," "dissolves," "wipes," and "keys." Switchers used for production or at master control don't just switch at the vertical interval. Most can mix two synchronous video signals together. The percentage of video from either source can be varied either manually via a "fader" bar (the lever used in the movie *Star Wars* on the Death Star to blow up Princess Leia's planet was a fader bar on a Grass Valley 1600 production switcher) or in many cases by automatic means. The percentage can be varied from 0 to 100 percent for one source while the other source goes from 100 to 0 percent. This process was originally called a "mix" but now is generally known as a "dissolve." Many switchers are also able to switch many times per field. The rate at which they switch used to be determined solely by pattern generators, but many now also use lookup tables stored in memory. These patterns allow the switcher to perform wipes. When a switch from one source to another happens during the vertical interval, it is a simple cut between sources. If it happens not only during the vertical interval but again somewhere during the field back to the original source, then it is a "vertical wipe." If the switching alternates between two sources once per line, it is a "horizontal wipe.".

  As a result, switching can occur based on an operator making a new selection via a button on the control panel, or operation of a fader bar, or based on a pattern generator in the case of a wipe. However, switching can also be based on video levels, and this is known as *keying*. There are three common types of keying: self, external, and chroma. *Self-keying* occurs when levels above a certain threshold for a selected video signal are used to replace video from another source. The video above the level threshold to be keyed becomes the switching pattern generator. From an operational standpoint the process is often referred to as *layering,* but it is technically still a switching operation. The switching occurs between the original video, usually called the background video, and the key video. *External keying* occurs when a device, such as a character generator or DVE, provides a separate keying signal along with the regular video source. The key signal can be sometimes referred to as the hole cutter, and the regular video the "fill" signal. Once again, though, this is really a switching operation with the hole cutter taking the place of the pattern generator.

Two types of key switching operations are available. One is known as a *nonlinear key,* which is what we just described, that is, either the video fill or the video being layered on top is switched between—they are never mixed. A newer approach is the *linear key,* where the key video and background video are not directly switched between. The switching occurs between the background video and a "mixer." The mixer bases the percentage of background to key video, based on the level of the hole cutter key signal. This process provides the keys with an element of transparency.

A close relative of the video key is the chroma key. In a *chroma key* the separate red, green, and blue (RGB) primary signals from a source are available to the switcher. This is usually accomplished by feeding these three signals from the source, such as a studio camera, into the switcher. Many switchers have options that will take the color difference signals, in the case of an analog or digital component, or a complete composite signal in and decode it into the primary colors. Chroma keying in its basic form is different from regular video keying in that the source itself, commonly a studio camera, uses its own levels to create the pattern for the video to be overlaid instead of the overlaid video generating the pattern. A common example is the meteorologist standing in front of a blue screen. The background video is the composite (or complete component) picture out of the camera. However, the three separate RGB signals are also going to the switcher from the camera. A knob on the switcher control panel is turned to select the primary color (or mix of two of the three colors) that is to be used. In the case of a blue screen, the keying threshold is then adjusted so that blue over a certain threshold out of the camera creates the switching pattern. The switcher then switches between the composite/component camera video and the overlaid video, which is usually some sort of computer graphic.

It should be noted here that there are many specialized boxes that perform only the chroma keying function. Producing realistic chroma keys is part science and part art. One of the basic problems is caused by light contamination from the blue (green is commonly used also) screen background and the subject in front of the blue screen. This is the cause of much "tearing" seen at the edge of the key. Many specialized boxes have proprietary circuitry for addressing the problem. Many assume that a transition, and not an abrupt switch, is needed at the switching points. Another problem that is now being addressed is that the overlaid video and the background (opposite of what you would expect—the overlaid video for our weather example would be the computer graphic that appears behind the meteorologist) both appear in focus. This creates a lack of depth of field, which looks artificial. So some dedicated chroma key boxes now soften up the overlaid (or actual background) video so that it looks out of focus.

Some of these specialized chroma key boxes are now tied to workstations from a number of vendors like Silicon Graphics and HP to produce what are commonly called *virtual sets*. These systems replace carpenters, painters, and lots of space with computer graphics animators and software programmers. Instead of a news set, on-air talent sits in a sea of blue or green while the set is keyed around them. The high-powered computers know where the talent is in space (or where in the set) and what the cameras are shooting. The workstations control the video generation directly or through a dedicated chroma key box. Methods for determining camera position have become fairly ingenious. In one approach multiple infrared sensors look at geometric shapes mounted on top of the camera, while in another a sensor camera is mounted vertically to the main camera looking at geometric patterns in the lighting grid.

As we briefly mentioned in Sec. 9.4.2, timing pulses are generally removed near the input of a digital switcher. The serial video must be converted to parallel for digital processing in the switcher so each input video word will consist of 10 parallel lines. When a TRS comes along (a string of three words = 0 $\times$3FF, 0$\times$000, 0$\times$000), all the parallel lines are high for the first word and then all go low for the second word. Imagine the inductive kick and strain on the switcher's power supply when all input lines go high (240 on a 24-input switcher) and then low at the same time. This is what would happen if all 24 inputs were timed to within one clock of each other. In order to prevent this from occurring, all TRSs are stripped off.

Digital component words, a Cb color difference sample, a luminance sample, a Cr color difference, and another luminance are processed sequentially. However, the two color difference signals are cosited spatially with each other and with one of the luminance samples. It is required that they be buffered in time. Since this is the case, it is not much harder to provide more buffering for the incoming samples. As in the case of the TBC, ring buffering can be instituted so that exact input timing of each source is not required. Digital switchers generally have $\pm\frac{1}{2}$-line timing windows. If things are out of the window, things either have to be delayed for the straggler, or sync gens can be used to advanced the timing of the straggler.

One other aspect of switchers should be touched on before we move on, that is, that most large switchers are composed of smaller switchers. In large switchers these smaller switchers are referred to as *mix/effects (M/E) banks*. In the past a single M/E bank could only perform one of the four types of switching at a time. Now most can perform more than one, sometimes all types at once. These M/E banks used to be fixed on how one cascaded into another. M/E1 could go M/E2 or M/E3, but M/Es 2 and 3 could not be used in M/E1, etc. Now most large switchers have architecture that allow any M/E bank to be fed into any other M/E bank. This used to cause problems in the analog domain as the propagation delay would increase as a particular video source went through additional M/Es. Each M/E would add some delay. Originally, when switchers had a fix order for cascading through the M/E, the varying delays were fixed.

In fact, every M/E bank had what was called a reentry section if an input that was directly selected on an M/E bus was used versus using the output of a M/E bus upstream. The reentry provided the necessary delay to match the delay from upstream M/E busses. Why would you need to worry about timing of the individual M/Es and not just the switcher output? Because often multiple program feeds are needed out of one switcher. This happens frequently in remote situations where one switcher is feeding multiple clients. Some clients might not want all the graphics or might even require a different set of graphics. If the operator on an analog switcher is changing the order of the cascaded M/E banks, the timing out of the various banks would change, providing grief for the client of that output. Digital switchers often circumvent this because either the digital switcher upstream has built-in frame syncs that delay everything by a field or frame or the input of the client's digital switcher has a wide timing window.

This prompts one more concern in the digital realm. Frame syncs and devices with built-in frame syncs will delay the video versus the audio one video frame each time. It doesn't take long before a noticeable lip sync problem occurs. Many organizations limit the amount of frame syncs in a path to 2. A number of vendors offer frame syncs that delay the audio as well. It is also possible to buy stand-alone audio delays.

9.4.3 Routing

Compared to high-tech computers and DVEs, routing switchers rarely command much attention. However, they are a critical part of nearly every facility's infrastructure. A functional and flexible routing system is the key to minimizing errors and improving overall efficiency. Today, as facilities migrate toward digital, they find that many of their analog sources tend to migrate with them. Even facilities that have the luxury of starting from scratch find that the concept of an all-digital facility is usually just that, a great concept but not quite obtainable.

Why is that? Well, one obvious answer is that instead of yesterday's digital islands, there are now analog islands, especially where there is little or no economic benefit from going digital. For instance, it is still economically advantageous in situations that call for numerous small video monitors. Many people that do studio camera setup and matching still find good old NTSC to be the best way to monitor their results. Also, audio is probably the biggest reason for keeping analog signals floating around a digital facility. In many instances analog is more cost effective because analog mixers and processing equipment are still less expensive than their digital counterparts. Many digital facilities mix and route audio in the analog domain and only digitize and imbed the audio into the serial bit stream as it approaches the transmission path out of the facility.

9.4.3.1 Multiformat facilities. Today's facilities need to install routers that work with a variety of signal formats, including analog NTSC video, analog audio, SDI digital video (either SMPTE 244M or SMPTE 259M), and AES digital audio. In addition, routing time code and machine control signals is common. Although a separate router matrix is typically required for each signal type, it is unacceptable for each matrix to require its own set of control panels. In nearly all modern routing systems, when dealing with more than one matrix of crosspoints, it is possible to have one router control all of the various router matrices involved. In today's digital facility, the SDI router typically controls all of the other installed routers because the primary format is now serial digital interface (SDI), SMPTE 259M, and all signals are usually controlled in conjunction with how the SDI signals are routed. That said, there are times when signals must be split away from the main SDI signal. So, how should you think about and cope with various routers handling different format signals (Fig. 9.10)?

The easiest way to visualize and understand the problems and the opportunities presented by this situation is to think of the overall router system in a multiformat facility as a three-dimensional object such as a stack of square (or rectangular) waffles. Each waffle represents a type of format that is being routed. One waffle could represent AES audio routing, another could represent analog audio routing, and the third might be the analog NTSC routing waffle, and finally, on top, the SDI routing waffle. Visually, the waffle analogy plays well in the two-dimensional aspects of routing. The waffle's gridiron pattern represents the router's X-Y switching matrix. Inputs come on the Y axis and are switched or connected to outputs going out the X axis. As we move from one waffle up to the next, we are traveling in the Z axis. Hence, our three-dimensional object. No two waffles in the stack need to be the same size; they can each be different, and these various sizes can be in any position in the stack. SDI sources and destinations in a newly constructed digital facility generally outnumber analog or AES digital audio sources. Often, the analog islands left over in a digital facility preclude the need for a large analog matrix. Now, instead of calling them waffles, let's call them levels. The router matrix for each format is called a *level*. For instance, there could be an SDI level, an analog NTSC level, an AES level, an analog audio level, and possibly more.

9.4.3.2 Organizing and planning a multilevel system. With the various levels identified, it is important to understand how to organize and use these resources. The initial point of confusion usually occurs when planning the input and output usage of a new multilevel routing system (Fig. 9.11). As an example, let's consider a 32 × 32 SDI router, a 16 × 16 AES router, and a 16 × 16 analog audio router. The SDI router is assigned as level 1, the AES router is assigned level 2, and the analog audio router is assigned level 3. The output of a VTR (VTR 1) is connected to input number 1 of each of the three routers. The VTR's SDI output goes to the level 1 router input number 1, with the AES output going to the level 2 router input number 1, and the analog audio output going to the

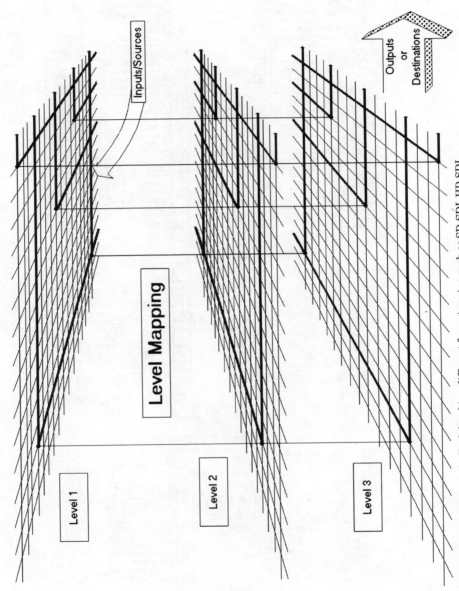

Inputs/Sources

Outputs
or
Destinations

Level Mapping

Level 1

Level 2

Level 3

Figure 9.10 3D router stack. Each level is a different format router, such as SD-SDI, HD-SDI, AES, time code/machine control, analog audio, or video.

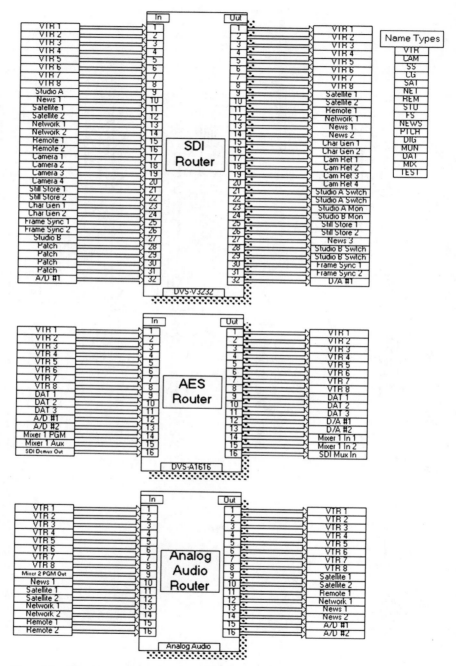

Figure 9.11   Three routers, each a separate level in the routing stack.

level 3 router input number 1. Suppose we connect seven more VTRs in the same manner into inputs 2 through 8. For the most part, when the inputs from the VTRs are switched to various outputs via matrix crosspoint changes, all three levels should switch in unison, a process called *level mapping.* Commanding that input 6 be fed to output 13 causes that to happen on all levels. This is the simplest approach to setting up a multilevel routing system. However, it can greatly hamper the system's flexibility. If only one level needs to be switched or "broken away," a router control panel with that capability is needed.

Going back to our example, let's connect two still stores to the SDI router level at inputs 21 and 22. Audio usually isn't associated with a still store, and wasting these four inputs into the analog audio router is not desirable. Notice that the same inputs are not found at all three levels. Level mapping does not work in cases where nonrelated resources are at the same inputs but on different levels (Fig. 9.12).

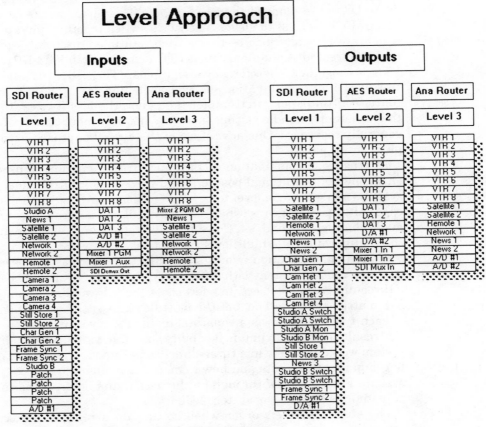

Figure 9.12 Another way to look at the level approach.

A second approach, entirely contrary to the level approach, is the virtual approach. Most routing systems today allow an input and/or output to be assigned to a different place in the matrix. This means that because VTR 1 is physically connected to router input 1 on all levels, the router could be told that VTR 1 is virtually at input position 101 or 201 or 389. Virtual input 389 would then be mapped to physical input 1. So anytime we "talk" VTR 389 with the router, it knows we are really talking about the input physically arriving at input 1. The input and output on each level can be assigned to a separate virtual placeholder. For instance, SDI (level 1) input 1 could be assigned to VTR 101, AES (level 2) input 1 could be assigned to VTR 201, and analog audio (level 3) input 1 could be assigned to VTR 301. These three virtual labels can be separately assigned to control panels so that VTR 101, 201, and 301 all appear. Selecting VTR 101 would only change the SDI level, while selecting VTR 201 would only change the NTSC level, etc. (Fig. 9.13).

In the case of the still stores and DAT machines, they could be virtually mapped away from each other, placing still stores at virtual inputs 21 and 22, while DATs are mapped to virtual inputs 41 and 42. In the case of virtually mapped VTRs, if all the levels needed to switch together, phantoms or salvos, which are basically macros that switch multiple crosspoints simultaneously, could be created. A phantom called VTR 1 could switch VTR 101, VTR 201, and VTR 301 together. Therefore, by assigning a VTR 1, 101, 201, and 301 to a control panel, any combination of level switching for VTR 1 is available. For router control panels, the simpler the panel, the less it costs. Control panels that inherently handle various levels are not only more complex to operate but generally cost more. One advantage of the virtual approach is the use of simpler control panels.

In practice, a combination of level and virtual mapping is used, depending on the mix of operational positions and the type of control panels used. Some systems allow a single level to be split or folded into two virtual levels. This is especially advantageous when one physical router frame is handling AES audio channels 1/2 and 3/4. Considerable thought and planning are needed for the configuration of a multilevel router (Fig. 9.14).

Many implementation efforts start with the router as the base. As already stated, a routing system with different types of routers should be thought of as a three-dimensional object. In a few cases the old two-dimensional model of an X-Y matrix switch will do, but the most flexibility can be obtained by thinking of each type of router as a separate, distinct Z layer or level. To help ensure the result allows maximum flexibility and functionality, visualize what the system would look like in a three-dimensional stack. Then determine how best to group inputs/outputs and how to map the available resources to the router system. Finally, follow through on the design and be prepared to make a few adjustments based on operator feedback.

The steps and their order when setting up a router from scratch follow. Interdependencies carry on from early steps to later steps.

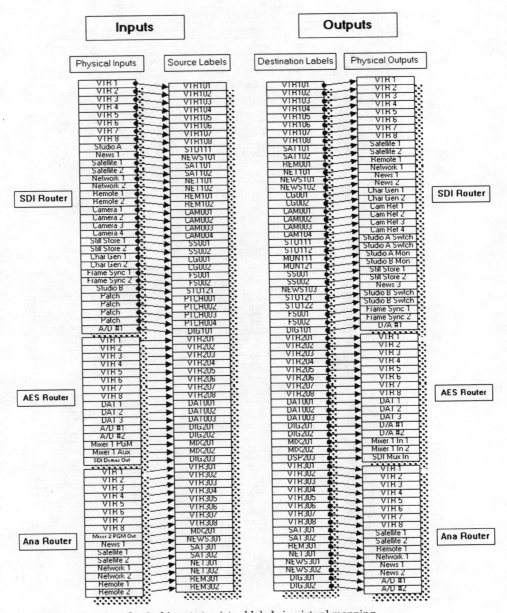

Figure 9.13   Mapping physical inputs to virtual labels in virtual mapping.

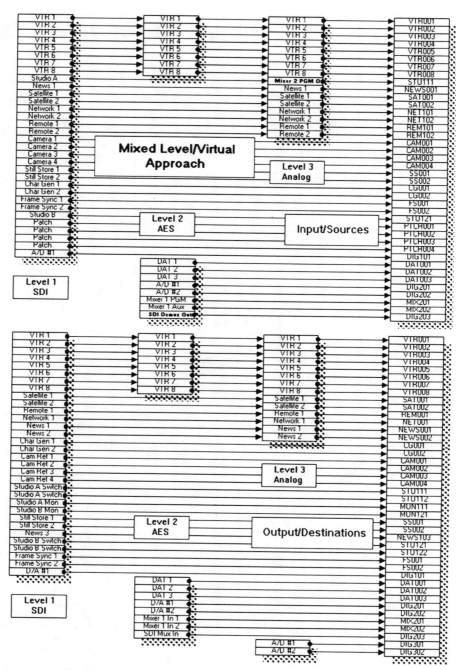

Figure 9.14 Combination of level and virtual mapping.

1. Define the router space. This space consists of not just the physical I/O but the entire naming space. Various systems usually have some upper limit on the size.

2. Organize the various physical matrices into the space they occupy, for example, on a particular level. Also define the different levels and the names that the levels will have.

3. Identify all input/output signals on all levels. In most systems you have to assign each to a category. Category names identify the input or output type such as VTR, CAM, etc. Some systems have restraints on name length and the number of category names. Many of the restraints revolve around the display limitations of control panels. A common point of confusion is that the same category name can be used for both inputs and outputs.

4. Determine how many virtual names or labels you need to use. This is often the toughest step to understand conceptually, but it is the step where virtual and level mapping are actually assigned. If you are going to completely level-map all inputs and outputs, you need to assign enough names to cover the matrix with the most inputs and the one with the most outputs. As an example, if you had a 64 by 48 matrix with a 32 by 32 level-mapped below it, your name or label list would be 64 inputs and 48 output names. If total virtual mapping is to be used, then you need as many virtual names as the sum of the inputs and outputs of all the physical matrices. If you have four 64 by 32 matrices, then you would need 256 virtual input and 128 output names or labels.

5. Virtual meets reality as each name or label is assigned to a particular matrix's input or output. Most systems create separate input and output name or label lists. So the actually physical number assigned will be input or output, depending on the list you are in. A name can be assigned to more than one level. As an example, a VTR could have inputs or outputs on digital and analog video and audio layers, along with time code and machine control layers. However, an actual physical input or output can only be used once. Two different labels can't be assigned to the same input or output, but you can find the same name assigned to a particular output and a particular input, say VTR1.

6. Assign any needed phantoms or salvos. These are macros that most routers allow so that one button push changes multiple crosspoints. They are defined and given a name. Thereafter, they can be assigned to control panels like any of the virtual labels.

7. Next the control panels are set up and programmed. In some systems the system controller is given the address of each panel to be polled on a common bus, and input and output labels are assigned to control panel pushbuttons. On other systems the devices that comprise the control network must be entered, and the types of panels that are available. Some systems have intelligent control panels that tell the system what input to select on what crosspoint; some panels only provide information about what key on the panel was pressed and leave it to the system control to determine what the operator is requesting. In either case a file must be created for each panel. In the intelligent panels the file resides in the panel, although the system controller might

also keep a copy of it. In the dumb control panels the file is only kept by the system controller.

8. Decide the inputs that won't be available at selected outputs. This is usually a job protection feature so that the "Playboy" channel or a VHS machine with other "racy" material can't accidentally make it to air.

9. Set up any "tie lines" or "path finding" lines needed. Most systems can send an output on one level to the input on a different level, usually to get analog video to a digital output or vice versa. Obviously, there must be an A/D decoder or D/A encoder between the analog and digital matrices. To limit the amount of these needed, program the router so that a few outputs of the analog router go to decoders and a few outputs of the digital router go to encoders. Many times it is also necessary to specify which inputs on each level can go to which outputs on the other. The router will automatically find an available tie line and use it. If all tie lines are taken, then the path won't be made. So a careful investigation on peak tie line demand must be undertaken. Another concern is that often only the panel that has requested the tie line path can release the path, although most systems allow the system setup software to override the "protected" path (Fig. 9.15). Finally, two approaches are possible in terms of setup. On some systems you can change settings on the fly. As soon as you make the change, it

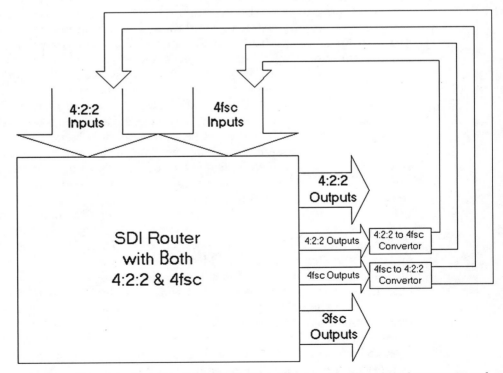

Figure 9.15 Here, one physical matrix is divided into two virtual levels, a digital component and a composite level. Tie lines (through convertors) connect the two levels.

takes effect. On others you modify a database file that then must be compiled and loaded on the system.

From a hardware standpoint there are three types of routers:

1. *Wideband analog switchers.* These treat digital signals as analog, and most use high-frequency equalizers. They degrade the digital waveform with every pass through the router, and usually fall out of SMPTE spec after a few

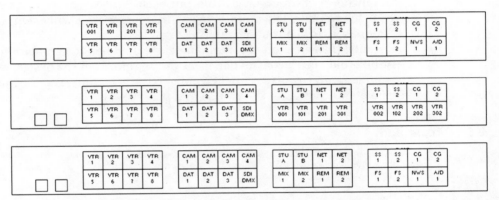

Figure 9.16 Simplest control panel approach. These panels control one router output each. The available inputs are labeled on each button according to the three-level example.

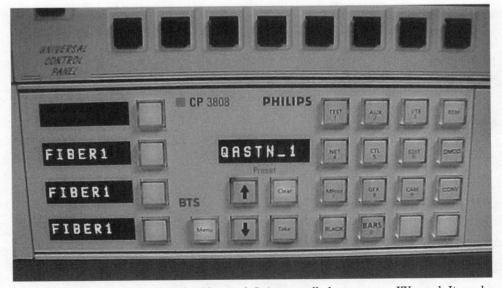

Figure 9.17 Here is an extremely flexible panel. It is generally known as an *XY* panel. It can be programmed to switch any crosspoint, which means it can connect any input to any output.

passes. However, these switchers can usually pass a wide range of bit rates. Some wideband analog routers with bandwidths of 100 MHz can pass digital composite signals, although not very reliably and only with one pass.

2. *Digital switchers with reclockers at every output.*   Some of these systems will automatically lock to the required bit rate.

3. *Digital switchers with reclockers at every input and output.*   Some of these systems can also automatically relock to different bit rates. This approach allows routers with large matrices to be built. They are very good at eliminating high-frequency jitter, but reclocking does not eliminate the low-frequency wander. The buildup of low-frequency jitter will eventually limit the absolute size of the router matrix.

Finally, the selection of control panels should be given much thought. Many just mix and match control panels at the beginning of the design without much thought. Router control panels can be extremely simple and intuitive or very flexible and confusing (Figs. 9.16 and 9.17).

9.5   Bibliography

Boston, Jim, "Multilevel Routers," *Broadcast Engineering*, December 1997, p. 52.
Robin, Michael, "Routing Switchers: The Mezzanine Approach," *Broadcast Engineering,* December 1998, p. 36.

Storage

Now we are going to examine how video/audio is stored. Storage is needed so that, unlike in television's earliest days, material doesn't have to be produced in real time. Today, storage means one of two things—storage on linear video-tape or on the disk drives of a video/audio server. Server-based storage is rapidly displacing tape in many applications, but tape continues to make sense in a number of applications. We will examine the history and technology of VTRs first.

10.1 The VTR

It has been said many times that the only universal tape standard we will ever see was the first standard—the quadruplex VTR. It used 2-in tape, and was called a "quad" because it used four rotating heads, as we shall see shortly. It recorded composite video and a single analog audio channel, although it had a lower-quality "cue" channel. It took many seconds to play back "locked" video and produced no still picture in playback. Originally, the machines could only record black-and-white video and they couldn't be locked to the house. This means that effects, such as dissolves, wipes, and keys (which weren't possible when VTRs first appeared), could not be done if a VTR was involved. Eventually, electronics was added that allowed the machines to be "genlocked" to other equipment. However, additional electronics had to be added to make these machines truly synchronous with other equipment. These shortcomings were not because of the format but because of limitations in the technology of the day (see Figs. 10.1 to 10.4).

Later, another analog composite format was introduced—the type C VTR. SMPTE developed three helical formats. *Helical* meant a VTR that had "slant" or diagonal video tracks laid down on tape. Ampex invented type A. Sony developed a machine closely resembling the type A. The compromise between Ampex's and Sony's formats led to the type C. The type B was a machine from

Figure 10.1 RCA's first videotape recorder—the TRT-1. The first VTR—the VR-1000—was introduced by Ampex in 1956 (see Fig. 2.6).

Bosch that used 1-in tape but laid down video tracks that more closely resembled a quad than either Ampex or Sony. The development of the frame sync allowed this VTR to have many of the same attributes that type A and C machines introduced, namely, slow motion and reverse playback. Type A and C machines also allowed "confidence" record. These machines had separate record and playback heads. The playback head followed behind the record head and played back the freshly recorded video, allowing the operator to see,

(a)

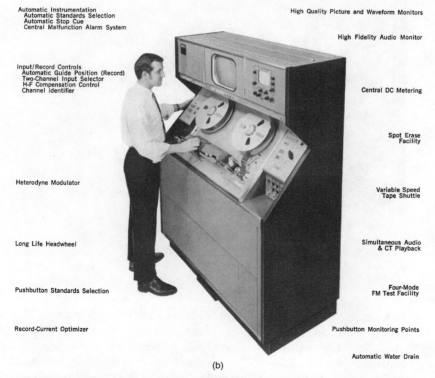

Automatic Instrumentation
  Automatic Standards Selection
  Automatic Stop Cue
  Central Malfunction Alarm System

High Quality Picture and Waveform Monitors

High Fidelity Audio Monitor

Input/Record Controls
  Automatic Guide Position (Record)
  Two-Channel Input Selector
  H-F Compensation Control
  Channel Identifier

Central DC Metering

Spot Erase
Facility

Heterodyne Modulator

Variable Speed
Tape Shuttle

Long Life Headwheel

Simultaneous Audio
& CT Playback

Pushbutton Standards Selection

Four-Mode
FM Test Facility

Record-Current Optimizer

Pushbutton Monitoring Points

Automatic Water Drain

(b)

Figure 10.2 (*a*) Ampex VR-1200 loaded with all the servo and video processing options (*courtesy of Tim Stoffel*). (*b*) RCA's second-generation competition, the TR-70, surrounded by marketing hype.

Figure 10.3   Ampex's third-generation AVR-3.

and have confidence, that the VTR was actually recording. Since the heads literally dug into the tape, a common malady of quad VTRs was the "head" clog. No, this was not a plumbing problem on a ship, but a problem that occurred when a piece of the videotape's oxide coating became adhered to, or wedged onto, the head, effectively stopping the recording of video information (Fig. 10.5).

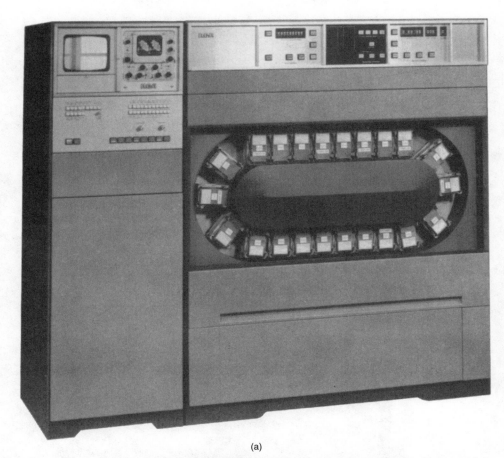

(a)

Figure 10.4 (*a*) RCA introduced the first self-loading, dual-transport quad VTR geared toward the automatic playback of commercials and other short-segment tapes—the TCR-100. (*b*) Ampex's response to the TCR-100 was the ACR-25. It was technically more advanced than the TCR-100, and made extensive use of pneumatic systems to load and guide tape movement (*courtesy of Tom Matthies*). (*c*) Control panel of ACR-25 (*courtesy of Tom Matthies*).

Yet another format that appeared and made electric news gathering (ENG) possible was the Umatic (Sony's trade name) color under VTR. As we will see shortly, although it facilitated ENG, it was hardly a high-quality format. The intended replacement for the 1-in type C VTR was the $\frac{1}{2}$-in component machines. Next came the digital machines, some composite but most component. Initially, the digital machines didn't use compression, but they were extremely expensive. Most later machines resorted to compression to help hold down costs. Now HD is on the storage scene and tape formats have not diminished, but they have actually proliferated. We will look at the VTR's history and then the current tape formats.

(b)

(c)

Figure 10.4   (*Continued*)

Figure 10.5 Late-generation Ampex C format VTRs. C format VTRs are still found in extensive use in television stations and production houses even today.

10.1.1   History

The problem with recording video on tape has always been the high frequencies needed in video. With digital video the problem increases a hundredfold. Even analog composite video was extremely tough for early videotape recorder designers to overcome since analog video frequencies, which are much higher in frequency than analog audio frequencies, produced much shorter wavelengths than audio. This meant that the head gaps used for the video head would have to be much smaller than the gap on a corresponding audio head. The use of much higher tape-to-head speeds was also needed to resolve these higher frequencies. In 1955 the BCC went on the air with its VERA (Vision Electronic Recording Apparatus) video recorder. This machine was a longitudinal recorder, that is, the video was laid down the length of the tape as on an audio recorder. Most VTRs that followed used, and still use, a "transversal" type system. Video tracks are either laid across the tape or at least at a steep angle across the tape. These methods greatly conserve tape usage. The VERA consumed tape at 200 in/s, so that a reel 5 ft in diameter was required to hold $\frac{1}{2}$ h of programming.

10.1.1.1   The quad VTR. Ampex had a better way. The method was known as quadruplex (quad) recording. This system used four video heads on a "drum" which transversed almost directly across the tape as a capstan pulled the tape by. This meant that as each head traveled across the tape, it only laid down approximately 17 lines of video. Thus, it took about 16 tracks to record one field of video. The tape was pulled past the heads at 15 in/s, and the "head" drum rotated at 240 r/min, which meant that the head drum made four complete revolutions for every field recorded. Thus, it took about 16 tracks to record one field of video. In addition, they could not make a clean edit. Any editing literally was like film. The tape was physically spliced together. The machines did not play back synchronized to the rest of the "house." This meant that when the machine was put on the air, it caused a vertical roll to occur on monitors and TV receivers. Also, the machine could not be used as a source in effects. Nothing could be "supered" (the key wasn't around yet and neither was the "wipe") over them.

In the 10 years that followed, the quad machine was brought to maturity. The machines were made to record color (the "high band" of FM recording), and options were available to allow electronic assembly and insert editing. Ampex developed Intersync, Amtec, and Colortec (RCA called it Pixlock, ATC, and CATC). Intersync forced the drum and capstan servos to lock to external reference signals (see Secs. 10.1.2.2, 10.1.2.9, and 10.1.2.10). This meant that the video playback from these VTRs was as close to being "locked" or in sync with the other TV signals in a TV station, production house, or remote truck as servos could mechanically be, within 1 μs. The video was brought the rest of the way into exact synchronization by electronic means. The Amtec system eliminated errors to a point where a black-and-white picture looked correct. Colortec removed the rest of the errors so that a color TV picture would look OK. (See Secs. 10.1.2.3 and 10.1.2.11.)

10.1.1.2 The cart machine. The next big step in quad VTRs was the introduction of the cart machine—VTRs with two tape decks that incorporated a magazine or carousel that could hold about two dozen "carts." These carts held about 3 min of 2-in tape. The carts facilitated the automatic loading and unloading of the tapes. The two tape decks were essentially the same as on older quad machines, except for features necessary to accommodate automatic loading and unloading. However, they had the same format (or footprint) as older quad machines, which means information was laid down on the 2-in tape the same way. A tape recorded on an older 2-in machine could be respooled into a cart and played back on one of these cart machines. Only two basic models were built—RCA's TCR-100 (Fig. 10.4a) was introduced first around 1971 and the more advanced Ampex ACR-25 (Fig. 10.4b) was introduced about a year later.

10.1.1.3 The helical VTR. While the quad machine was the mainstay of the broadcast industry (about 18,000 were sold worldwide and it is said that this was the only universal VTR standard we will ever see), the "industrial" TV industry needed a tape machine that was much more cost justifiable than the quad. Thus, many companies (including Ampex) worked on a VTR based on a helical format. Today all VTRs are probably descendants of these early helical machines. The major problem with early helical machines was that the scanners were never built with much accuracy.

What is the helical format? This format lays a whole field of video down on each track across the tape. Each track is at an approximate angle of 30° from the edge of a tape. This is accomplished by having the tape wrap around the drum or scanner as it traverses through the tape path from the supply reel to the take-up reel. As the tape is pulled past the scanner, it is forced to travel slightly up the scanner as it passes it. This process creates the helical video track. Due to the shallow angle of the video track across the tape, the video track containing one video field might be quite long (over 12 in) even on a $\frac{1}{2}$- or 1-in-wide tape. Before the development of computer-controlled machining, the guides to control the tape movement past the scanner were never very exact. Thus, the video track laid down on tape by one helical machine might not match the guide on another helical machine, and would thus not be able to be played back or at least played back very well. This problem was referred to as an "interchange" problem. The fact that many of these machines only needed one head for video recording and playback (instead of a quad machine's four) made them very attractive though. These machines couldn't play back color until "color under" techniques were developed, or the TBC became available (see Secs. 10.1.2.3 and 10.1.2.11). A company called IVC was the first to employ color under techniques with a color under machine. Their machines used 1-in tape, and had a cumbersome tape path. IVC had what it referred to as an alpha-type wrap around the head. It was called alpha because the tape path around the scanner resembled the Greek character alpha. The tape entering the scanner passed under the tape leaving the scanner. Thus, the tape wrapped around the entire 360° of the scanner.

Sony overcame the complicated tape loading of helical machines with the automatic-loading Umatic format at the end of the 1960s. Not that this machine had a simple tape path, it's just that the machine took care of loading and unloading of tape. The tape wrap around the Umatic scanner is referred to as an omega wrap because the tape path around the scanner resembled the Greek character omega. The tape does not completely wrap around the scanner. The tape was $\frac{3}{4}$ in wide and housed in a cassette. This machine also used color under recording techniques. This format was the first to actually make the recording and playback of color video a possibility in the home or office. Early Umatic machines even came with TV tuners so off-the-air signals from broadcast TV could be directly recorded. This machine spelled the demise of the IVC 1-in format and severely curtailed other 1-in formats.

10.1.1.4 TBCs. In the early 1970s a major VTR development occurred. For the first time electronic digital (and computer) techniques were employed on video. This development became the time-base corrector (see Sec. 10.1.2.11). The TBC could take video with all its time-base (or velocity) errors, which were caused by the fact the scanner and capstan never exactly duplicated the conditions found when recording or matched the playback video to "house" video, and make the video exactly match the external video reference. The TBC does what Amtec and Colortec do in quad machines; TBC matches the output of the VTR through synchronization to external sources (the VTR must vertically lock to the external source on its own).

10.1.1.5 ENG. Around 1975 many broadcasters discovered that if they took portable Umatic machines which were then becoming available and used the new portable cameras, they could cover news much more easily and economically than with 16-mm film used at the time. With the advent of low-priced editors, the ENG boom was born. Sony and JVC soon responded with Umatic machines specifically for broadcasters. It is estimated that approximately 1.5 million Umatics have been sold to date.

10.1.1.6 The consumer formats. Although the original Umatic machine appeared around 1970, its price tag aimed it more at the industrial user than the home consumer. Additionally, it had a maximum recording time of 1 h and used fairly expensive tape, which limited its success in the average home. However, development was pushing the "home" VTR (or VCR as it is generally known) closer to the living room. Toshiba and Sanyo offered a system called the V-CORD I in 1974 and the V-CORD II in 1976. One-hour recording was standard, but 2 h were possible if used in the "skip" mode. In this mode every other field was not recorded. This format never caught on.

Sony announced the Betamax format in 1975. It was the first format to feature video azimuth recording. This system had basic 1-h record/playback capability, but it could be extended to 2 h if the guard bands between video tracks were eliminated. Azimuth recording takes advantage of what is known as azimuth loss (a disadvantage in audio recording). The higher the frequency, the higher the signal loss from any difference in the angle between the record and the play heads. Thus, heads in this machine are not perpendicular; the opposite horizontal line is reversed on one of the video heads when it is recorded. Since the chroma phase would normally be reversed every other line, the end result is that one video head has chroma phase reversal from line to line. The other video head records chroma normally (chroma phase reversal from one line to the next). On playback the color information from the video head that was recorded normally is now reversed every other line, so that there is no chroma phase reversal from line to line. The end result is that chroma played back from one head is of all the same phase, while chroma from the other head is of the opposite phase. This tends to cancel crosstalk from one video track (or head) to another. Like the Umatic VTR, the Betamax system uses the color under system of recording. Thus, the chroma information is recorded at a lower frequency than the monochrome information. In a color under system the chroma information uses the luminance signal as a "bias" to impress its information. Companies originally in the Betamax camp included Sony,

Toshiba, and Sanyo. Organizations putting their name on Beta machines were Zenith, Sears, Pioneer, and AIWA. This consumer format came in second to the VHS format in consumer's hearts. It is arguable that if Sony were not pushing the Betamax transport for its broadcast equipment (Betacart), Sony would have let this format die.

In 1976 Matsushita introduced a single-head machine that was sold in the United States as the Quasar VR-1000 or the "great time machine." This system gained brief acceptance, but eventually quit the competition. Also in the autumn of 1976 JVC introduced the VHS format. This system had two video record/playback heads on a scanner drum 62 mm in diameter. This system used coplanar cassettes (supply and take-up reels mounted side by side) like the Umatic and Beta formats before it. This machine used azimuth recording techniques like the Betamax (heads have ±7° offsets). The process VHS machines use for chroma crosstalk cancellation is to advance chroma information by 90° every horizontal line on one head during recording, while delaying chroma information by 90° each line on the other video head. During playback, this process is reversed. Companies originally in the VHS camp were Matsushita (Panasonic), JVC, and Hitachi. Companies putting their name on VHS machines were Ward, Sylvania, Curtis Mathis, MGA, Penney's, Sharp, and Akai. This format emerged as the clear winner in the contest for the consumer dollar. The distant second place winner was the Betamax format. Evidence of this fact comes from Sony, which eventually announced that it would start manufacturing VHS machines. One of the big advantages of this format is that it has a simple tape path. While the Betamax's tape path resembles the Umatic tape path, the VHS tape path resembles an "M" (hence it has been called an M wrap). In 1978 Toshiba developed a longitudinal VTR with 1-h playing time, which could be extended to 2 h. It had a head-to-tape speed of 6 m/s and had 220 tracks down the tape.

10.1.1.7 The C format. Around 1977 Ampex developed the AST head. Along with computer-controlled machining technology, the AST head allowed the perfection of the 1-in machine. The format that Ampex used for laying information down on tape came to be known as the A format. This machine was the first machine to allow broadcast quality in slow motion and up to 2 times normal playback of video. Even though the head-to-tape writing speed was much lower than quad machines (1006 in/s), the video quality was considered better than on quads. What especially helped 1-in machines was the fact that there was no "banding" because of different video heads reproducing different segments of the video field as in quads. (See Secs. 10.1.2.1, 10.1.2.9, and 10.1.2.10.)

At the same time, Sony was working on its own format for a 1-in VTR. In a rare show of cooperation the industry, Ampex, and Sony got together and agreed on a common format for 1-in machines. This format is called type C, and was introduced in 1978. Type C (or simply C) machines had all the attributes of Ampex's A format machine. All these machines require a TBC as an integral part of the system if they are to be used as playback machines. The overall cost for a 1-in machine was less than a 2-in (or quad) machine. One hour of 1-in tape cost less than 2-in tape, plus the 1-in machine was physically smaller than the quad machines they were replacing. These attributes made the 1-in machine the mainstay of broadcast recording and many industrial productions as well.

10.1.1.8 The B format. One additional format was marketed, the type B VTR. This machine was really only marketed by one company, Bosch. It was a 1-in machine that recorded segmented tracks of video for each field (much like a quad). Its quality rivaled or surpassed type C machines, but it had a serious drawback. Due to segmented tracks of video for a field, it required a still store and a complicated capstan servo to perform slow, fast, or stop motion. This machine was popular for a while in Europe but few were sold in the United States.

10.1.1.9 The M format. Around 1980 RCA, along with Panasonic, announced the "hawkeye" VTR. This machine used the consumer VHS tape transport, but that is where the similarity between the two ended. This format also became known as the M format because the tape path in a VHS transport resembled an "M." It was the first of the component VTRs (see Sec. 10.1.2.5), and was small and had much better quality than Umatic VTRs, mostly because it didn't have the color jitter that Umatics have. The M format never got very far, and was supplanted by the M-II format in 1985.

The M-II format was a joint effort between the Japanese Broadcasting Company (NHK) and Matsushita (Panasonic). The M-II format uses a cassette slightly larger than the home VHS cassette. Cassettes can hold from 10 to 90 min of tape. Portable versions of this VTR hold 10- or 20-min versions. The M-II uses a special thin (metal particle) tape designed for optimum performance in M-II machines. This new tape affords an automatic 7-dB increase in the signal-to-noise ratio realized, so that the M-II can record luminance up to 5.5 MHz, have a K factor of 2 percent, and have a signal-to-noise ratio of 50 dB. Due to this special tape, a new transport had to be devised—one not too different than the Beta or Umatic paths. There are 10 heads on an M-II scanner—a pair each for record and playback for each of the luminance (Y) and chroma (C) channels, another four for the "AST" or autotracking of both the Y and C channels, and a flying erase head for the Y and C channels. The M-II format uses a two-track analog recording format—one for luminance and one for chrominance.

Because of all the heads on the scanner, these machines offer the same nice feature that 1-in machines offer, that is, confidence recording. These machines allow the recording of four audio tracks, two of which are conventional longitudinal tracks with Dolby and the other two of which are combined with the luminance and chroma tracks.

10.1.1.10 The Betacam format. M-II was definitely not the only $\frac{1}{2}$-in component format. In 1982 Sony introduced the Betacam, which was based on its consumer Beta transport. In 1987 Sony beefed up the Betacam with a format called Betacam-SP. This format is compatible with the older Betacam, but the SP version uses the same type of metal tape that M-II uses. The Betacam-SP has four audio tracks (Betacam had two). The two newest tracks FM record the audio for increased audio quality on those tracks. The studio versions of the Betacam-SP can accept a normal Beta cassette (which holds 30 min of programming) or a larger cassette that can hold up to 90 min of programming.

10.1.1.11 Early format wars. Of the major networks, only NBC decided to go with M-II. NBC bought $50 million worth of M-II machines by the mid-1990s. It had over 1000 units.

Many people maintain that over multiple generations $1/2$-in VTRs never equalled 1-in VTRs. It has been estimated that applications where five generations or more are needed account for 6 percent of VTR usage. However, other than this, $1/2$-in machines can edit and perform slow, fast, and stop motion as well as 1-in VTRs. The 1-in VTR offers better K factor specs, which equates to less ringing. Also, the amount of moiré is less in 1-in than $1/2$-in formats. This shows up as an edge pattern at low frequencies and a blurring of the high frequencies. The bandwidth of a 1-in machine is 5 MHz wide, with 4.5 MHz for Beta-SP and 5.5 MHz for M-II.

In an effort to "soup up" the 1-in format a few years ago Ampex introduced the Zeus Processor which was a TBC. Ampex claimed that this allowed broadcast quality of 1-in recordings down to the twenty-third generation. This TBC also provides great suppression of jitter and interfield motion artifacts and provides superb manipulation of slow and stop motion playback. At $30,000 the Zeus was expensive. It was estimated that at their peak there were 35,000 type C VTRs around the world.

To add to the analog tape format confusion Panasonic introduced the S-VHS (super VHS) VTR in 1987. This machine used the same VHS transport as consumer units, but it used component video techniques. This system had specs which claimed to produce 400 lines of horizontal resolution and a signal-to-noise ratio of about 47 dB. Although S-VHS horizontal resolution was quite good, chroma artifacts during playback compared to those found in Umatic VTRs.

10.1.2 Technology

Now we will examine some of the technology that made analog VTRs possible.

10.1.2.1 AST heads. AST was an Ampex Corporation term. Sony called theirs dynamic tracking (DT) heads. Ampex perfected the system. An AST head is a video playback head sandwiched between two piezoelectric crystals. When a high voltage is applied to the crystals, they bend. This moves the video head side to side in relation to its normal trajectory down a video track. What use is this? Well, when a tape is played at any speed other than the speed at which it was recorded (such as slow motion), the geometry or relationship between the video track as laid down during recording and the video head playing the tape back is lost. This is because the video track is at a different angle to the edge of the tape than it would be because the tape speed is not the same as it was during record. Thus, in slow motion the video head might start off transversing the center of the video track but will then wander off the track and onto the adjacent track. "Guard-band" noise will wander through the picture. The AST head will bend to stay with the track, eliminating guard-band noise. The AST servo controls the AST head. In order to stay with the video track, the AST servo "dithers" the head slightly from side to side of the track to make sure it stays near the center of the track it is trying to follow.

10.1.2.2 Capstan. The capstan is the steel shaft in the tape path that, along with a pinch roller, pulls tape past the video heads. The capstan controls the speed of the tape. Its main job is to align the video tracks so that they fall in position to have the video heads scan them. If the phase of the capstan is wrong, the video heads will miss the video tracks and guard-band noise will occur. The capstan is controlled by incoming video when recording, and it lays down a record of its movement during record. This record is called the *control track*. During playback, the control track is recovered and compared to an internal or external reference to recreate the exact speed of the tape as encountered in record. Adjusting the tracking knob on a VTR controls the phase of the capstan (tape speed–to–video head relationship) during playback.

10.1.2.3 Color under. This system of video recording allows tapes to be played back straight to a TV receiver or monitor without any correction of tape velocity errors. What are velocity errors? Velocity errors are errors caused by the fact that, upon playback, the video heads scanning video tracks on the tape and the capstan system which is pulling tape through the system vary minutely in speed as compared to the conditions and speed history encountered during recording. These errors do not affect black-and-white information much (picture width in extreme instances), but this problem greatly affects chroma or color information. Color information is derived by looking at a specific frequency (3.58 MHz) and comparing the phase of information found at that frequency with a known standard (color burst). When a VTR plays back a tape and the video heads are at the wrong speed as they scan the video tracks on the tape, all information is played back at a slightly wrong frequency (and phase). This "screws up" all chroma information. To get around this problem, a process known as *heterodyning* is applied. Heterodyning is the process of "beating" or combining two different frequencies. What happens when you do this? The two frequencies are combined along with their sums and differences. If you were to separate the chroma information from the luminance information and you "beat" the 3,580,000-Hz chroma signal with a 4,270,000-Hz reference signal, as one of the by-products you would get a 688,000-Hz (the difference between 4,270,000-Hz and 3,580,000-Hz color information signals) signal with the color information on it. If the luminance information were recorded on a frequency of 2,000,000 to 4,000,000 Hz along with the 688,000-Hz signal, the chroma information would be below the luminance information in the frequency spectrum. Thus, color under. Now on playback if you took the 688,000-Hz color information signal off the tape (with its velocity errors) and "beat" it with a 3,580,000-Hz reference signal, one of the resultant signals would be a 4,270,000-Hz signal containing the color information from the tape (with all its velocity errors). If during a second step you now beat this newly created 4,270,000-Hz signal with a 688,000-Hz chroma signal derived from the chroma information signal from the tape (again with all the velocity errors), one of the resultant signals would be a steady 3,580,000-Hz chroma signal. A monitor or TV receiver would display locked chroma. In reality the luminance would still have its velocity errors but the viewer would not notice them. Umatic and Betamax machines use the 688,000-Hz frequency as the recorded color signal. VHS machines use 629,000 Hz.

10.1.2.4 Comb filter. The comb filter is a circuit that is widely used in VTRs for separating chroma (or color) information from luminance (or black-and-white) information. Video in goes to a balanced bridge and to a one-horizontal-line delay, then on to another node of the balanced bridge. The comb filter works because the luminance information usually changes very little from one line to the next and chroma information would be 180° out of phase from one line to the next (if the chroma information for the two adjacent lines were the same). The luminance is obtained between the one-line filter and the bridge and chroma out another node of the bridge.

10.1.2.5 Component. Component is a recording method used by most of today's VTRs. Generally, the chroma or color information is separated from the luminance or black-and-white information. In the case of most analog component VTRs the chroma after separation is demodulated along the B-Y and R-Y axes. These chroma signals are then time-compressed so that both signals can be combined into a signal channel and occupy the same amount of time. This signal is then FM recorded as the chrominance channel. The luminance information is recorded as the other video signal on the tape. Both the chroma and luminance signals have horizontal sync and a reference burst signal (for phase alignment and jitter correction) incorporated on them.

10.1.2.6 Editing. Today all VTR editing is electronic. Two main types are available—insert and assemble editing. *Assemble editing* is where new video and control track (see Sec. 10.1.2.2) is laid down starting at the beginning of the editing. The end of the edit will have a hole in the video. Insert editing only lays down new video (and audio if desired), but uses the old control track. It leaves no hole in the video information at the end of the edit.

10.1.2.7 Frequency modulation (FM). FM was used to lay video information on tape in analog VTRs only. Digital video bit rates are too high to use a modulation carrier. As we will see later, the bit stream itself, along with channel coding, is used to align magnetic domains on digital videotape. FM is the shifting, or sweeping, of a main frequency in concert with an input signal. In VTRs the input video is made to "deviate" a carrier off its center frequency. The most negative portion of the video signal (sync) drives the carrier or frequency to the lowest possible frequency allowed by the system. The most positive portion of the signal ("white" in the video) drives the carrier to the highest frequency the system allows. FM is needed to record and play back video in VTRs because heads can be made to handle a range of about 10 octaves. Audio, which covers the range from 20 to 20,000 Hz, is 10 octaves wide. A head can handle that range. However, video, which ranges from 30 to 4,500,000 Hz, covers 18 octaves, which is too wide for any head to handle. To get around this the video is made to "deviate" a carrier, around 3,000,000 Hz in many systems, ±1,000,000 Hz. This would mean that the frequency range employed would cover 2,000,000 to 4,000,000 Hz, which equates to only 1 octave, well within the bandwidth of any head. The electronic circuit that performs the deviation in frequency is called a *voltage-controlled oscillator* (VCO). This circuit is built around a component called the *varactor diode*. This diode changes its capacitance as a function of the voltage applied across it. (Also see Sec. 10.1.2.8.)

10.1.2.8 Heads. Heads are the devices that actually impart or recover information from the tape. On videotape recorders there are generally heads that record and play back video, heads that are used to erase information from the tape, heads that record and reproduce audio, and heads that record and play back control information. Some video playback heads are special in that external forces affect their operation (see Sec. 10.1.2.1). The output of a head in terms of voltage depends on the number of turns of wire that comprise the head, multiplied by the rate of change of the magnetic flux on the tape. With everything else constant, the output from a head will double every time the frequency doubles (up 1 octave). This doubling continues until the frequency wavelength equals the gap in the tip of each head. Ideally, the gap in a video head should be about 0.0001 in. Video heads, as a very general rule of thumb, last approximately 1000 h of use. They generally require quite a bit of current to drive them; it is not uncommon for over 50 percent of the total current draw of a VTR in record to be used for driving the record heads. In fact, the tape is required to dissipate the heat that the record heads generate. (See Sec. 10.1.2.7 for additional information.)

In recent years scanner heads have endured a lot of controversy. It has been claimed that head life has dropped, and it is not unusual to have heads go a few thousand hours before replacement while manufacturers often warranty heads for 500 or maybe 1000 h. Many users are also confused as to when a head needs replacement. Early tape machines had predictable head-tip protrusion and head-gap depth. Thus, simple measurement of the tip protrusion or height yielded information about where a particular head was in its life cycle. This is generally not true today. Head-gap depth can vary as much as 20 percent or more, and the bottom of the head gap (the distance above the scanner surface) can vary widely. Thus, simply measuring head-tip protrusion will usually not yield much about the usable head life left. The only dependable way is to optically look at the head tip from the side with a microscope.

However, even this can confuse the uninitiated. Head wear is not linear. Often 80 percent of a head's life is from the bottom 20 percent of the head tip. Heads often will not wear uniformly, setting up turbulence in the tape that affects the performance of the following heads. This leads to an extremely gray area of tape performance. Over the years batches of tape from various manufacturers have, at times, varied between being too abrasive and not abrasive enough. Tape has a lubricant in the magnetic media that usually takes a few passes to reach peak effectiveness. Thus, even on correctly formulated tape, head wear is maximum the first few passes. If there is not enough lubricant or if the magnetic media are simply too coarse, heads will wear fast and possibly unevenly. However, if the tape is not abrasive enough, problems can occur, as the heads will clog easily and develop various coatings, with the result that the tape will not help clean the heads. The backing for the magnetic media, the Mylar, can sometimes have different pliability characteristics that will affect head wear. Sometimes changes in brands of tape will cause the heads to undergo new contouring to conform to the characteristics of the new tape. This can sometimes accelerate head wear.

These tape attributes have led manufacturers to stress the use of cleaning tapes in recent years. The cleaning tape used to be looked down on, but its ability to remove films and stains from the head surface, along with minor head recontouring, often requires its use.

10.1.2.9 Scanner. The scanner is the circular "drum" that contains the video head(s). Scanners contain one video head (plus record, flying erase, and three sync heads) on 1-in VTRs. VHS and Beta machines have two video heads (some models also have additional video heads for slow speed, or search, of flying erase heads). Umatics generally have two video heads and two flying erase heads. The tape is pulled past the video heads at a slant or angle from the trajectory of the video heads. Thus, video tracks are slanted across the tape. The scanner spins at 60 r/s. This means one video head will lay one complete video field down on one video track, while the other head will lay down the next track. In record the scanner is controlled by incoming video. In playback the scanner is made to spin at the right speed by comparing a tach signal generated by the spinning scanner to a 60-Hz reference. The phase of the scanner (such that the video heads start scanning the video track at the proper time) is controlled by comparing off-tape video (the vertical interval) with reference video (its vertical interval). The flying erase heads mentioned earlier precede the video heads generally by 120°. These heads are used for editing. They allow exact pinpointing of a new recording's start. Some machines have a separate record head. In C type 1-in machines the record head allows a head to record video while at the same time allowing a head 120° behind it on the scanner (playback head) to immediately play back what was just laid down. This lets you confirm that the record heads are working, and is known as confidence recording. Thus, the separate playback head is sometimes called the confidence head.

10.1.2.10 Servos. There are generally three, and sometimes four, servos in a VTR. These are known as feedback or closed systems. A servo takes outside control information and combines it with information coming from the device to be controlled and uses it to effect control of the device being controlled. Servos can be mechanical or electric. Servos in VTRs control the scanner and the capstan, both of which are electronic. Some consumer machines combine these two servos. Most machines have servos to control supply and take-up reels. In consumer and most Umatic machines they are mechanical. In industrial and broadcast machines they are electronic. A fourth servo in many broadcast machines is the AST or DT servo, which is electric. (See Sec. 10.1.2.1 for information on this.)

10.1.2.11 TBCs. Early TBCs relied on analog techniques, which used varactors to vary the RC time constants of the video path. Later TBCs all relied on digital technology. A TBC is used to do what its name implies, correct time-base errors imparted into the video on playback. Since the scanner servo is a mechanical system, it will vary slightly in speed on playback, which means the video information will vary in frequency as well. This variation normally causes only minor errors in the black-and-white portion of the video, but it causes major errors in the color information, since the color information is contained as phase differences in a single frequency (3,580,000 Hz). If tape speed is off at all, not only will the phase information in this color frequency be off but also the color frequency might not be seen as the right frequency at all and thus not recognized as color information. Other black-and-white frequencies close to this color reference frequency might be mistaken as the color frequency. The TBC samples incoming frequency at either 3 or 4 times the rate of the color reference signal (3,580,000 Hz). These samples are converted to digital bytes of information (either 8 or 9 bits wide) and stored in computer memory. External reference signals then control the reading out of the data bytes from memory. These bytes are then reconverted to analog video and fed out of the TBC. Thus, video goes into the TBC at a rate determined by the output rate of the VTR, but comes out of the TBC at a rate determined by the external reference signals from the "house." Early TBCs had expensive, large analog-to-digital and digital-to-analog conversion systems and limited amounts of memory, usually only enough to handle three to six horizontal lines of video. As memory has become much more compact and cheaper, many TBCs today can store an entire frame of video.

10.1.2.13 Analog VTR specifications. Here are the analog formats compared to a few digital formats to convey why digital VTRs are quickly replacing analog VTRs.

Effective video head writing speeds

Quad	38.1 m/s
Type C	25.5 m/s
Umatic	5.2 m/s
Betamax	4.0 m/s (fast speed)
VHS	3.3 m/s (fast speed)
D-2	27.4 m/s

K factors

Type C	1 percent
Umatic-SP	Less than 3 percent
Betacam	2 percent
M-II	Less than 2 percent
D-2	Less than 1 percent

Video bandwidth

Type C	5.0 MHz
Umatic	3.3 MHz
Umatic-SP	4.3 MHz
Betacam	4.5 MHz
M-II	5.5 MHz
S-VHS	5.0 MHz
D-1	5.5 MHz
D-2	6.0 MHz

Video signal-to-noise ratio

Type C	49 dB
Umatic	47 dB
Umatic-SP	47 dB
M-II	49 dB
S-VHS	47 dB
D-1	56 dB
D-2	54 dB

Differential gain

Type C	2 percent
Umatic-SP	Less than 3 percent
M-II	Less than 2 percent
D-2	Less than 2 percent

Differential phase

Type C	2°
Umatic-SP	Less than 3°
M-II	Less than 2°
D-2	Less than 1°

Luminance/chrominance delay

Type C	20 ns
Umatic-SP	Less than 25 ns
M-II	20 ns
D-2	Less than 10 ns

Moiré

Type C	40 dB
D-2	0 dB

Video track widths

Betacam	86 μm
M-II	40 μm

10.1.3 Current formats

Today more tape formats are available than ever. Although SMPTE has sanctioned many formats in the last 10 years, most are pushed by only one or two manufacturers. Each brings some technological slant on how to build a VTR. However, there appear to be more offerings than the market can absorb. Digital VTRs can eliminate the effects of multiple generations and dropouts due to tape wear. Some of the formats use various levels of compression and some don't use any. The noncompressed formats tend to be more expensive than their compressed counterparts.

Digital VTRs process video and audio differently than analog VTRs. First, once the video and audio data are inside the VTR (many digital VTRs convert analog video and audio to digital first), many VTRs will first compress the video. Most use JPEG compression techniques, although MPEG compression is beginning to be used. Once the video is compressed, it is thereafter treated like data. Error-correction code (ECC) is added to the video. The resulting blocks of data are then shuffled to ensure that large dropouts do not wipe out contiguous blocks of data but instead blocks from various areas of a field. This allows enough recovered data on playback to use the ECC to totally reconstruct the lost data. After block shuffling, the audio data are often multiplexed into the shuffled video bit stream. Audio data are usually compressed in time to fit within the time frame provided the audio. Now the combined data have another layer of ECC added to the stream. This second ECC is known as the inner ECC, while the first ECC is known as the outer ECC. From here the data go to a channel coder. The channel coder takes the place of the FM modulator in an analog VTR. The purpose of the channel coder is to create enough edges in the final bit stream going to the record head to make the bit stream self-clocking and to conform to the frequency spectrum allowed by the record/play heads. Also, dc components in the signal must be eliminated due to signal path, head, and tape limitations.

When played back, either correction or concealment can repair digital video that has missing data. Correction occurs when the missing data's value is determined using ECC. The video data coming out of the VTR are exactly the same as the data fed in. Concealment occurs when too many data are lost for the ECC to be able to correct. In this case concealment circuitry looks at data spatially around what was lost and replaces that data through interpolation. This technique is possible because of the shuffling of spatial data before the data were laid down on tape. Because concealment techniques are not possible with the audio, the audio is generally multiplexed in with the video, so that the audio data end up in the data track such that they are recorded in the middle of the tape. Tape damage is less likely to occur at the tape's center than along the tape's edges.

Digital VTRs have the same general servo requirements as analog VTRs, that is, scanners (drum with record/play heads) must rotate at the correct velocity and phase so that the record/play head rides down the center of the data track on the tape. This means that the tape must be pulled at the correct speed past the scanner by the capstan. Also, the reel servos must produce the correct amount of tape tension so that the tape is not distorted by incorrect hold-back tension coming out of the supply reel.

Information density laid on tape is much greater for digital VTRs than for analog. The *track pitch* (distance from the center of one helical tape track to the next) is extremely small. It ranges from 32 μm for Betacam SX down to 15 μm for DV-CAM. The DV-Pro series machines are at 18 μm, while most of the rest are between 20 and 22 μm. As a comparison, analog Betacam had a track pitch of 80.5 μm. The digital VTR designer is faced with a number of trade-offs. Narrow track pitches require more robust ECC and channel coding schemes. Narrower track pitches have a lower signal-to-noise ratio than desired for analog operation but playback is unaffected as long as it is above the digital error cliff. To ensure correct tracking with narrow tracks, some VTRs lay pilot tones at predetermined locations in the track that are used by the VTR's capstan servo, along with traditional control track pulses, to center head travel down each track.

Track pitch can be increased by increasing tape speed, but then thinner tape will have to be used to allow the same amount of recording time as comparable size cassettes. Scanner speed can be increased to pack more data onto the tape, but mechanical complexity and power consumption would increase. Speed, and thus power and noise, needs to be as low as possible in the field scanner. In fact, Sony's digital Betacam field acquisition camcorder uses a slower scanner angular velocity than its studio counterpart. This is accomplished by time-expanding the data laid down in the field and recording multiple tracks on tape simultaneously. The trade-offs result in a final minimum wavelength laid down on tape that is between 0.5 and 1 μm.

Many digital VTRs can also play back legacy analog tapes. D5 can optionally play back D3, digital Betacam and SX can optionally play back analog Betacam-SP, D9 can play back S-VHS, and DV-CAM and DV-Pro can play consumer DV. In some cases the scanners are different sizes, such as digital Betacam and analog Betacam. As a result of powerful TBCs and DT head technology, these differences are overcome. Often head rotation in the digital VTR is not the same as the television field rate any more. Digital Betacam's drum rotation rate during digital playback (analog playback is still ~30 Hz) is 90 Hz. Sony's SX has a drum rotation speed of ~75 r/min, while DV-CAM's is as high as 150 r/min. Many, such as digital Betacam once again, do not lay down an entire field in one track. Digital Betacam has two heads side by side with different azimuths that require three passes (six total tracks) to record one field of data. D5, which lays down data at more than double the rate of digital Betacam, uses four side-by-side heads and three passes (four for 625/50), for a total of 12 tracks/field (16 for 625/50). Both formats have two sets of record and playback heads on opposite sides of the scanner. To make three swipes across the tape per field requires scanner rotation rates of 90 Hz.

Digital recorders bring other new tricks to the table. One is called *preread*. Preread allows a single digital VTR to both play back and record at the same time. The play head reads data off the tape, which is fed as audio and video out of the VTR. This video can be manipulated and immediately sent back to the VTR to be recorded. This process will allow effects such as layering to be done with a single digital machine. Some digital VTRs can also record and play back video at rates faster than real time.

Some VTRs that perform compression will allow a few lines in the vertical interval to stay uncompressed so that closed captioning and test signals are recorded without interference. A few will allow most of the vertical interval to be passed noncompressed.

10.1.3.1 D1. The D1 was the first widely used digital VTR. There was limited experimentation with digital type C VTRs but these never made it to market. Both D1 and D2 are 8-bit devices and both use cassettes that are 19 mm ($^3/_4$ in). D1 was the first of three digital formats that used 19-mm tape. D2 and Ampex's DCT format also use the same width tape. D1 machines can play back/record both 625 and 525 because both have the same sample rates under digital component. Sony introduced the first D1 machine (DVR-1000) in 1987. This VTR accepted CCIR-601 component video. In a nutshell this is video comprised of a series of digital samples. This signal is made up of data bytes (8 bits wide) arriving at a rate of 27 Mbytes/s. Half the samples contain luminance information; the other half are one of two chroma difference signals. There are four luminance samples for every two of each chroma difference signal. Hence this is also known as 4:2:2. Luminance is sampled at a 13.5-MHz rate. Each chroma signal is sampled at a 6.75-MHz rate. This 6.75-MHz sampling rate for the chroma means that a chroma bandwidth of 3.37 MHz is possible, allowing high-quality chroma keying. The D1 machine uses 19-mm ($^3/_4$-in) tape in a cassette. The tape uses a cobalt-doped gamma ferric oxide coating. A cassette is capable of holding 76 min of program material [94 min if thin (13-mm) tape is used]. This VTR directly records the digital 4:2:2 signal as digital information on the tape. The result is excellent video quality and very little quality degradation over multiple generations. A major disadvantage to this machine is that it is very expensive. It listed around $140,000.

The D1 VTR had a fixed upper drum with a headwheel that rotated through a slot in the drum cylinder. The headwheel contained all the rotating heads. It was replaced as a complete unit. D1 records four channels of audio at 16 bits of sample resolution at 48 kHz. The video S/N ratio is 56 dB, with a K factor of <1 percent. Since it is a component device, differential phase/gain and Y/C delay values are not issues. Moiré is zero.

10.1.3.2   D2.

The same year that Sony offered its first D1 VTR, Ampex offered the D2 format. This format handles video much like a TBC. It takes incoming analog video and converts it to 8-bit digital bytes at a rate 4 times that of color subcarrier. It then records these bytes on tape. On playback it reverses the process and outputs regular analog video. Its major advantage was that its cost was not much more than high-end type C machines. D2 was much cheaper than D1. As a result, D2 greatly outsold D1. However, over time other VTRs offered component recording, and these machines were more reasonably priced. Composite VTRs are losing out now to component VTRs. The fact that all high definition is component will only hasten the retirement of the composite VTR. In 1988 Sony offered a D2 machine. Ampex and Sony both supported D2 development. Since 4Fsc is on different frequencies for PAL and NTSC, there are PAL and NTSC versions of D2 and D3.

The D2 format also uses 19-mm ($^3/_4$-in) cassettes, much like the D1 VTRs. The magnetic medium is different in that it uses a high-coercivity metal-particle tape. D2 is much more conservative in its tape usage because D2 tape speed was 131.7 mm/s versus 286.6 mm/s for D1 and also because D2 tracks have no guardbands. A cassette can hold up to 208 min of program material. Since the D2 composite format had a bit rate of 144 Mbits/versus 270 Mbits/s for D1, everything could run slower and, therefore, more cheaply than D1 VTRs. This machine can perform slow motion and still playback. Ampex claimed that it produced broadcast quality video out down to 20 generations. D2 VTRs consumed less power and weighed less than D1. As with all the current tape formats, with the exception of D1, D2 had a rotating upper drum that contained the heads. Some D2 machines allowed the individual replacement of those heads.

Video S/N is slightly worse than D1 at 54 dB. It matches the D1 K factor at <1 percent. Since it is digital, this composite machine beats analog composite in differential phase (<2 percent) and differential phase ($<1°$). Like the D1 machine it has zero moiré.

10.1.3.3   D3.

D3 was Panasonic's answer to D2. It is also a composite (4Fsc) device but it used 13-mm ($^1/_2$-in) tape and the same magnetic media as D2. The D3 VTR is an 8-bit VTR.

Power consumption and machine weight were greatly reduced as compared to D1 or D2. Tape speed was further reduced to 83.9 mm/s. This was slower than even analog Betacam, which pulled tape at 118.6 mm/s.

Note: There is no D4 as the number 4 carries some unlucky implications in some Asian languages.

10.1.3.4 D5. D5 is Panasonic's answer to D1, a component system that uses 13-mm ($^1/_2$-in) tape, the same as D3. In fact, the transport is the same used on the D3. This machine recorded 10-bit video using no compression. The data rate to tape is 288 Mbits/s. The D5 can play back D3 tapes.

D5 pulls tape at 167.3 mm/s. It weighs 80 percent more and consumes 50 percent more power than D3. D5 can also be used to record HD. To do this, an MPEG compression/decompression unit is electrically placed around the D5 VTR. SDTI is sent and received from the D5 VTR. This external box accepts HD 4:2:2 directly at a luminance rate of 74.25 Mbytes/s and each chroma difference signal at half the luminance, or 37.125 Mbytes each. This equals the SMPTE 292M 1080i bit rate of 1.5 Gbits/s. Since SDTI has a payload of 207 Mbits/s, the external box performs at (1.5 Gbits/s at 207 Mbits/s) just over 7:1.

10.1.3.5 D6. The D6 format is the only full bit rate HD VTR proposed so far. It was developed by BTS and Toshiba and uses 19-mm cassettes mechanically identical to D1 and D2, but with a different tape formulation. This format would lay down eight data tracks (one cluster) per pass of the record heads. Each track would be able to handle 150 Mbits/s, resulting in an aggregate bit rate of 1.2 Gbits/s. This corresponds to the 1.25 Gbits/s necessary for the active video region of 1920H × 1080i HD. To make room for these tracks the tape speed would be very high, 497.4 mm/s, 30 percent faster than the 2-in quad VTR's tape speed (381 mm/s). Track pitch would be 21 mm, with the pitch between clusters at 176 mm.

10.1.3.6 D7 DVCPro. The D7 DVCPro format is based on the consumer DV format, but it is beefed up and intended for ENG applications, according to Panasonic. Many mid-sized markets will likely be tempted to use it as a production format due to its relatively low cost. This format comes in a number of flavors generally based on the rate at which data are recorded onto tape. All three types use 6.35-mm-wide metal-particle tape. Regular DVCPro uses a 4:1:1 sampling structure at 8 bits. It pulls tape at 33.8 mm/s and lays data down on tape at 25 Mbits/s. It uses 5:1 JPEG intraframe compression. This is a machine that can play back tape at a speed faster than real time (4 times). It records two 16-bit audio channels sampled at 48 kHz. DVCPro can play back DV-CAM tapes.

DVCPro 50 lays down data at 50 Mbits/s but at double the tape speed of DVCPro (67.6 mm/s). It has a 4:2:2 video sampling structure at 8 bits. It also uses JPEG intraframe compression but has only a 3.3:1 sampling ratio. It records four 20-bit audio channels, also at 48 kHz.

Panasonic currently has a 100-Mbits/s speed to tape VTR in the works (DVCPro 100). It is intended for the HD market. Currently, none of the DV-based VTRs (including DV-CAM) has preread capability.

10.1.3.7 D9 digital-S. The D9 digital-S format is based on JVC's S-VHS format. In fact, a D9 (or digital-S as it is also known) machine can play back S-VHS tapes. D9 is intended by JVC to be a general-purpose format. Many facilities currently using S-VHS might find this a natural progression into digital VTRs. Its sampling structure is 4:2:2 at 8 bits. It also records four 16-bit audio channels at 48 kHz. These machines can play back at a rate of 2 times real speed. Tape speed is 57.7 mm/s. It uses a 3.3:1 JPEG compression scheme and records data on tape at 50 Mbits/s. It can record up to 124 min on a cassette, and has preread capability.

10.1.3.8 Digital Betacam. Digital Betacam uses a mild 2:1 JPEG compression scheme. It records 4:2:2 component video at 10 bits. Data rate to tape is 127.8 Mbits/s. Digital Betacam uses $1/2$-in tape, and moves it at 96.7 mm/s. It is fairly power stingy and light compared to its competition. This VTR has preread capability. It records four 20-bit audio channels sampled at 48 kHz. As an option, this machine can play back analog Betacam tapes. This machine is intended by Sony to handle the upper-end production chores in a facility.

10.1.3.9 DCT. Like Digital Betacam, Ampex's digital component technology (DCT) VTR uses 2:1 but it records 8-bit video. DCT uses 19-mm-wide tape at a rate of 131.7 mm/s (146.5 mm/s for 625/50), so 3-h recordings are possible. Ampex claimed that recordings down to 30 generations were possible. After D1 and then D2, DCT along with D5 are the most power-hungry digital formats. DCT machines are heavier than most other VTRs as well. The tape transport, though, might be the most elegant and easiest to service. It uses pinch rollerless and frictionless air-lubricated tape guides. This was the first VTR to sport a floppy drive to facilitate upgrades.

10.1.3.10 Betacam SX. The Betacam SX is the first "hybrid" VTR as it includes a disk drive for video/audio besides the standard VTR tape transport. It, therefore, has both linear VTR and nonlinear editor attributes. The base system has two hard drives (RAID-3 protected) that store 90 min of material. Over 6 h of external hard-drive storage can be added. Material on tape can be copied onto the hard disk at a rate up to 4 times real time. It uses MPEG-2 4:2:2@ML (I and B frames only—GOP of 2) 10:1 compression. The video is sampled at 4:2:2 at 8 bits. It also records four audio channels sampled at 16 bits. SX passes the majority of the vertical interval through uncompressed. This format is intended by Sony for ENG applications.

Data are laid down on tape at 18 Mbits/s. This low bit rate is intended so as to let SX data move between machines through DS-3 type communications channels. Tape speed is 59.6 mm/s, and SX uses $1/2$-in tape. The data rate and tape speed allow for a wide-track pitch, 32 μm. This wide-track pitch also allows SX to eliminate the dynamic tracking head. Two staggered rotary heads cover an area wider than the track. The head receiving the strongest rf signal is used. Another neat feature of this machine is that it is possible to record and play back separate material simultaneously. Faster than real time playout can be achieved via tape playout but not via the hard disk.

10.1.3.11 HDCAM. The HDCAM format is from Sony and it is the first portable HD recorder. It uses $\frac{1}{2}$-in metal-particle tape. Tape speed is 96.7 mm/s, and up to 124 min of HD material can be recorded. Like the SX format before it, much of the mechanics and electronics are based on digital Betacam. The bit rate to tape is 140 Mbits/s. It samples video at 3:1:1 at 8 bits. The actual luminance sample rate is 50.625 Mybtes/s; each chroma difference channel is 16.875 Mbytes/s. This produces an uncompressed bit rate of 675 Mbits/s. Therefore, compression (675/140) is 4:1. It records four channels of 20-bit audio sampled at 48 kHz.

A current debate is ongoing between the HD-D5 and HDCAM video processing approaches. While HDCAM subsamples the video up front, its compression ratio is less. HD-D5 does not subsample up front, but it has a greater compression ratio.

10.1.3.12 DV-CAM. DV-CAM is Sony's DV entrant. It samples 4:1:1 at 8 bits and lays data onto tape at 25 Mbits/s. Compression is JPEG at 5:1. It pulls tape at 28.2 mm/s, and uses $\frac{1}{2}$-in metal-evaporated tape. Sony intended this format for the industrial user. DV-CAM cannot play back DVCPro because its track pitch is not as wide as DVCPro's, but it can play back DV-consumer (track pitches in DV-consumer are 10 mm, while DV-CAM's are 15 mm and DVCPro's are 18 mm).

10.1.3.13 DV-consumer. The DV-consumer machines are available from a number of manufacturers and generally record up to 3 h on the same tape used by DV-CAM. Track pitch is much less than the professional formats, 10 μm for standard play and only 7.7 μm for extended play. Tape speed is also much lower, 18.8 mm/s. Sampling is 4:1:1 at 8 bits and it uses JPEG 5:1 compression. JVC introduced a dockable VTR to portable cameras that allows regular DV recordings that can then be played back on either DC-CAM or DV-Pro.

10.1.3.14 Data recorders. Many in the industry have clamored for simple tape-based bit buckets, or a tape transport that accepts data via SCSI or Fibre Channel interfaces. These machines couldn't use any lossless compression, so they would have to be devices with high recordable data rates. Noncompressed video VTRs are often referred to as "rasterized bit buckets." Video transports such as D1 or D5, which have high throughput, have been given SCSI interfaces to handle data rates as high as 23.8 Mbytes/s for D5 decks and 19.0 Mbytes/s for D1. This compares favorably with many RAID throughputs. Cassettes that would be used to store 2 h of video could instead hold approximately 125 Gbytes of data. Other transports, such as Sony's digital Betacam, have been transformed into bit buckets. These data recorders will continue to gain importance as bit streams such as MPEG, which has no relationship to the classic television field, need to be achieved off video servers. If the tape transport is treated as simply a bit bucket, format issues at the I/O

Figure 10.6 A cross between tape- and disk-based technologies, this data recorder, an offline storage device, stores files not currently needed on a video server on data tape. It can hold hundreds of hours, or, more exactly, terabytes, of multimedia data.

of the transport will no longer matter as everything is simply data wrapped in SCSI and maybe Fibre Channel or P1394 wrappers. The only format issues will be how that data are organized and laid down on tape (Fig. 10.6).

10.2 The Video Server

Anyone contemplating the design of a facility today will most likely base his or her design around one or two basic building blocks. In recent years the router has been the cornerstone of facility design. Now the video server has evolved to a point where it will be the centerpiece of many new installations. Not that the router is no longer a central piece of infrastructure; the server has recently matured to the point where it makes sense to organize some facilities' workflow around it. The video server and the router are often tightly intertwined with one another in that a router functionally wraps a server, or set of servers, around it. That router is often controlled by the same system that controls the server.

Many technologies have come together to make today's multimedia server possible. Some of these are the same advances that have pushed us into DTV, mainly the rapid progress by the compression and microprocessor/DSP industries. However, the strides that the disk drive manufacturers have made are just as important in the physics of producing drives with greatly increased data density. A lot of the progress in that area is due to the breakthroughs in READ/WRITE head technology. The drive vendors have also put DSP to use to help support the breakthroughs realized in head design. Finally, the RAID has made the server viable along with the accompanying improvement in server and disk drive interface methods.

10.2.1   Tape versus disk

This argument is still very much valid. It is well known that disk-based editing can be much quicker than tape because of the nonlinear techniques that can be used. In addition, large storage capacities are becoming more affordable. The fact that the average spot inventory in a medium-sized station is under 20 h has now made servers viable for most broadcasters. However, often the way processes are handled and the pattern of workflow must change to accommodate the integration of a server into the plant. Depending on system architecture, it is possible to end up with a server system where one catastrophic event could extinguish the entire commercial spot or news story inventory. Losing one drive in a RAID and not catching it before a second drive quits is one such scenario. Many systems maintain library databases completely separate from the data stored on the RAID. Losing that database will, in some instances, render the still intact RAID data unusable.

Short of a fire it's almost impossible to lose an entire tape library. However, tape means videotape machines and often larger systems that house the VTRs to automate playback. VTRs require maintenance and often the replacement of very expensive parts. However, the cost of storage per bit of data, while dropping rapidly for disk, still favors tape storage. In short, tape has its applications and so does the server. Most will agree that tape will slowly diminish in scope while disk-based storage will grow.

This change is evidenced by two approaches to the server and tape marriage. The first instance involves server systems that use disk storage for what can be referred to as near-term storage, that is, news stories that will play today or this week or commercial spots that are still in today's or future program logs. For long-term storage data would be offloaded from the near-term disk storage to data recorders. Data recorders can achieve sustained data transfer rates of over 15 Mbytes/s using SCSI interfaces. Many of the data recorders in use today have evolved from their video recorder brethren. In fact, many use the same tape transport and servo circuitry.

A second coupling of tape and disk is the so-called hybrid VTRs. These machines integrate a regular VTR with an internal disk drive. A number of these machines can be combined to effect a larger virtual server. These machines have video/audio connectivity through normal television routing paths, but they are also connected via a computer network such as Ethernet. Users can browse the content on the various VTRs and request that copies of material be sent between machines. The material being copied goes through normal video and audio paths, while the control commands are sent over the computer network in use. The video and audio material can often be sent at rates faster than real time using the SDTI digital video format or the ASI data format.

10.2.2   Server applications

Like DTV, servers mean different things to different people. Servers can be used as simple replacements for the VTR, and these are generally known as *digital disk recorders* (DDRs). Many mimic the VTR from both an operational and interface standpoint, and have RS-422 interfaces that talk some of the common VTR protocols. These systems are good for things that play out often. They also find use in time-shift applications. This ability is a welcome relief to many Mountain time zone affiliates that must provide their own network delay. Many network affiliates also use this capability to delay network fare, noticeably at 11:30 p.m. so that they can extend the local newscast. The amount of storage, along with the compression rate, will determine how much time shift is available.

From simple VTR chameleons many DDRs morph into more ambitious creatures. Many DDRs can produce multiple playout streams. While some have internal storage, many have external storage, including RAIDs. Some DDRs even have SCSI, Fibre Channel, and Ethernet connectivity. Since current HD DDRs generally have only a single record/playout channel, they are often referred to as glorified VTRs.

Frequently, the delineation between the DDR and the full-blown server is indiscernible. Systems that store and manage commercial and news story inventories are generally thought of as servers, but server systems can consist of a combination of DDRs and one or more servers to form large systems. This is the case with a number of vendors who offer news and enterprise servers. To some users the DDR is a server. Most stand-alone DDRs used for commercial spot playback are referred to by their owners as servers. A DDR used in an edit bay is most likely referred to as a nonlinear editor. However, that DDR could be a part of a larger server system. Generally, large, higher-capacity DDRs that have client DDRs are properly referred to as servers.

The very term video server is rapidly becoming obsolete. First, they not only deliver video but also audio. Some can be made to act as still stores. However, additional development is occurring that will result in these devices being called multimedia servers. Currently, a number of vendors have file transfer and sharing protocols that allow not only the sharing of video/audio but data about that video/audio. These data about video/audio are known as metadata. The protocols break the file into objects that can be manipulated individually. These objects can be split up and operated on at different workstations simultaneously. With the introduction of MPEG-4, this evolution will continue. MPEG-4 also breaks video scenes and audio passages into objects. No longer is compression based solely on spatial and temporal relationships; instead the baseband scene is considered a collection of objects. Objects could be a talking head, the background wall, a chart, etc. Most objects in a scene will still have spatial and temporal compression performed on them, but that compression will be done without reference to the other objects. At the receiving end, these objects will be reassociated and layered back together.

10.2.3 Server architecture

Figure 10.7 depicts the most common server architecture.

The encoding/decoding subsystems (see Fig. 10.8) often occupy the same card or box. Each pair is usually called a channel. The storage control subsystem is the gatekeeper of the actual storage system, which is usually some form of RAID (array). A control system, usually a PC or workstation, sits on top of the storage control box and usually the encoders/decoders. Some systems have enough separate control tasks and user applications occurring simultaneously that multiple networked PCs are required. Connection between the PC and the storage control box can be anything from Ethernet to RS-422. Compressed data to and from the encoder/decoder is often SDTI or ASI. Connection between the storage control and the RAID array is usually SCSI, but Fibre Channel is gaining favor. Even if Fibre Channel is used to connect to the RAID array, internally the RAID controller to individual disk drives that comprise the array is usually still connected with SCSI.

Although some servers can handle uncompressed video streams, most apply either JPEG or MPEG compression schemes into and decompression out of the server. In addition to the fact that uncompressed video consumes great amounts of disk space, it also consumes internal server bandwidth. The storage system can only write to and read from the disk at a finite rate.

A number of elements define the bandwidth or throughput of a server system. The first is the compression ratio. JPEG is still used in many systems

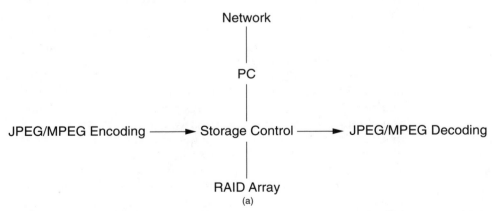

Figure 10.7 (*a*) High-level server architecture. (*b*) Schematic diagram of audio/video server (Grass Valley Group's Profile XP media platform). (*c*) Major components of audio/video server (Grass Valley Group's Profile XP media platform).

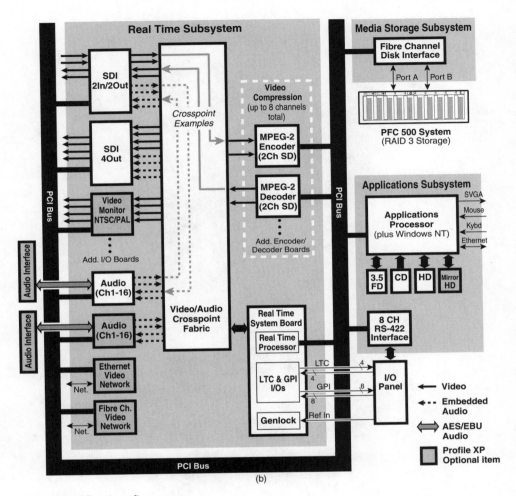

Figure 10.7 *(Continued)*

because it allows for editing on every frame since, in essence, every JPEG frame is an MPEG I frame. MPEG is used in many servers where editing is not an issue, such as spot playback servers. MPEG provides a four- or fivefold compression efficiency over JPEG. The more the bit rate can be knocked down via compression, the less reading and writing have to be done to the storage system. Lower internal server bit rates mean that more clients or consumers of server I/O capacity can be served simultaneously. Many systems let you select the trade-off to be made between compression (video quality) and the number of I/Os in use.

All I/O into the server generally ends up being time multiplexed for transmission to and from the storage system. This means that internally video must be moved at a rate faster than real time. Until recently, individual disk drives

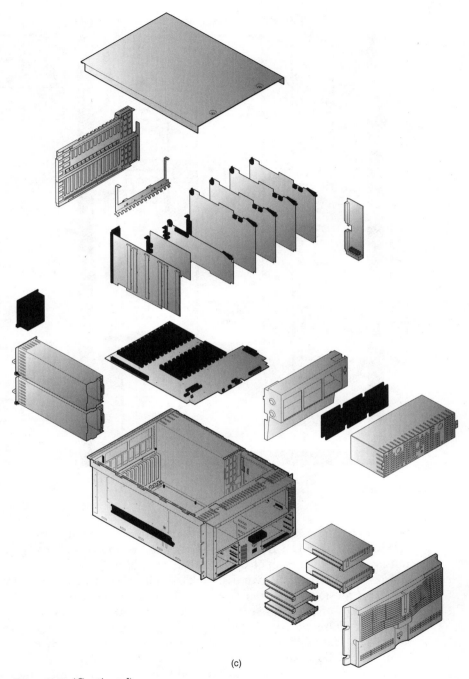

(c)

Figure 10.7 (*Continued*)

Figure 10.8 Here is an SD baseband (270 Mbits/s) disk-based server with external SDTI encoder and decoder so that it can also record 1.5-Gbit/s HD video.

could handle between 5 and 10 Mbytes/s of I/O throughput. Drives will soon be available that can handle rates close to 50 Mbytes/s. RAIDs allow throughput higher than for individual drives, but as drives are added to an array, a point of diminishing returns is reached. RAID throughput using SCSI often maxes out at around 25 Mbytes/s. As the throughput need is increased, additional complete RAIDs have to be added. Currently, the most common system for tying multiple RAIDs into one large server storage system is via Fibre Channel. The RAID redundancy scheme in use will tie up some bandwidth. Internal available server bandwidth is often 30 Mbytes/s or higher; some are above 50 Mbytes/s.

A number of other elements conspire to limit throughput. Most disk interfaces have transmission overhead. The most common, SCSI, has a 20 percent overhead. One-fifth of the bytes sent over a SCSI bus are for handshaking and control. If Fibre Channel is used, the overhead is even greater because Fibre Channel is essentially carrying SCSI traffic, albeit serially and at a much higher rate. Fibre Channel also uses 8-bit/10-bit channel coding, which further shrinks throughput by 25 percent. This is partially why 1-Gbit Fibre Channel seldom has actual throughput higher than 400 Mbits/s.

10.2.3.1 Video formats. Many servers still accept NTSC video, but most immediately convert the video into one of the digital component formats. Once in the digital component domain, many systems reduce the amount of succeeding compression needed by subsampling the chroma. While SMPTE 125M, which specifies the standard 4:2:2 digital baseband component video in use throughout the plant, undersamples chroma by one-half versus the luminance, this subsampling is only in the horizontal direction. To further reduce bit rates, many servers also undersample chroma in the vertical axis as well. Hence, many systems use 4:2:0 sampling structures, which immediately reduces the active video bit rate by 25 percent. To save on bandwidth and disk storage, most servers only record the active video portions of the signal. The ratio 4:2:2 has 207.4 Mbits/s of active video data, while 4:2:0 reduces that rate to 155.5 Mbits/s.

10.2.3.2 Compression formats. Once chroma subsampling has taken place, compression is usually performed. The three main compression standards all use DCT, where video samples in the time or spatial domain are taken to the frequency domain. While this alone doesn't provide any compression, the next step does. The frequency components generated, of which there may be more than the original spatial samples, can be scaled or multiplied by some value or scaling coefficient. The value of this coefficient determines the amount of compression. By scaling the frequency coefficients generated by the DCT process, many of them essentially become extremely small. The small ones are thrown away, and this is lossy compression. We throw away some high-frequency information, but hopefully not enough to notice. The next step in the process is to order the coefficients in such a way that lossless processes, such as run length and Huffman coding, can be applied, further reducing the bit rate. This is essentially what JPEG and MPEG do when they produce I frames. JPEG-type compression up to 4:1 is considered essentially transparent. Video out of the server will virtually look the same as video into the server. Compression ratios of 8:1 or even 10:1 will compare with the best analog studio VTRs at a single generation. At 10:1 compression, we have 20.7 Mbits/s for 4:2:2 video and 15.6 Mbits/s for 4:2:0. MPEG takes this process one step further by adding temporal compression to the spatial compression that JPEG employed. Anchor I frames are sent only so often, interspersed with frames that only have information concerning motion and compressed information about pixels that have changed values. There are two types of these, B and P frames. There is also a hierarchy of dependence among the frames. P frames rely *only* on previous I or P frames. B frames rely on preceding and succeeding I and P frames. Due to this interdependence, the order of generation and the actual order of intended display of these frames are not the same. This means that encoding takes place over many frames. Thus, MPEG encoders are more expensive than JPEG encoders. These groups of I, B, and P frames are organized as groups of pictures (GOPs). Every time an I frame is sent, a new GOP starts. Some MPEG-2 profiles support all three types of frames, while some others only support I and P frames. The MPEG 4:2:2 profile does allow for only I frames to be sent. Long GOPs (a common one is 15) allow for higher levels of compression, while shorter ones achieve less compression but allow easier editing because edit-friendly I frames occur more often. Some systems use GOPs of 2, sending alternate I and P frames.

10.2.3.3 File formats. Some lower-end systems store the video as GIF, TIFF, or EPS. Some organize JPEG or MPEG data into larger file structures such as AVI, QuickTime, or OMF file systems. Some systems will treat each "clip" as a separate file, while some systems create a single large file that holds all "clips." The second method is used for several reasons. The first reason is that some systems are organized using a legacy VTR program management style, an approach where all RAIDs are treated as one virtual drive. Time code values are assigned to the entire disk storage area. The time code is broken down into blocks, with lengths such as 10 or 15 s. A 1-s clip would consume an entire block. This approach allows no clip or program fragmentation. Enough consecutive free blocks must be available to accommodate a new clip being inserted into a vacated area. The second reason this approach is used is that it requires a simpler file structure, video and audio are easier to synchronize, and the file is not fragmented so the average disk seek time is reduced, resulting in a higher system throughput. While SCSI drives might sustain transfer rates below 1 Mbyte/s with reads of random blocks only 1K long, transfer rates of 8 MBytes/s would be realized with 64K-long sequential block reads. Even if the read cycle were 64K blocks long, sustained transfer rates of over 2 Mbytes/s would be unlikely if the blocks were randomly read. Regardless of which file format is taken, a single- or multifile approach, many systems use external PC/workstations to maintain a database for managing the file system on the RAID. As mentioned earlier, if this database is lost or becomes corrupted, all programming could still be lost. As a result, many systems and experienced users will install a second RAID system, one to hold the file database.

10.2.3.4 Control systems. Large server systems can require fairly elaborate methods for controlling the entire process (Fig. 10.9). Simple servers usually have a PC or workstation that controls encoding/decoding parameters, file management chores, and the operator/user interface. As mentioned earlier, many user interfaces provide the illusion that the server system is a bank of VTRs. Larger systems can have other assets to manage than just the operation of a DDR. Many larger systems actually have multiple servers to break up the workload and to organize the workflow. To tie these servers together, routers on many levels are sometimes employed. Video, audio, time code, and even machine control routers are often employed to integrate multiple servers and editing workstations into a single system. In addition to controlling video and audio paths many servers rely on more than one network topology. Some also use Fibre Channel for transfer of video/audio between servers, while using Ethernet for control purposes. Ethernet is often used for sending the metadata that describe a clip.

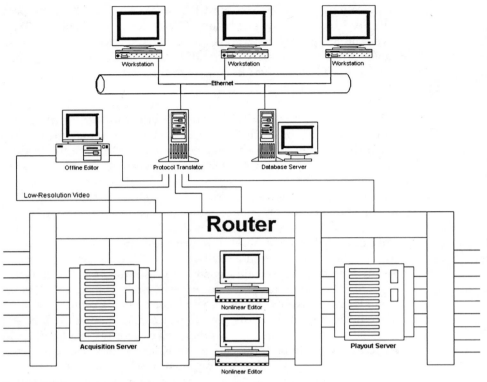

Figure 10.9 Large server system for news programming.

10.2.3.5 Acquisition. Large server systems are frequently divided into three subsystems, based on the way work is processed in a facility, often specifically a news facility. The first phase is the gathering or acquisition of program material. Material is recorded and an operator usually enters some metadata about the clip being recorded. These metadata are usually stored in a database that is common to the entire system and not just the acquisition server. The acquisition server usually has equal or even greater input than output capability. Bandwidth management usually gives the input greater priority than the output to ensure that live incoming material is not interrupted. Clients requesting material from the acquisition server might experience anything from faster to real-time downloading to much slower than real time, depending on the needs of the acquisition server's input streams. Many servers have Fibre Channel or Ethernet connection with other servers. The Fibre Channel connectivity is usually to enhance and expand SCSI properties. Some believe that Fibre Channel will be used in the future to link television devices inside the plant, while ATM over Sonet will be used outside the plant. Others think that Fibre Channel will continue to be a storage link only. In addition to using normal video and audio paths to link servers together, many systems can send video/audio files to other servers via Ethernet. TCP/IP servers can currently ship data between them at rates up to 1 Mbyte/s. This is slower than real time but allows multiple servers to be tied together without expensive video/audio routers. Clients requesting material from the acquisition server would be considered part of the production or manipulation subsystem.

10.2.3.6 Editing. In a newsroom system, editing is the department where the news stories are edited. There are two uniquely different approaches to this phase, offline or online editing. Offline editing has two approaches. The first is the use of shadow servers. Shadow servers record everything the acquisition server sees incoming, but at a much higher compression rate. These greatly reduced bit-rate images are then sent via computer LAN, say gigabit Ethernet as an example, to client workstations requesting the material. The video/audio quality is good enough for the editor at the workstation to generate a type of edit decision list (EDL) describing scene cuts and effects. When the editor is done, the control system uses the EDL to transfer the material out of the acquisition server to the transmission or playout server in the order the editor requested. This "finished" story might be sent at a rate slower or faster than real time between the acquisition server and the transmission server. In the second offline method a client workstation directly manipulates the acquisition server to view high-quality video and audio. The local workstation still builds an EDL that is used as before to end up with a finished piece residing in the transmission server. The most common method of editing in large systems, though, is online editing. A client workstation requests that a copy of the acquired video/audio be sent to the workstation. The workstation generates a finished story and ships the piece to the transmission server itself. These workstations are usually full-blown nonlinear editors.

10.2.3.7 Transmission. The transmission server is a symmetrical copy of the acquisition server. It gives priority to playout, not to incoming feeds from production clients or the acquisition server. To make these three subsystems work together, a layer of control above the three must manage the flow control between the three subsystems.

10.2.4 Server storage

The individual server actually consists of a number of subcomponents. Disk storage is usually a separate external device. Almost all storage means a stand-alone RAID of some type (Fig. 10.10).

10.2.4.1 Disk interfaces. Up until recently, internal PC busses were not capable of moving the data required for video applications internally through a computer or workstation. Today, even with new bus technology, which allows much higher throughputs, that bandwidth must still be shared with internal PC housekeeping and operating system traffic. Servers that seem to house

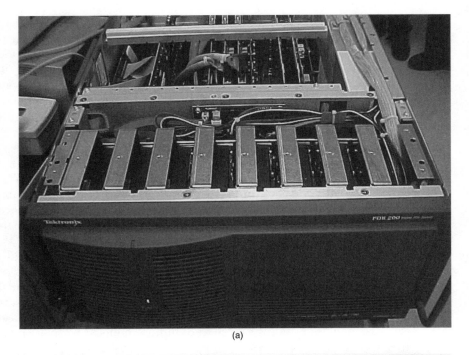

(a)

(b)

Figure 10.10 (*a*) Internal RAID storage in a video server. (*b*) Same video server with external RAID.

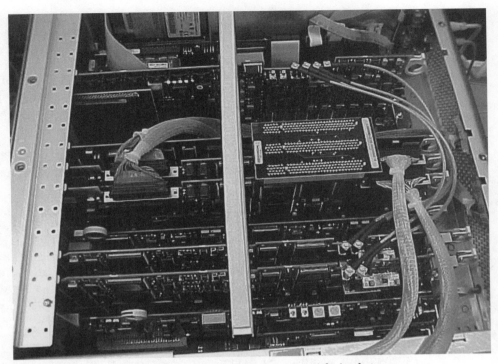

Figure 10.11 An example of over-the-top architecture, center right in photo.

video/audio storage internal to the box usually have separate storage for the PC use and the video/audio data. "Over-the-top" architectures are often used to get I/O to and from the internal video/audio RAID to peripheral cards plugged into the PC's expansion slots. This is a separate bus, not one of the PC's data buses, that moves data from I/O to storage. Most video server systems have stand-alone encoders/decoders that communicate directly with the storage control system, which in turn communicates with the actual storage subsystem. A number of interfaces are used to connect to the actual RAID array (Fig. 10.11).

Integrated device electronics (IDE).   The most common interface used inside PCs is known as IDE, also known as the AT bus attachment (ATA). IDE replaced the ST-506 controller used in the original IBM PC. IDE moved the disk-control function from the ST-506 ISA bus card onto the drive itself, so that data caches could be incorporated on the drive, something SCSI drives were already doing. A major advantage of IDE is that it doesn't require a software driver as it is supported in the PC's BIOS. However, the IDE interface can only support two drives, and maximum data transfer rates were approximately 3 Mbytes/s. IDE had no inherent limitation on disk size (more than 100 Gbytes); IDE's limitations are due to the lowest common denominators between IDE and BIOS/DOS. Most BIOS imposed a 528-Mbyte limitation. Early BIOS allowed 63 sectors/track, while IDE allows 225. Although the BIOS allowed 255 heads, IDE made allowance for only 16. The real limitation, though, was the number of cylinders. IDE allowed for 65,536, while the BIOS only allowed for 1024. New BIOS have solved this by equating drive physical cylinder, head, and sector addresses to logical block addresses, like SCSI does. Today, 70 percent of all drives built use enhanced IDE (EIDE). EIDE can handle drives up to 8.4 Gbytes in size. Many EIDE interface boards now handle four drives. As a result, volume EIDE drives cost less than drives with other interfaces when the size of the drive is less than 1.2 Gbytes. Operating systems like NetWare or Unix never had this problem as they directly read the drive's parameter table and bypassed BIOS. A limitation that keeps EIDE/IDE an internal interconnection solution is that its cables are limited to 18 in. Another reason is that IDE/EIDE data transfers are still below 20 Mbits/s.

SCSI. Known as the small computer system interface, this is the most common interconnection method used between a RAID and the rest of the server system. This is a parallel interface, with most implementations using 68-pin cables. If differential drivers are used, 16 lines are for data and 18 are for control. SCSI comes in a number of flavors, with SCSI-3 and ultrawide SCSI the most common. Both come in narrow (1-byte transfer at a time) and wide (2-byte transfer at a time) versions. SCSI-3 has a 10-MHz clock, while ultrawide has a 20-MHz clock. Therefore, the ultrawide SCSI interface can transfer ($2 \times 20E6$) 40 Mbytes/s. Of course, this rate is not normally maintained because of SCSI handshaking and control overhead. In fact, the highest throughput of an SCSI device with no mechanical limitations is about 38 Mbytes/s. Due to the physics and mechanics of disk drives, the maximum output of most current drives isn't much over 10 Mbytes/s, and many are closer to 5 Mbytes/s. SCSI is organized around initiators and targets. In a server system the storage controller is the initiator and the RAID is the target. SCSI-3 and ultrawide SCSI allow one initiator to converse with up to 63 targets (an earlier version allowed only eight). Actually, multiple initiators can talk to the same 63 targets, which is what happens when a server has multiple RAIDs. Often there is a two-tiered SCSI tree. The server's storage controller acts as an initiator in communication with one or more RAID controllers acting as targets via SCSI, while the RAID controllers inside the RAID act as initiators and carry on separate conversations with their individual drives, which are internal RAID targets.

Fibre Channel. Fibre Channel is both a storage and a network solution, but most implementations to date are for storage. The physical layer can be either coax or fiber. Two Fibre Channel (FC) topologies are available, the arbitrated loop (AL) and the switched channel (S). The arbitrated loop is similar to a token ring LAN approach. At any given time, one device has the loop to send data to another device. When the ring is free, devices arbitrate to use the loop. A set of rules governs who wins the right to use the loop. As with most things in life, some devices play by the rules and some don't. With FC-AL the loop bandwidth is shared by all devices on the bus. In switched FC-S Fibre Channel uses a switch, much like a router in television, to set up connectivity from one device to another. At any given time, one device talks to one other device, but the two devices have all the available bandwidth to communicate. The throughput is higher but the system needs expensive switching hardware to make this work. Looking physically at the boxes involved in a Fibre Channel installation won't usually tell you which topology is in use. Many FC-AL installations have every FC node connected to a concentrator, which provides what appears to be a "star" topology, but in reality the loop is electrically preserved. The concentrator is used to shunt around FC devices that fail, thus preserving a closed loop. For storage applications FC is usually used to carry SCSI commands. It can be thought of as a serial implementation of SCSI. The SCSI commands and data are wrapped in FC packets. Except for managing the transport and flow of FC packets, FC has no command device control language of its own. FC can also be used to carry TCP/IP packets and can thus be used as a LAN. There are server systems that use FC in both configurations.

SSA. SSA is a point-to-point store-and-forward technology developed and used mainly by IBM in their drives. SSA allows strings of attached peripherals to transfer data simultaneously between each other and to a host. SSA uses a dual-ring topology for redundancy if one ring fails. Each node or device on the SSA ring has four ports, two for each ring. Both rings allow for 40 Mbytes/s through a node. SSA also usually carries SCSI protocol.

P1394. Also known as Firewire, P1394 is intended as a storage and peripheral linking topology. It also can carry SCSI commands, known as serial bus protocol. There are already camcorders, CD-ROM drives, and hard drives that have P1394 interfaces. This means that video can be offloaded from the camera to the hard disk with no intervention from a PC. Currently, P1394 runs at 200 Mbits/s but much higher speeds are on the way. The P1394 cable consists of two copper pairs, one for data and clocking and the other for power. P1394 will probably never be a network application because the cable between nodes is limited to 4.5 m. P1394 divides its time between two modes, one geared to transmitting data and the other to transmitting video/audio. The data mode is asynchronous where one device sends data to another node. About 20 percent of P1394 bandwidth is devoted to this opportunistic transfer of data; the rest of the bandwidth is used to send video/audio. No handshaking or flow control is used. One device simply sends the video/audio data isochronously to any and all devices interested. No devices acknowledge its receipt and there is no retransmission of the data.

FDDI. This has a token ring structure but usually fiber is used instead of copper. It allows large-area networks with hundreds of workstations or nodes. It also allows for a second redundant ring like SSA. Interestingly, FDDI has an isochronous mode like P1394 for the deterministic transfer of data such as video.

SDTI. SDTI is the same data format as SMPTE 125M, except that video data are replaced with MPEG data. The TRSs, SAV and EAV, are sent as usual.

Instead of 10-bit words, the payload is 8-bit bytes, with the other 2 bits used to ensure that data values reserved for only TRSs are not sent. The ancillary data space between the EAV and SAV signals contains data format information and error checking. SDTI also has a destination address sent during ancillary space time so SDTI devices downstream know for whom the data are intended. A number of manufacturers use this as a mezzanine compression layer. Since it has proper TRSs, SD digital component devices will pass it with no problems as long as it does not process the active video, as such production switchers and proc amps would be off limits. The data pattern can even be displayed on an SDI monitor. Several systems use SDTI between the encoders/decoders and the RAID. A related approach is ASI, which has the same SD digital component bit rate, 270 Mbits/s, but no TRSs. The ASI signal can be passed through any SD component device that does not require TRS signals. Most digital routers do not look at TRS signals.

HIPPI. HIPPI is a high-speed point-to-point connection technology that operates up to 800 or even 1600 Mbits/s over cables 25 m or less. It uses 4-byte-wide data transfers and, as such, has clocks of 25 or 50 MHz. Like SCSI, it allows one connection at a time. This technology has not found use in television as a storage connection, but some devices, such as telecines, are using it to transmit video data to computer workstations.

10.2.4.2 Disk drives. The typical 7200-r/min hard drive has an average access time of 13 ms, and transfer rates of approximately 6 Mbits/s.

10.2.4.3 MTBF. Disk mean times between failures average 400,000 to 500,000 h, and some actually boost close to 1,000,000 h. These predictions are for populations of drives, not individual drives. One thousand drives, each operating for 1000 h before one fails, will produce an MTBF of 1,000,000 h, but this doesn't necessarily predict what one drive operating for 50,000 h might do.

10.2.4.4 RAIDs. The RAID is the piece of equipment that makes today's server possible. The acronym RAID is said to mean either redundant array of independent drives or redundant array of inexpensive drives. The concept of the RAID array was developed at the University of California (Berkeley) in 1987. The RAID can provide protection against a single drive failure that might eliminate some or all of the stored data on the array. It does this by generating extra data that are stored and that can be used to determine the data that resided on the failed drive. This usually takes the form of parity data, but the form that the protection scheme can take can be instituted in a number of ways. The various schemes are commonly referred to as RAID levels. The different RAID schemes offer various levels of data bandwidth and protection. RAIDs can become quite large and expensive. Commercial RAIDs containing as many as 192 drives, offering 4 Tbytes of storage, are available and cost well over $500,000.

Most RAIDs in use today allow drives that have failed and been replaced to be rebuilt online. That means that while the RAID is in use the lost data are regenerated and loaded onto the new drive. This wasn't always the case. Depending on the size of the drive, most can be rebuilt in several hours offline. Online background rebuilding can take many hours. Some RAIDs can be unforgiving with operational mistakes. On some if you take a drive offline, the only way to put it back in service is to perform a rebuild of that drive. Also, it should be noted that if for some reason a drive is taken offline to pull it from the array, it can take 30 s for it to spin completely down and park its heads.

As drives now have spindle revolutions per minute generally above 7K and some new ones at 10K, generated heat is higher then ever. Many RAIDs have backup fans to ensure adequate cooling in the face of fan failure. Since the cooling fans are mechanical (just like the drives themselves), they will eventually fail; it's just a matter of when. Most RAIDs provide overtemperature indications, often more than one warning—the first indication is a warning and the second results in automatic shutdown, but many server vendors override the shutdown overtemp. Drives running above 50°C undergo an unwanted stress test. Another interesting subtlety in some RAIDs is that the internal SCSI bus is often activity terminated by the last drive in the chain. If that occurs, the activities of all drives in that chain will be disrupted if the end drive is turned off or removed.

RAID levels 0, 1, and 2. RAID level 0 actually offers no protection at all. What it does do is provide for high bandwidth by allowing multiple READ/WRITES simultaneously as data are spread to all disks in the array. However, if any disk fails, all data in the array are lost.

RAID 1 has been around for a while. RAID 1 is simply disk mirroring. All data written to one disk are simultaneously written to a second disk. RAID 1 provides the highest level of protection but at the highest cost. This approach quickly becomes too expensive for video servers, but RAID 1 is often used for audio.

RAID levels 2 and higher offer different schemes for using parity data. RAID 2 is not used because its error-correction scheme has been incorporated into most drives.

RAID level 3. RAID level 3 strips data across all the disks in an array, 1 byte at a time. There is a drive dedicated to storing ECC (parity). RAID 3 is good for applications that transfer large amounts of sequentially stored data (such as video). This method allows for higher data transfer rates because the data transfer of each disk is available in parallel. Data stripping is what permits multiple I/O streams from a single array, as data can be quickly read out and cached so READs for another stream can commence. Depending on data rates (based on compression ratios), a single RAID is generally able to support around four I/O channels. RAID 3 can lose a single disk with no data loss and still provide high bandwidth even after a disk has failed. RAID 3 is the most common method for protecting video data storage.

RAID levels 4 and 5.   RAID levels 4 and 5 strip entire blocks of data across disks in an array. One block is stored on one disk, the next block on the next disk. This arrangement is good for randomly distributed data such as might be found in a database, where access to individual data blocks is much quicker, but actually large data burst transfer rates are lower than RAID 3. RAID 5 shines in applications with frequent transfers of small data files. RAID 4 is the same as RAID 0 since it doesn't provide for data protection; hence it is not used. RAID 5 uses and distributes ECC data over all the disks. RAID 5 provides a higher level of security than RAID 3 but at the cost of system bandwidth. It allows faster READ than WRITE operations because as the data are spread out as blocks instead of bytes, it takes longer to access those data. Many server systems use RAID 3 to protect video data and RAID 5 to protect the servers operating system, applications, and database software.

10.2.5   Other media

Two types of optical storage are available—magneto-optical (MO) and compact disk (CD). MO uses a rewritable, removable disk cartridge which can be written to on both sides. Writing to disk is done with a combination of laser and magnetics. MO playback is accomplished using a polarized light that is beamed at the magnetically polarized surface of the disk. The phase angle of the reflected light will vary according to the disk surface's magnetic polarization. The CD uses light only. The technology behind the CD has become extremely defined and, like magnetic disk drives, the CD has reached commodity status.

10.2.5.1   CDs.   The CD was developed by Philips and Sony and this technology is the predominate optical storage method. The two inventors license it to others. Licenses must comply with a set of specifications for audio CDs known as the Red Book, which specifies one song per track. Physically, there is only one track on a CD, but each program, most likely containing a song, is also referred to as a track. These tracks are comprised of sectors. Each sector is 3234 bytes, which equates to $\frac{1}{75}$ s. A sector is also known as a block. Actually, only 2352 bytes are available for program material. Error-detection and -correction code use 784 bytes. There are also 98 control bytes, which indicate whether the program data are music or data, the running time, and synchronization words. There are also ninety-six 6-bit user-definable words. An initial thought for these words was for graphics. Very few actually use these words. There are 4500 blocks of data per minute. A track can be as short as 300 blocks (4 s) or as long as 325,000 blocks (72 min).

CDs can hold up to 99 tracks. A later set of specs for the computer CD-read-only memory (CD-ROM) industry was created, and this book of specs is known as the Yellow Book. Although the Red Book contains error-correction specifications, the Yellow Book allows for more bytes per sector to ensure a higher degree of data integrity. This overhead cuts down by 10 percent the amount of actual data on the CD. There are a number of other standards related to CDs. There is an Orange Book for MO and CD-write once (CD-WO), a Green Book for CD-interactive (CD-I) that addresses the problems of synchronizing separate audio and data tracks on the same disk, and the White Book which sets the video CD standard. The White Book uses ISO MPEG-1. These agreed upon standards are allowing CD usage to continue to grow.

Digital audio CDs are capable of holding 74 min of audio, which is equivalent to 650 Mbytes of storage. Actually, most CDs only have approximately 553 Mbytes of real storage due to the fact that the outer region of the disk is more susceptible to defects and that some drives have trouble reading the longer tracks. The CD is 4.75 in (120 mm) in diameter and 1.2 mm thick, and has a polycarbonate plastic surface. Underneath this is a layer of aluminum or gold, which is backed by protective lacquer onto which the label is attached.

The polycarbonate plastic surface has pits etched into it. Areas where there are no pits are known as *lands*. Reflected laser light that shines on the pits and lands varies as the CD track passes by. Every variation represents a "1." CDs use 8- to 14-channel coding. MO disks and phono records spin at a constant angular velocity (CAV); therefore, the data density is lower at outer tracks than at the inner tracks. CDs spin at constant linear velocity (CLV). Disks spin faster at inner tracks than at outer tracks. To meet the Red Book spec of a sustained data rate of 150 kbytes/s, the spin rate is 530 r/min at the inner spiral and approximately 200 r/min at the outermost spiral. CLV hurts data access slightly as disk velocity must be changed if data access is separated by many tracks.

The digital audio Red Book standard describes the CD media at the physical level, players at a machine level, and the decoding of PCM information of the CD track. The Red Book specifies 16-bit quantizing with 44.1-kHz sampling.

The Yellow Book permits two modes of storing data. Mode 1 is specified for computer data, while mode 2 specifies the method for storing compressed audio, graphics, and motion graphics, also known as video. The block sector length, 2366 bytes, is different from the Red Book. Both mode 1 and mode 2 have 12 sync and 4 header bytes. Mode 1 has 318 error-detection/correction (EDC/ECC) bytes and 2048 user bytes. Mode 2 has no EDC/ECC bytes and devotes all 2366 bytes for user data. Mode 2 is not used.

The Green Book standard allows for the interleave of data and compressed audio and video on the same track. Closely related is the White Book, which specifies the use of MPEG-1 video. As a result of MPEG-1 requirements, playback rates of 187 kbytes/s (1.5 Mbits/s) are specified. Use of MPEG-2 is not feasible based on the Green Book standard. MPEG-2 players need a 4- to 10-Mbit/s rate, impling a rate at least 4 to 8 times the player rate. However, this also means that the 74-min playing time would decrease to 9 to 18 min.

Writeable CDs are now available. These CDs use an organic dye as the recording medium. There is a layer of gold over the dye so the gold will not corrode as a result of contact with the dye. When recording, the laser burns the dye to create mounds or lands, which reflects the reflectivity of the gold. A standard CD player sees the mound as if it were a land.

10.2.5.2 DVD. The new disk that is remaking the optical storage scene is the DVD. The acronym is said to mean two things—either digital versatile disk or digital video disk. While it is the same physical size as a CD, its storage is much greater, 4.7 to 17 Gbytes/disk versus 682 Mbytes for the CD. The DVD is a boon for multimedia developers combining audio, graphics, and video into the playback experience. DVD will most likely replace the use of VHS tape. The DVD will cost about one-third the cost to manufacture versus VHS.

There are a number of physical variations to the DVD. The simplest version is quite similar to the CD, but instead of the normal protective layer with a label attached on a CD, the DVD disk has a second polycarbonate substrate layer. Two-layer DVDs use this second layer, but this layer is still necessary on single-layer DVDs because the DVD laser expects to find data at a specific depth on both single- and dual-sided disks. Even with the disks the same size, the simplest DVD holds 4.7 Gbytes. The DVD accomplishes this data density by using a shorter wavelength laser than does the CD. While the CD laser is a 780-nm infrared laser, the DVD laser is a 635- to 650-nm red laser. The maximum pit length on a CD is 0.834 μm, while it is only 0.4 μm on a DVD. DVDs also have less data overhead than CDs do.

The simplest DVD disk is a single-sided single-layer disk. With DVD, it is possible to add a second polycarbonate layer over the first layer. The top polycarbonate layer is backed with a semireflective layer. Below this top layer is the normal layer. The DVD laser is made to focus on either the top semireflective layer or the bottom reflective layer, which nearly doubles the storage capacity of the DVD to 8.5 Gbytes. DVDs can be two-sided as well. A two-sided DVD disk has double the storage of the single-sided version, that is, 9.46 Gbytes. These two-sided disks can be dual-layered too. A two-sided dual-layer disk has a 17-Gbyte data capability.

All DVDs use the universal disk format (UDF), which is a subset of the ISO 13346 standard for data exchange for file management. Device-operating systems need drivers that recognize UDF. Although the executable programs are still platform-specific, at least the data are portable. ISO 9660 is the CD-ROM file standard. Some early DVD disks have both file structures pointing to the same data.

DVD and CD use a concatenation of two codes for channel coding and error correction, namely, eight to fourteen modulation (EFM) and cross-interleaved Reed-Solomon code (CIRC). EFM is used for transforming the digital bit stream into a sequence of binary symbols, called channel bits, which are suitable for storage on the disk. EFM is a dc-free run-length–limited (RLL) code. The number of sequential-like symbols in a binary sequence is known as *run length*. RLL code is used to prevent intersymbol interference when the sequence is transmitted over a bandwidth-limited channel and to ensure an adequate frequency of transition for recovery at the receive end. CIRC is used for detection and correction of erroneously received data. Cross interleave allows errors to be masked if correction is not possible. CIRC actually happens before EFM. It takes user data and arranges the data into arrays with error-correction code in each row and additionally in each column.

DVD uses sectors of 2418 bytes; 2048 are actual user bytes with 52 sync, 16 header, and 302 error code bytes. The 2418 bytes are channel coded to 16-bit words. Thus, 2048 data bytes are translated to 4836 bytes, which means that there are 2.36 channel bits for every user bit. This is more efficient than CD.

10.2.5.3 MO. MO technology uses a laser to heat a point on the disk to approximately 200°C. At that temperature, an inductive head can change the magnetic polarity of that point. During playback a polarized light source bounced off the surface of the MO disk will change its phase angle, depending on the magnetized state of the disk's surface. It is a metal-based disk.

MO disks spin at constant angular velocity (CAV), which means that the disk's revolutions per minute are the same whether the READ/WRITE heads are at the inner or outer track on the disk. Thus, the data density is greater at outer tracks than at inner tracks. The MO has two READ/WRITE heads—one on top and one on the bottom of the disk. When the upper head is at the outer-most track, the lower head is at the inner most track. In this situation the upper head outputs 80 Mbits/s, while the lower head outputs 43 Mbits/s. When combined, the two heads have a 123-Mbit/s data rate. As the upper head moves inward, the lower head moves outward. Thus, the data rate of the upper head is dropping while the lower head's data rate is going up. Together, their summed data rate is always 123 Mbits/s. The MO disk is 300 mm in diameter and spins at 1800 r/min.

10.3 Bibliography

Boston, Jim, "Video Servers," *Broadcast Engineering*, May 1999, p. 92.

Processing Digital Audio

In many ways digital audio is actually tougher than digital video. Since digital audio is sampled at rates far below video, the ears are able to detect imperfections in the audio that go unnoticed in the video streaming by at much higher data rates. High-quality audio samples require 20 or even 24 bits, whereas comparable video requires only 10 bits. In addition, errors in the video can be masked if the corrections cannot be made, whereas masking audio errors is problematic.

The other problem with digital versus analog audio is that the transition in technology between the two is so great. With analog video fairly sophisticated processing and test equipment is needed, but analog audio does not nearly need the same complexity. Bandwidth was much narrower for audio, and mixing and switching requirements were few. Now with digital audio the digital bit streams are actually more complex than their video counterparts. The audio bandwidth requirements are now equal to what analog video used to be. Switching between digital audio sources is no longer trivial if that switch is without artifacts. In addition, mixing digital audio signals can require complex DSP rivaling that applied to video.

Until recently, the cost-to-benefit ratio for digital audio was nowhere near what analog audio provided. That is changing, but digital audio equipment is generally still larger than its analog counterpart, it consumes more power, and often more boxes are required to accomplish the same task. Manipulation that could be done with analog audio often can't be used when digital audio is implemented. Like digital video, digital audio shines when multiple passes of the signal or data into and out of storage are required.

11.1 Analog Audio Primer

Initial audio acquisition devices, a fancy way of saying microphones, are all analog in nature. They work by using sound pressure waves to modulate an inductor and generate minute voltages or to modulate a capacitor which varies the amount of current allowed through the mic by an outside source. The first type is known as an inductive mic, and the second is known as a condenser mic. Condenser mics are generally smaller than inductive mics. Power to condenser mics is usually supplied by installed batteries or by an external source. Many audio mixers provide the required power, and this is known as *phantom power*. Thus, the mixer acts as a power supply. Many intercom systems used in television operate in much the same way, with the intercom power supply providing the power down the microphone cable to power the various intercom stations connected together in a daisy chain fashion.

11.1.1 Audio sources

The signal levels from a microphone are often on the order of -60 dBm (0 dBm = 0.775 V into 600 Ω), in other words, 0.775 V with a current flow of 1.3 mA (-60 dBm represents a value equal to one-one thousandth of the value at 0 dBm). With levels so low, many elements can interfere with these audio signals. Sixty-cycle power, or hum as it is called, is the most common. That is why professional audio is usually balanced. The object is to get what is captured by the mic to the audio-processing and mixing circuitry as intact as possible.

11.1.2 Impedance

Originally, the most important attribute for delivering as much power from a source (say a microphone) to a receiver (say an ear piece on a headset) was for both to have the same impedance. The transmitter (mic) and the receiver (ear piece) should have the same impedance. Additionally, the cable that connects the two must have a similar impedance.

11.1.2.1 Impedance matching.

Many microphones have a transmit impedance of 150 Ω. Common microphone cables have 110-Ω impedance, while the inputs to a lot of analog audio equipment are 600 Ω. Thus, a microphone would be slightly mismatched from the cable to which it is connected, and the cable mismatched from the input to the audio processor/mixer. Wherever an impedance mismatch occurs, some of the energy or power is reflected back toward the source. So some reflection occurs between the mic and the cable, and some more reflection occurs between the cable and the processor input. In addition, the power reflected back from the processor input has a small amount reflected again when it arrives back at the microphone, this time back again toward the processor input. Thus, not all the energy traveled from the microphone to the processor and some arrived late because of reflections.

With video, this impedance mismatch shows up as "ghosting." NTSC color subcarrier has a wavelength of approximately 60 m in most high-grade coax, which means that for the ghosting to be pronounced, the cable runs will have to be fairly long. An audio tone at 10K has a wavelength of 30 km. Cables will have to be much longer than normally encountered for reflections to be heard. So for the telephone company, long-run impedance matching was imperative.

11.1.2.2   Power transfer.   The other important reason for impedance matching was to achieve the maximum transfer of power. When the transmitter, the transfer medium (the cable), and the receiver all had the same impedance, then one-half the transmitted power would be delivered to the receiver and one-half would be dissipated in the line. Any other combination of impedances would result in less than one-half the power being delivered to the receiver. With early communications equipment, power was necessary to make the receive device work. More specifically, it was the current component of the power that usually needed it; the more, the better.

11.1.2.3   Low Z to high Z interconnection.   Today, most solid-state devices require very little current for operation. It is the voltage impressed across the input of the solid-state device that drives it. So many analog audio facilities found in television today use devices with high-impedance inputs, generally around 10 kΩ. This is done to maximize the voltage drop across the input. Higher voltage drops mean higher input voltages, and, in turn, higher signal-to-noise ratios—generally double, or 6 dB. This gross impedance mismatch results in higher reflections. However, as we just mentioned, the line lengths are generally such that the reflections are not noticed. So the convention today is for the transmitter to have a low-impedance output and the receiver to have a very high impedance input.

11.1.3   Audio patching

Audio patching allows some added flexibility not found with the matched impedance methods. Since a low-impedance transmitter now drives a high-impedance receiver, multiple receivers can be driven at once by a single transmitter, and this is known as *bridging* (multiple receivers being fed by a single source). If all inputs to receivers are high impedance, this means that you can always bridge more than one receiver across any source.

11.1.3.1   Patch-panel formats.   Many patch panels today allow bridging of more than one receiver across a single source. Proper patch-panel wiring has signals coming from sources arriving at the top half of the jackfield and signals going to inputs on the bottom half of the patch panel. A "normal" patch (no patch cord inserted into the upper or lower halves of the jackfield) internally connects the signal arriving at the top of the jackfield with the line going to the input of the device wired to the bottom of the jackfield. Inserting a patch cord into the upper or lower half of the jackfield would interrupt that path. A patch cord is inserted on the upper half of the jackfield to route the arriving signal to a different input that plugs the other end of the cord into the desired bottom half of a different jackfield. This arrangement is known as a *full normalized jackfield.*

The low- to high-audio impedance scheme allows implementation of what is called a *half normalized jackfield.* This jackfield is internally wired so that if a patch cord is connected to the top half of the jackfield, the path to the lower half of the jackfield is not broken. The path from the source to the normal receiver is still connected. However, the patch cord can now be plugged into a different lower half of a different jackfield to bridge this second receiver across the same source as the normal receiver. This one source can feed to receivers, which is especially helpful for monitoring a source. *Note: This process is only possible with analog audio; it is not applicable for digital audio.*

Patch panels provide two important services for video, audio, machine control, etc.: (1) to provide flexibility in the plant and (2) to troubleshoot paths that are in trouble and to patch around that trouble spot.

11.1.4 Analog audio levels

As a point of reference, we will briefly look at analog audio levels because very shortly we will equate these to digital audio levels. We mentioned earlier that microphones commonly output audio at very low levels. Thus, signals generally below -20 dBm are referred to as low-level signals. High-level audio signals generally are in the 0-dBm or higher range. Common levels today that equate to 0 VU on audio level meters are either $+4$ or $+8$ dBm. The two different levels stem partially from the fact that many unbalanced audio signals were set to $+4$, which meant that balanced signals with power in two conductors instead of one would have nearly double the $+4$ power (Fig. 11.1).

Figure 11.1 Modulation meters that relate modulation percentages to power levels.

11.2   Analog-to-Digital Conversion

As with digital video and digital transmission, digital audio is extremely robust, but it can be degraded and broken. In many applications it makes good sense but not in all cases. As people who have built digital plants already know, going digital doesn't eliminate analog. It usually just adds additional layers of fabric to a broadcasting or production facility. If for no other reason than microphones, a facility will continue to have an analog audio layer.

11.2.1   Sampling rate

This is a quick review of material in Chap. 4. The sampling rate must be above the Nyquist rate to prevent aliasing. The higher above the sample rate, the better as it lowers the noise floor; too high, though, and the bit rate, along with system complexity, increases.

11.2.2   Quantization resolution

Resolution also is important to the signal-to-noise ratio. The higher the sample rate, the lower the effective noise floor. Each additional bit in the sample can mean a 6-dB lowering of the theoretical noise floor.

11.2.3   Digital audio levels

As with all digital attributes, signals that are higher than they should be do not gradually degrade as they push against the amplitude ceiling of the path they are in. A digital signal is unaffected by any distortion until its amplitude is such that all bits must be high to describe it. Then, any additional amplitude is simply clipped.

To graphically show the user exactly how much headroom is left, most digital audio meters do not revolve around the 0 VU point. Instead, most display the standard level as −20 dB, with the top level labeled 0 (Fig. 11.2). The display in Fig. 11.3 shows you that you have 20 dB of headroom at normal levels.

Additionally, not all digital meters try to mimic the ballistics that analog meters have. The ones that do generally accomplish this by digital integration, that is, by taking multiple samples and finding their average value. The more samples averaged, the more the integration. Some meters do not try to approximate analog meters at all.

11.3   Digital Audio Formats

The common digital audio formats were covered in Chap. 6: AES-3, SMPTE 272M (embedding AES digital audio into SDI video), and SMPTE 276M (AES through coax). A number of other digital formats for moving audio data are also available.

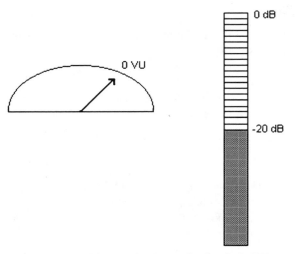

Figure 11.2 Digital equivalent to analog level of 0 VU is −20 dB.

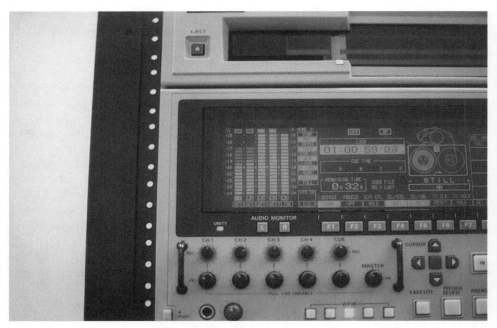

Figure 11.3 Four-channel VTR with digital audio meters.

11.3.1 Pulse-code modulation (PCM)

There are a number of PCM interconnects. PCM refers to the process of taking analog amplitudes and converting them to a number value. These all convey data much like AES but using different serial data formats and often multiple conductors for multiple channels and for separate clock signals.

11.3.2 Multichannel audio-digital interface (MADI)

The MADI standard allows 56 simultaneous digital audio channels to be sent down a single 75-Ω coax. A separate cable for synchronization is required. Each multiplexed channel is essentially an AES/EBU subframe. All the subframes must be at the same sample rate. MADI uses a four- to five-channel code at 100 Mbits/s. The actual payload data rate is 25 Mbits/s.

11.4 Embedding Audio in the Video

Early on in the industry's history people came to realize that the time it took for a television receiver to perform a retrace and start scanning the next line was an inefficient use of the signal. Sure, horizontal synchronization and color phase reference information were sent during that time. However, even with fairly wide horizontal sync pulses and nearly a dozen subcarrier cycles during the blanking period, over 30 percent of blanking time was void of content. Compared to the horizontal blanking, the vertical interval could be thought of as a payload wasteland. Of course, early receivers and monitors needed all those equalizing pulses and vertical serration pulse trains for the analog synchronization circuitry to work. Many horizontal and vertical sync circuits worked by integrating the sync pulse(s), and this process requires time and, thus, fairly wide sync pulses. In addition, due to physics, it does take some amount of time for vertical and horizontal retrace to happen. Therefore, while the horizontal retrace sequence took 17 percent of a horizontal line, the vertical sequence took approximately 20 lines.

It wasn't long before test signals, closed captioning, other types of data, and control signals were found inhabiting the vertical interval, hidden from the casual viewer. Today, digital audio is a predominant inhabitant of many digital video signal streams. It not only occupies portions of the vertical interval but also the horizontal interval. In fact, the vast majority of the audio data found in an SMPTE 259M signal stream resides in the horizontal, not the vertical, blanking intervals. There is enough room in a digital component bit stream for eight AES audio pairs (that's 16 individual channels), although digital composite only has room for two channel pairs. We are going to look at how we get audio data into the SMPTE 259M bit stream (Fig. 11.4). Since most audio sources still produce analog audio signals, the first step is to get analog audio into the digital domain. Analog audio signals need to be quantized into a digital number sequence. The two parameters of most concern during this process are sample rate and sample resolution.

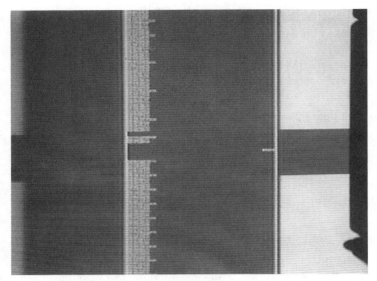

Figure 11.4 Digital audio mapped into horizontal blanking space of a digital component video signal. (Also see Fig. 6.4.)

11.4.1 Mapping AES to SDI

While most audio sources start life as analog signals, not all do, and this can cause some problems. Why? Because the sample rate of various equipment might be different. While most professional equipment has settled on a sample rate of 48 kHz, most consumer equipment has not. For example, compact disks have sample rates of 44.1 kHz. The EBU, along with the BBC, has used sample rates as low as 32 kHz.

The SMPTE 272 specification, which describes embedding AES/EBU audio into ancillary data space in an SMPTE 259 bit stream, allows for sample rates from 32 to 48 kHz. This means that you can have various pairs of data embedded in the same SDI video bit stream with different sample rates. As already stated, since most professional equipment is sampled at 48 kHz, that is what we will concentrate on here.

Be aware, though, that if a plant has digital audio signals with different clock rates, often the only reliable way to get from one bit rate to another is to go back to the analog domain and then back to the new bit rate. Even if done in the digital domain, sample rate conversion is not a totally "transparent" operation.

11.4.2 Monitoring embedded audio

Another common area of misunderstanding is when test equipment, such as waveform monitors, displays the ancillary data space along with the active video. It must be understood that the ancillary data displayed on most test equipment cannot be used for any critical measurements, only as an indication that the data is present. Why?

Let's start with the TRS signals that are displayed. TRS stands for timing reference signals. The timing reference signal is comprised of four data words. The first three are unique values (not allowed in active video). They are 3FF, 000, and 000. The fourth word is known as the "XYZ" word, and it indicates which field you are in, whether this is an EAV or SAV sequence, and whether you are in the vertical interval or not.

The slew rate for going from 3FF to 000 (all bits ON to all OFF) is a method for getting outside the bandwidth of normal SDI video. In essence, to pass this signal intact would require a bandwidth of 27E6/2, or 13 MHz. The vertical bandwidth of most waveform displays is not nearly that wide. Therefore, much ringing and pre- and overshoots will be apparent when looking at these pulses. (See Figs. 4.4 and 6.2.)

The slew rate also depends on the parade mode you are looking at on the waveform monitor. If all three channels are selected (Y, Cb, and Cr), you should be aware of what that display is really showing you. Just as horizontally you travel in time across the waveform display, the Y, Cb, and Cr lines displayed each scan are one line delayed from the component in front. What this means is that if you select line 20 for display (with all three channels selected), the Y line is actually line 20, but Cb is line 21 and Cr is line 22.

The SAV pulse shown with the Y signal is actually the XYZ word (it will have four different amplitude values, depending on whether it is the vertical interval or field 1 or 2). SAV associated with Cb is the 3FF word, and SAV with Cr is one of the 000 words.

Why? Because if this data space was actually video, Cb happens where 3FF occurs (maximum positive chroma value), Cr happens where the second 000 occurs (0 chroma value), and Y occurs twice—once at the first 000 and secondly where the XYZ word occurs. It gets even more confusing with EAV if ancillary data are present because ancillary data begin right after the XYZ word.

These data values present slew rates to the vertical deflection circuits of most displays that are way out of band as well. Therefore, the TRS ringing contains ringing from the ancillary data also mixed in.

The only way to look at ancillary data in detail is with a device that displays actual ancillary data. Back in Fig. 6.7 the array of data shows the EAV string in the top row, with an audio data packet starting in the second row.

11.5 Audio-Video Delay

Many things conspire in a digital facility to separate the audio from its associated video in time. Frame syncs, DVEs, and some TBC's delay video behind the audio are just a few examples. Some networks set limits on the number of frame syncs allowed in a backhaul path (feed from remote location back to the network facilities). Vendors are available who offer audio delay boxes to compensate for the video delay. People generally can more accept the delay of sound with respect to video than vice versa. Sound travels at a slower rate than light, and up to a point this is a natural occurrence. However, current systems tend to delay video more than audio. The EBU recommends a maximum sound advance of 40 ms and a sound delay of 60 ms. Almost all format conversions are bound to cause A/V delays.

Due to the temporal compression nature of MPEG, conversion of video into an MPEG stream can take many frames. Time stamping in MPEG is supposed to eliminate timing problems in that realm. MPEG audio processing generally takes less time than its video counterpart. The AC-3 encoder encodes time code so that it can time stamp the AC-3 PES to match the video PES. The AC-3 encoder also has additional time delay, up to 200 ms, that can be added. This delay can be manually set or controlled externally.

11.6   Fighting Clicks and Pops

Another common area of confusion is what are commonly known as "pops" and "clicks." These can occur any time two nonsynchronized audio sources are switched. Simple switching, such as occurs in most routers, causes nonsynchronous source changes which, at best, will cause mutes, and, at worst, large irritating pops. This is a problem with both AES streams and AES streams embedded in SDI streams. Although video routers will switch on line 10 in the vertical interval, ensuring video continuity, embedded audio in that video will normally not have an AES block boundary at that point.

The resultant switched output stream will have a new bit stream which has no phase relationship with the stream it replaced, which means the receiver will need its local phase-locked loop to reacquire and lock to the new bit stream. This usually won't happen until the next AES block boundary. The amount of interruption is usually determined by the AES device receiving the embedded or deembedded stream.

11.6.1   Synchronization

Just like video has always been, digital audio is now processed much more seamlessly if AES sources are synchronized to each other. Almost all AES streams used in television carry analog audio that was sampled at 48 kHz. AES streams that contain different sampling rates will not work together. Often, the AES stream will have to be taken back to analog and resampled to convert from one sampling rate to another.

11.6.1.1   Frequency synchronization.   In addition to the sampling rate, the AES streams must produce blocks at the same frequency. If they don't, pops and clicks will occur occasionally if nonsynchronized audio is mixed as buffers over/underflow. To ensure correct synchronization, a few pieces of equipment take video reference and produce an internal AES reference. A better method is to use an AES reference stream to lock various AES streams together.

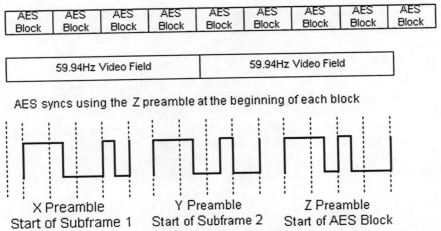

Figure 11.5 The top part of this graphic shows that digital video and audio boundaries do not occur together. The bottom shows the preambles' various AES components.

11.6.1.2 Phase synchronization. Almost all digital equipment uses buffers. As such, most equipment can handle AES streams that are frequency locked but not phase locked. Buffers can be made of various lengths, which allows nonphased inputs to be aligned so that AES block starts are in phase (Fig. 11.5).

11.6.2 VTR maintenance

In digital VTRs, whether or not the audio is embedded in the video data stream when it enters the VTR, it is laid down on tape as if it were part of the video stream, so that any problems that plague video also affect the audio. In addition, error concealment, when error correction has failed, is much less noticeable in the video than in the audio. Something as simple as mistracking can show up as pops and clicks.

The tape path should be cleaned often to prevent VTR-induced audio problems. The guide surfaces and edges where guide and ceramic meet should be checked for contamination. The rabbit guide around the scanner should be cleaned with a cleaning stick to remove any contamination. The cleaning roller should be checked because dirty rollers will contaminate the video heads. Many machines have a tape scraper that removes contaminants from the tape surface. It is recommended that the scraping edge be replaced on a regular basis. If muting occurs, change the blade earlier. The tape path should be checked to make sure the tape is being guided correctly. The rf waveform should be checked to ensure that it is collapsing evenly and uniformly at both ends. Any breathing and nonuniformity at the same time can result in audio mutes. If the machine is not tracked correctly when editing, high errors can cause muting. Low tip projection can result in additional muting.

The environment can have a big impact on audio muting. A dirty environment will add to muting. Low humidity can cause buildup of material on the video heads, which can lead to mutes. Occasional use of a cleaning cassette can reduce video-head contamination, which may improve on audio muting.

11.6.3   Digital-to-analog clicks

Finally, it should be mentioned that once an analog audio signal has been turned into a discrete number sequence, digital attributes can follow it back into the analog domain (Fig. 11.6). If two signals that were digital signals are converted back to analog, and they are switched back and forth, pops and clicks can still occur because of the discrete nature (even though it is analog again) of the signal. A series of impulses each represented by sine^2 (sine-squared) functions are output. Each of these functions, which can also be thought of as pulses, is a sample.

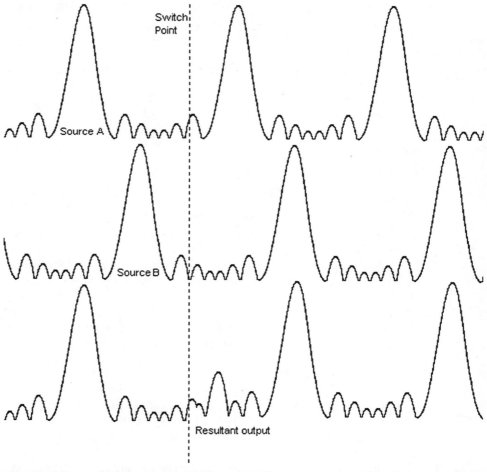

Figure 11.6 Two audio streams converted from digital back to analog and the effect of a resultant switch (bottom).

The sine-squared pulses occur because the D/A output does not have infinite bandwidth, and they null at the center of each preceding and succeeding sample. Thus, they do not affect a continuous pulse train from a particular D/A; they do become a problem if one D/A's output is replaced (switched) by another's output.

11.7 Multiple Audio Channels

Historically, audio has received less attention than it should. A single audio channel, monaural or mono as it is known, has dominated throughout the history of television broadcasting. Many stations have added stereo transmission capability, but most programs pass through broadcast facilities as a single channel of audio no matter how many audio channels were initially produced. Most VTRs and video servers today offer four channels of audio. Digital component serial video can, in theory, carry 16 channels of audio. With the introduction of DTV, multiple channels of audio will be the order from now on.

With AES digital audio one audio cable can now carry two audio channels, but that capability is also possible in the analog domain.

11.7.1 Dolby surround

Since it is hard to deal with more than two channels of audio in a plant (routing, etc.), Dolby created a "surround" analog encoder. It takes the four channels and encodes them into two channels (total left and total right). The home Pro-Logic decoder will decode the signal into left, center, right, and surround channels.

A Dolby surround encoder requires a pair of analog audio paths. This system lacks some of the features of Dolby E, including the low-frequency effects channel. Also, it can only handle four channels, so six-channel audio must be mixed into four channels (left, right, center, and mono surround) first. If the encoder is set to the 2/0 mode and the surround mode parameters are correctly set, the receiver can use these settings to decode a left and right analog audio pair into four separate channels. In-phase information on the left half of the pair is a summation of the center and the left channel, with the out-of-phase information being the mono surround channel which is subtracted from the total signal. The right half of the pair has the center and the right channel's sum as the in-phase information, while the surround information is the out-of-phase signal. This system has the surround signal as the out-of-phase signal in both halves of the stereo pair, but its polarity is reversed on the left side. This four-channel encoded audio can be mixed like any other audio as long as interchannel balance and phase are maintained. Over 31 million Dolby surround decoders are in use by consumers. This system might be a way for broadcasters to pass multichannel audio through the plant. The Dolby surround pair would be fed to an encoder like any other stereo pair, but the metadata also fed into the encoder would instruct the Dolby receiver on how to decode the audio data.

11.7.2   Dolby Pro-Logic

Dolby Pro-Logic decoders decode matrix-encoded surround material. The decoder has four channels: left, right, center, and surround. The decoder delays the surround channel slightly to mask front channel leakage into the rear surround channel. This process works because of the Haas effect, which states if two similar sounds arrive at slightly different times, the listener will not perceive the later sound. To stop rear surround sounds from leaking into the front channels due to audio head azimuth error, the surround audio is run through a low-pass filter.

11.7.3   Dolby E

Dolby E, or editable, is intended to be the digital equivalent of Dolby surround in the plant. It uses a less aggressive compression scheme than AC-3. AC-3 is intended for final delivery to the viewer. Dolby E can be considered an intermediate-level compression (or mezzanine compression) scheme, much as SDTI (SMPTE 305) is for video. Dolby E allows up to four AES pairs (eight audio channels) in one AES pair. It uses an extremely mild 4:1 compression ratio, which allows Dolby E compression to be cascaded or concatenated 8 to 10 times. Dolby E is also intended for distribution applications. The transport is via AES protocol down coax or twisted-pair. This allows existing AES infrastructures to handle 5.1 surround sound. Any effects that must be performed, such as a remix of any sort, require that Dolby E be taken back to AES baseband (Fig. 11.7).

Dolby's system uses the nature of human hearing and psychoacoustic phenomena to compress audio with what is called a perceptual audio coding system. The parts of the audio most perceptible to human hearing are given the highest resolution, and thus encoded with the highest number of bits. Fewer bits are allocated to masked, or nonaudible, portions of the signal.

With multiple channels, phase between these channels will become important. There are devices that vectorally compare channels to each other, and it will be important to understand what these devices are conveying (Fig. 11.8).

11.7.4   AC-3

ATSC uses Dolby AC-3 (along with the American DVD). It is designed for final delivery (transmission), and provides up to six channels in one stream. It has automatic down mixing at the receiver, dialog normalization, and dynamic range control.

AC-3 is also known as Dolby digital or Dolby D. It was developed in 1992 to allow 35-mm theatrical prints to carry multichannel sound. It is also the DTV audio system used. This standard is spelled out in ATSC's document A/52 entitled "Digital Audio Compression Standard." This standard is intended to be used in the final stage of delivery to the consumer. It is meant as a transmis-

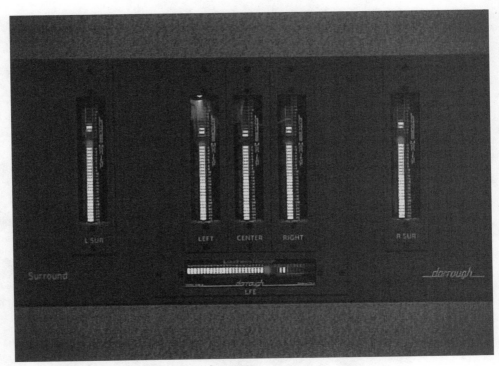

Figure 11.7 Metering required for Dolby surround.

sion distribution system to the consumer, and is not intended for use in multiple encoding and decoding applications. Artifacts start to show up after two or three AC-3 encodes/decodes. The Dolby system takes six channels of 20-bit audio sampled at 48 kHz, which would create a single composite 6-Mbit/s bit stream, and reduces this stream to 384 kbits/s, a reduction of 15:1. This reduction is acceptable because it is done for the final link from the broadcaster to the viewer's home, and thus no additional generations of coding should occur. At the 384-kbit/s rate, this system is not meant to be used upstream for distribution or program generation. If this system had to be used upstream, testing has shown that, with a data rate of 640 kbits/s, a few encode/decode generations could be done transparently. Or if only two channels are sent at 384 kbits/s, then 10 or more encode/decode generations could be done.

AC-3 achieves this coding efficiency by taking advantage of limitations in human audio perception, which is known as a psychoacoustic compression technique. Humans hear in the range of 20 Hz to 20 kHz. Signals are inaudible if their sound pressure level is below the threshold of quiet, which varies by 60 dB over this frequency range. At 2 Hz, the threshold is at a sound pressure level of 60 dB; at 60 Hz, it is near 35 dB. It is lowest at approximately 1.5 kHz (near zero). At 5 kHz, it is about 10 dB, at 10 kHz, it is 20 dB, and at 20 kHz, it is at 40 dB. Spectral masking causes louder tones to mask the

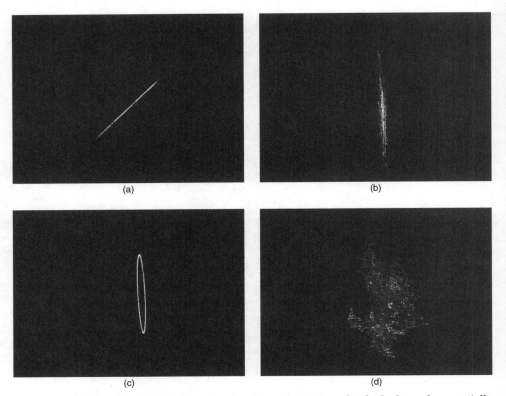

Figure 11.8 (*a*) Two channels of stereo tone in phase. (*b*) Mono audio (both channels essentially have the same audio information). (*c*) Mono tone with two channels slightly out of phase. (*d*) Stereo program with lots of left/right separation.

presence of nearby, softer tones. Temporal masking causes louder sounds to mask the presence of softer sounds immediately before or after the occurrence of the louder sound.

AC-3 breaks the audio frequency spectrum into small bands. It then counts on "masking" to reduce the audio data rate. Low-level sound frequencies that are near frequencies with high-level sounds are eliminated. This greatly lowers the bit rate. An AC-3 5.1-channel bit stream is less than a single channel of Red Book PCM audio. AC-3 typically is more efficient than Red Book PCM by a 12:1 margin. Masking can lead to interesting effects if the bit rate is set too low. Often the first to go is background noise. A sporting event with a sell-out crowd might sound like a sparsely attended event. A five-piece band might sound like it is a three-piece band. A singer might no longer have any instrumental accompaniment. Dolby encoders produce large audio delays of 70 to 160 ms.

An AC-3 encoder (Fig. 11.9) takes up to six PCM audio channels, metadata control information, reference, and time code, and produces an AC-3 coded bit

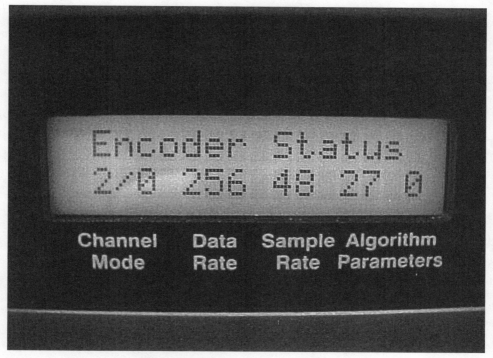

Figure 11.9 Front display of AC-3 encoder.

stream. An AC-3 decoder undoes the encoder process, but it also accepts information about the local setup.

AC-3 supports 32-, 47.1-, and 48-kHz sample rates. The AC-3 output bit stream data rate is 32 to 640 kbits/s. However, ATSC imposes a limit of 384 kbits/s on the AC-3 bit stream. The main and the associated services are limited to 512 kbits/s. Assuming 5.1 channels at 48 kHz at 20 bits, this is equivalent to a 13:1 compression.

AC-3 consumer decoders allow five ways to listen to an AC-3 soundtrack:

1. Full dynamic range 5.1 channel.
2. Reduced 5.1 dynamic range, which is intended for apartment dwellers or late-night listeners.
3. Two-channel Dolby surround encoded down mix, which may then be Dolby Pro-Logic decoded.
4. Normal two-channel stereo down mix.
5. Mono down mix.

AC-3 five-channel sound consists of left front (L), center front (C), right front (R), left back surround (LS), and right back surround (RS). LS and RS can be treated as a single rear channel (S).

ATSC AC-3 has a number of coding modes:

1/0	C	2/0	L, R	3/0	L, C, R
2/1	L, R, S	3/1	L, C, R, S		
2/2	L, R, SL, SR	3/2	L, C, R, SL, SR		

The low-frequency effect channel (X.1) can be added to any of the modes. AC-3 receivers can down mix the modes for whatever environment they are in—mono, stereo, 5.1, etc.

The .1 in 5.1 refers to a channel that carries only low-frequency effects audio. The 5 plus 0.1 means five full-bandwidth audio channels and a sixth effect channel—three front channels, two rear channels, and one effects channel would have the Dolby designation of 3/2L, or 3/2/.1.

AC-3 can carry surround sound audio which is handed off to a Pro-Logic decoder for four-channel decode. Many set-top boxes include an internal Pro-Logic decoder.

AC-3 encoders use two physical interconnection schemes:

AES: used like SDTI as a wrapper for 384-kbit/s AC-3 stream.

ASI: used as a wrapper for AC-3 elemental stream.

AC-3 ATSC service is divided into two service types, main and associated services. Main services consist of

Complete main (CM), that is, dialog, music, effects (normal service type).

Music and effects (ME), that is, dialog missing or separate.

Associated services consist of

Visually impaired (VI), that is, narrative description.

Hearing impaired (HI), that is, dialog may be processed for improved intelligibility.

Dialogue (D), which can provide dialog in multiple languages to accompany ME.

Commentary.

Voice over (VO).

Emergency (E).

With a 48-kHz sampling rate locked to a 27-MHz system clock:

$$CM < 384 \text{ kbits/s}$$

$$\text{Single channel} < 128 \text{ kbits/s}$$

Two-channel dialog < 192 kbits/s

Main + associated service (intended for simultaneous decoding)
< 512 kbits/s

A complete main service includes music, effects, and dialog. Each type of service could have its own program ID (PID) in the complete bit stream.

It should be noted that input phase and amplitude of the 5.1 input channels must be scrupulously maintained up to the AC-3 encoder. Associated channels must be encoded with the same time stamps in order to be decoded simultaneously.

Other audio compression techniques are MPEG-1 (layer-II) stereo audio encoding and MPEG-II (ISO) 1 to 5.1 (six-channel) encoding. These standards are used by DVB, the international DVD standard, and current DBS services (Direct TV, Echo Star).

11.7.5 Metadata

AC-3 has control data in its stream that tell downstream decoders how to process the audio data. These data about data are called metadata. This means that production, distribution, and final transmission facilities must be able to pass these metadata so they can be generated by the program producers.

AC-3 has automatic down mixing at the receiver, dialog normalization (dialnorm), and dynamic range control. Dialog normalization keeps the average loudness uniform between program segments but does allow loud audio elements through. This is done by determining the average dialog loudness and setting that to the same value regardless of peak loudness. This eliminates the need to change volume control between program segments or when changing channels (or media). If a listener feels the need to readjust the volume, the dialnorm value is not correct. A subjective method is to compare a program to a known reference and make them match (Dolby will provide examples of references). There are also objective tools, with integrating-type sound meters. Dialnorm is metadata.

Dynamic range control (DRC) keeps the loudness range under control and allows broadcasters to optimize the dynamic range (determines default). It also allows users to override the default to experience more dramatic, cinema-like audio. DRC is needed because listening conditions and product capabilities vary. In addition, down mixing in the AC-3 decoder needs peak protection since converting stereo into mono can result in a 6-dB increase, and converting five channels into stereo can result in an 11-dB increase. This is also metadata.

The metadata can be used to tell a mono, stereo, or a full-surround sound receiver how to mix the available audio channels. A mono mix will obviously be different from a stereo mix, etc. Metadata are part of the bit-stream information (BSI) sent in the header portion of the Dolby digital frame. BSI is used to tell the receiver the number and type of channels in use.

Facilities will need metadata pathways. A current problem is that there is currently no link for the networks and other distributors to send metadata to the affiliates, and even if there was, most broadcast facilities have no layer to handle it yet. Therefore, the local encoder must be kept in a suitable default mode.

Most metadata are static in nature. They only change with program format changes. An exception is the dynamic range data, which change in real time. Most of the data are set up during original production, while the dynamic range information must be set in postproduction, or adjusted during real-time monitoring at the master control. Generation of changes to metadata is known as *metadata authoring*. The channel status flag in the AES audio stream must be set to "1" to indicate that the AES stream is data and not a linear or baseband audio signal, so that nothing in the path tries to modify the data. Also, devices that the metadata pass through must be in unity, so no modification is attempted.

11.8 Bibliography

Boston, Jim, "SDI and Embedded Audio," *Broadcast Engineering*, September 1997, p. 46.
Dolby Laboratories, San Francisco, "Dolby Digital Broadcast Implementation Guidelines," 1991.
Hoffner, Randy, "Dealing with A/V Timing Disparity," *TV Technology,* November 30, 1998, p. 22.
Kapler, Robert, "Conversion Dominates at SMPTE," *TV Technology,* November 30, 1998, p. 8.
Moulton, Dave, "TV Sound: Mean Streets?" *TV Technology,* November 30, 1998, p. 32.
Pizzi, Skip, "Implementing Digital Audio," *Broadcast Engineering*, September 1997, p. 32.

12

DTV Philosophies, Strategies, and Implementation

As we all know, broadcast stations are wrestling with which entries in ATSC's table 3 to offer viewers. Looming over these considerations is the fact that Congress has let it be known that they want their HDTV. Network affiliated stations will naturally look to their networks for cues as to the digital infrastructure they will adopt. In almost all cases, HD plays some part in these plans.

The network HD distribution plans are

ABC.   720p60

CBS.   1080i60

NBC.   1080i60

Fox.   480p30 and occasionally 720p30

PBS.   1080i60 or multiple 480i60

As a result, new infrastructures will have to be added to most facilities. First, let's come to grips with the scope of a project like this. For those who have added digital infrastructure to an existing analog plant, you know that what you end up with is generally a digital layer over the existing analog layer. The common lament in this case is "if only we could start from scratch and build a digital only plant." However, those who have had the luxury of building a digital plant from scratch will tell you that, even then, you still end up with both analog and digital layers. One of the last bastions of live television is the "remote" or outside broadcast (OB) segment of the television industry. In many ways OB facilities are pioneers when it comes to implementing new technology. Anything that is lighter, smaller, adds versatility, or brings a new feature is a candidate for OB service. OB vans have evolved from NTSC to SD-SDI systems. Now HD-SDI vehicles are plying the interstates. The same layering of SD over NTSC holds true when building an SD digital truck from scratch. Most of you have already surmised what must happen when you build an HD truck. Yes, you need an analog NTSC layer, a SD-SDI (serial digital interface) layer, and an HD-SDI layer. You get to build three trucks in one.

For the uninitiated, let's look at why this is so. Although most sources have an HD path through the truck, not all sources have HD outputs yet. Almost all have SD-SDI, so some sources will also need paths to an HD upconverter. Additionally, just because you have an HD truck doesn't mean you still won't have SD customers. Therefore, many outputs of the truck will have to be downconverted back to SD, and up/downconverters are not cheap. They currently top out at over $100,000. As a result, only a limited number of these on a truck exist. This means that an SD digital router is required, along with an HD router. A remote truck today can find itself with upwards of 150 monitors onboard. Few of these usually need to be high-quality evaluation monitors, so it often makes economic sense to use lower-priced monitors (often with analog inputs) for less critical applications. Since most HD (and SD digital) equipment also has NTSC analog outputs, this infrastructure complete with its own router is usually desired. Surprisingly, the HD router matrix may be the smallest and the NTSC router matrix may be the largest. In some cases NTSC waveform monitoring still makes sense. Many video operators still find that camera balance and matching are easier to perform using NTSC test equipment than either SD or HD digital component test equipment because baselines are easier to quantize. In addition, the NTSC test equipment is much cheaper. Generally, the only paths that are solely in the HD layer are those tangential to the production switcher and DME/DVE.

The audio side of the truck has to undergo a similar stratification. The main audio layer today is AES, but often a sizable analog audio layer is present. The two audio layers will generally have their own routing matrices. A third level will probably evolve over time, as 5.1 surround sound, which requires six channels of audio, might have to be encoded into one or two AES channels because most VTRs today still accept only four separate audio feeds. This leads us back to the video. If a network distribution standard (network-to-affiliate feed) becomes dominant, this might extend back to the origination truck or the contribution feed, historically known as the "back haul." This would mean an ATSC-type signal might need to emanate from and be processed on the truck. Add a machine control layer to the truck and the result is eight layers of signals, each probably requiring routing capability.

Superficially, this is the fabric that needs to be woven into an HD truck, but the truck has to be useful and worthy of its clients. What amenities must be loaded into the van to accomplish this? Most first-rate trucks today must carry at least 12 cameras, but trucks carrying over 20 ply the interstates. A dozen VTRs onboard is not uncommon. Operational space and seating for several dozen people is not considered unreasonable. On top of this it is assumed that the mobile facility is extremely flexible and highly fault tolerant. Hence all the routing and patching on all the layers previously mentioned. Oh yeah, the interior has to be warm in the winter and, more important, cool in the summer.

As you would expect, everything about an SD-SDI truck is more than a NTSC truck, and everything in an HD-SDI truck is more than an SD-SDI truck. Let's start with price. To produce the facilities mentioned in the previous paragraph would cost $4 to $5 million on an NTSC truck. An SD digital equivalent would come in at $6 to $7 million. The HD example could easily approach nearly $10 million. Most big trucks on the road today have trailers which range in length from 48 to 54 ft, and have expandable sides to increase the interior "people" space. A ballpark figure of the weight for an empty trailer of this size is 40,000 lb. Most trucks this size will end up approaching the bridge weight limit in the United States of 80,000 lb. Generally, an NTSC truck could just make this weight and still carry all its assigned equipment. SD digital trucks have a much harder time accomplishing the same feat. Many trucks today travel in tandem with utility trucks or some of the operations or functions are off-loaded onto a secondary truck, often graphics or VTRs. Why? It just takes more stuff. While an NTSC truck might need 20 racks to house the needed equipment in the truck, an SD-SDI truck of similar capability could need up to a dozen more racks. This obviously cuts into the people space, hence the single, then double, and now triple expando trailer. The first trailers with single expando sides came out the curb side of the trailer. Next, a second expandable segment of the trailer was the area where the monitor wall would be in a production compartment turned lengthwise. The third expando trailer area is now out the back of the truck. Next comes the HD truck. It appears that an HD truck with similar capability to our NTSC and SD-SDI trucks could require up to 40 racks or more full of equipment. This new excess in equipment can make itself known in a number of other ways as well. While an NTSC truck's rack-mounted equipment might be in the 8000-lb range, the SD-SDI truck might weigh in at 10,000 lb, and the HD-SDI truck could tip the scales at over 14,000 lb. More equipment naturally means more power. However, digital SD and HD add an additional element. Component SD-SDI has a clock rate of 270 MHz. HD-SDI (all high-definition formats, along with all ATSC formats, are component) has a clock rate of up to 1.458 Gbits/s. High clock rates tend to force equipment to run in high-current-condition states a greater percentage of the time, which generally increases the power draw of that equipment. While an NTSC truck might consume 75 to 80 kW of power under full operation, the comparable SD-SDI truck would require in excess of 90 kW. The power requirements of the HD-SDI truck would probably consume at least 130 kW of service. Since an extremely small percentage of that consumed power in any facility actually ends up as useful signals, the generated heat is considerably higher in SD-SDI and HD-SDI trucks. These high-current requirements required on digital trucks mean that trucks no longer have the option of running on single phase. Most of these new trucks will require three-phase service. While 10 tons of cooling capability would suffice for the NTSC truck, 15 tons would be prudent on the SD-SDI version and 20 tons on the HD-SDI version. Actually, the environmental power requirements, lights and air conditioning, usually consume the bulk of the power. The hypothetical HD truck's tech power would probably be around 50 kW. The other +80 kW would be to maintain the internal environment.

There is another reason why the hurdle is raised when contemplating the HD facility. The energy content in an HD signal only half jokingly seems closer to light than dc. In an NTSC facility any energy running through your coax higher than 10 MHz was either noise or spurious harmonics. In an SD-SDI facility energy over 1 GHz down the coax was normal. Now with HD, energy approaching 4 GHz is desired. Why so high? SMPTE 259M and SMPTE 292M (SD and HD serial transmission standards) specify bit-shuffling algorithms that are designed to create lots of "edges," which are needed in a self-clocking system, which both of these are. In fact, enough edges are created so that, for all practical purposes, both signals can be considered square waves. Each bit cell, 1/270E6/s for SD-SDI and 1/1.5E9/s for HD-SDI, can be thought of as one-half the period of that square wave (see Chap. 4). What does it take to make square waves? From Fourier's teachings we know that signals with even symmetry, such as a square wave, require a fundamental sine wave and the odd harmonics to construct. Thus, component SD-SDI requires a 135-MHz fundamental sine wave and the odd harmonics. The most important harmonic is the third, which is 405 MHz for component SD-SDI. When the third harmonic's amplitude drops below 6 dB above the noise floor, the dreaded "error cliff" has been reached, as the SDI signal has effectively stopped being a square wave and has become a sine wave. This prohibits the serial receive circuitry from reliably detecting edges, and self-clocking at the receiver becomes problematic. How is HD-SDI different from SD in this regard? Multiply by 6 and you've got the answer. HD-SDI's fundamental is around 750 MHz. Its third harmonic is a mere 2.25 GHz, which will severely limit your choice in cabling. Keep in mind, though, this 6 times differential also equates to picture quality. Whereas an SMPTE 259M SD-SDI picture would have 691,200 active elements in a frame of video, SMPTE 292M HD-SDI has 4,147,200. This means a wide shot of the ballpark in HD will be nearly as engaging as sitting in the stands.

This segues into another challenge in building the HD-SDI facility—on wheels or on cement—cable size and weight. An interesting corollary is racked equipment weight versus overall cable weight. These weights usually are very close to each other. On a truck it is imperative to eliminate any unnecessary weight. One way to do this is to use the lightest and, therefore, usually the thinnest cable possible. The trade-off is that thinner cable tends to have greater losses as a function of frequency. Thin coax for NTSC tended not to work, as it rolled the chroma off. It seemed to hit its stride with SD-SDI, as it still tended to roll off higher frequencies much faster than lower frequencies, but the lengths used on trucks didn't cause the third harmonic in SD-SDI signals to attenuate to a level anywhere near problem levels. This is not so with HD-SDI. While SD-SDI might travel 600 ft down minicoax with no problem,

HD-SDI is limited to 200 ft down the same coax. So as far as trucks are concerned, the renaissance of minicable seems to be in the SD-SDI domain. As alluded to at the beginning of this paragraph, cable weights for NTSC, SD-SDI, and HD-SDI trucks should be in the 8000-, 10,000-, and 14,000-lb ranges in association with the weight of the racked equipment. Cable weight on an SD-SDI truck might break this rule due to the use of minicable. The cable size found on the truck is not only dictated by the flavor of video coursing through it. Audio has similar problems. If AES-3 audio is used over twisted-pair, it requires true 110-Ω cable because AES audio pumps energy into its cables comparable to NTSC video. Reflections at these frequencies become important. Cable that is 110 Ω tends to be larger than 600-Ω cable. With analog audio the upper frequency limit is typically 20 kHz. Even if cable with the wrong impedance is used, the reflections are such a small percentage of the overall path length that they are not noticeable. At AES frequencies, though, they would cause problems. This means that AES requires the right impedance. Therefore, larger-diameter cables must be used, which means that AES audio adds to the cable weight and space used on the truck.

AES adds complexity in several other ways. Large AES mixers can be thought of as audio routers with internal digital processing. This means that outputs are not hardwired to any particular purpose and can usually be specified to be either analog or AES. As a result, AES mixer outputs can sometimes be confusing as you try to decide what layer they belong in. AES mixers are serious DSP devices. They tend to be driven by elaborate computer-based systems, which means that UPS power systems need to be employed to ensure that power glitches do not become major audio events. Auxiliary analog mixing systems used as fail-safe backups should be employed on HD trucks. Equipment with AES interfaces also requires a reference timing signal—either AES's Word Clock or analog video reference. It is often customary to use the video reference in the AES mixer and to use the mixer's AES Word Clock reference to lock out the other equipment. AES allows flexibility not possible with analog audio; it allows audio to be embedded into the video. Up to 16 channels are supported, but the majority of equipment today uses only four channels, although eight-channel use is now beginning to be accommodated. However, embedding and unembedding this audio adds to a system's complexity. In addition, routing at the SDI level doesn't support embedded breakaway. AES routers tend to be smarter about the serial bit stream than SD/HD-SDI routers (AES routers often decode the bit stream, whereas most SDI routers don't). A final consideration is that, like HD test equipment, which is more expensive than SD-SDI or NTSC test equipment, AES test equipment is more expensive than analog audio equipment.

Cameras become a slightly different animal in the HD world. The triax camera no longer exists. Triax is basically coax with an additional shield. Although the losses are slightly less than with regular coax, after a few hundred feet the HD-SDI signal will have fallen apart. Also, as we discussed earlier, the HD-SDI signal will consume most of the available bandwidth of the cable. In triax camera systems we not only need to get HD video from the camera head to the camera control unit, we usually want to send control, return video, intercom, and often audio, not to mention power, down the same piece of copper. Triax just isn't going to work for HD. A special fiber-copper cable is needed. This cable has two single-mode fibers, two wires for camera control, and four conductors for power. Its diameter is approximately 0.36 in, and it comes in 50- or 250-m lengths. A 250-m cable weighs 57 lb. At least seven of these 250-m cables can be connected together, which means camera runs of over 1 mi are still possible.

Sync and test generators in the HD realm aren't yet as capable as their SD ancestors. It still might be necessary to use either an analog or digital SD sync generator as the absolute reference. One additional note about monitoring in this new world: Truck customers usually want the ability to view all the formats on a video and waveform monitor in three separate positions in the truck during the production as insurance that the signal is usable in all worlds. This multiformat monitoring is usually located at the TD/director, shading, and QC positions.

HD integration is like everything else we compared between HD-SDI, SD-SDI, and NTSC. It just takes more—more planning, more understanding, and more fortitude than the earlier SD formats. But the payoff, if we can enlist the viewer to subscribe to the advantages of high-definition viewing, will benefit both us and our end users. Well, as the Chinese adage goes, we are living in interesting times.

12.1 Understanding Your Facility's Business Plan

The engineering approach taken will obviously revolve around the business plan your facility's management adopts. Currently, three major business plans are under consideration by most broadcasters: high definition, multicasting, and interactive/datacasting.

All three approaches revolve around the use of the additional 6 MHz of bandwidth that comes with the newly allotted DTV channel. High-definition programming will consume a large portion of the DTV bandwidth but not all of it. As compression technology increases, the amount of bandwidth needed will be reduced. Intelligent encoding will also determine how much bandwidth is required. HD scenes of wide vistas with little motion won't require nearly the bandwidth that an HD sporting event will need. Since the most compelling applications for DTV currently revolve around HD, especially for the few consumers that have invested early in DTV, many stations will need to install HD levels on top of SD-SDI and their original analog video level.

The second approach is to multicast, that is, the broadcast facility will transmit more than one SD channel. There is a solid consensus by television management that multicasting does not mean the addition of personnel. Multicasting must be accomplished using the resources currently available, so that a facility that transmits multiple SD program streams probably won't have multiple master controls. It is most likely that only a single master control operator will be expected to manage multiple program streams in the future. Automation will become increasingly important as multicasting facilities come online.

Automation will most likely become much more important in all broadcast facilities. Just as switchers and routers (which, in turn, issue machine control commands) are controlled by automation systems, video preprocessors and ATSC encoders and multiplexers will need to be controlled in the DTV era. Things like noise filter selection and ATSC bit-stream bandwidth allocation will have to be under automation control as those types of settings will often have to be set to optimize the video/audio quality, depending on program content.

A variation to multicasting is to send a number of program streams which are actually related to the same program, for example, a telecast of a sporting event or even a regular program where the viewer gets to select his or her point of view. For sports events one subchannel could be for game play and another could be devoted just to replays or other isolated shots. For news one subchannel would be for the normal newscasts while another would be devoted to the top story or breaking news and another to weather or sports. This approach might ease the burden at master control as all main and subchannels would most likely go to commercial breaks at the same time.

The third current approach (others will certainly surface) revolves around data. Leftover bandwidth can be used to ship data to the home. However, the data will have to be such that they can be used by many consumers or where DTV can add value to the signal. DTV data bandwidth, even if the whole channel is devoted to data, is not wide enough to derive much revenue from commodity-type data. All DTV channel bandwidth devoted to data would yield the equivalent of 12 T1 channels. Even in the smallest television market, the revenue from 12 garden variety T1 circuits would not begin to replace normal advertising revenue. As a result, the data transmitted must be useful to many people or be such that customers will pay a premium for them.

The most common type of data now talked about for terrestrial transmission is Internet Web sites. Currently, systems are available that allow the gathering of Web pages from Web sites. These gathered pages, which consist of HTML code, Java applications, etc., are then inserted into the ATSC stream and transmitted. Customers of these data would most likely have a special card in their PC which would demod, demultiplex, and decode the data (Web pages). These Web pages would then be stored (or cached) on the consumer's hard drive. Currently, there are no DTV receivers that will deliver data to a port on the back of the set. When the customer points a browser to a cached Web site, the browser will use the site data stored on the hard drive, resulting in a much faster Internet experience. The broadcaster would continue to stream updated Web pages to the customer's PC. A broadcaster using this system would need a wide and reliable portal to the Internet.

Many believe the DTV killer application will be interactivity. Interactivity has been heralded as television's next big evolutionary step for the last 10 years. Maybe with the advent of DTV it will finally live up to its hype. ATSC streams could easily carry supplemental program data that could be called up by a viewer. The viewer's responses to the supplemental data, say information about ordering a pair of shoes during a shoe spot, would need to be sent via landlines. Possibly the whole transaction would need to be a landline since technology for secure transactions via the Internet is becoming mature while a conditional access infrastructure would be needed if any part of the transaction was over the airwaves.

Another type of data transport by broadcasters would be to provide specialized data for individual customers. It would have to be data that customers would be willing to pay a premium for. An example niche might be for customers who need bursts of data occasionally but not a dedicated leased line, much like the old dial-up data transfer. Customers with no landline access might find this service appealing.

12.1.1 Top-down report

The top-down report was a suggestion of the ATSC Implementation Committee to give stations some guidelines as to how to approach DTV implementation. It was purposely limited in scope and excluded production and postproduction issues. The report inventoried the various systems and the interfaces that could potentially exist in the DTV station. The report addresses the fact that some stations will have the luxury of building digital SD or HD plants that produce not only the ATSC stream but also the customary NTSC stream, while most stations will use their current analog infrastructure and encode its output into an ATSC stream.

The report defined the paths through a television plant in a number of areas and also among a number of planes. Areas included are I/O, routing, encoding and multiplexing, data services, and redistribution of signals (cable headends, uplinks, fiber, STL, etc.). Planes included video, audio, monitoring, control, and timing.

The report identified 21 separate operations done at master control and 14 done at production control. The operations identified as *master control* are

- Network feed
- Network news service
- Independent news service
- Local live feeds—studio, newsroom, ENG truck, satellite truck
- News programming/production studio
- Spots—national/local
- Station promos, PSAs, etc.
- Syndicated programming—legacy and digital

- Station branding and ID
- Pushback (DVE for weather, school closings, live news events)
- Video crawls
- Audio under/over
- Weather alert systems
- Automation control
- Multiple outgoing streams
- Audio issues—mono, LoRo, LtRt, 5.1
- Multiple channel broadcasts
- Encoder switching SD/HD
- Encoder switching data rates
- Traffic interface with reconciliation
- Live external closed caption

The operations identified as *production control* are

- News
- Remote—parades, elections, disasters, pool feeds
- Field acquisition
- Legacy formats—U-matic, Betacam, MII, Beta-SP, digital Betacam, DV, DVC Pro, Beta-SX, digital -S, D-1, D-2, D-3, D-5
- Live microwave/satellite
- Computers—control presentations, CG, system playbacks, etc.
- Captioning system/prompter
- Live—latency problems
- Transitions other than VI (DVE, CG, video mix/effects, chroma-key, virtual sets)
- DVE, CGs, and SS
- Weather graphics, weather radar
- Cameras
- Audio—mono, LoRo, LtRt, 5.1
- Live foldback to the field (mix minus)

Due to the potential for many different digital formats, the report suggests that a facility select a native format that other formats arriving from outside the facility are converted to. It also suggests that the station's native internal format is not necessarily the format that is encoded for transmission. Almost all DTV receivers will have native scan formats that might be different than

most production formats. Also, some stations will elect to do all internal processing in either SD or HD, and then upconvert the SD to HD or even down-
convert HD to SD. The station's native format will be influenced by many
things—network affiliation, group ownership philosophies, equipment cost,
and legacy equipment, along with DTV migration policy. It is suggested that
stations that start with SD and migrate to HD over time use SMPTE 305
(SDTI) as a migration bridge between SD and HD. The possibility that there
will eventually be an HD SDTI format could lead to HD baseband and SD
baseband routing using the same router. The report even suggests a single format at the 1.485-Gbit/s rate with all formats mapped to it. SD would simply
have many more words between EAV and SAV TRS than HD would.

Large markets might see HD native formats, while few, if any, small markets will. Some large market stations might process all video as HD, even if it
started life as SD. Most small market stations will, at most, pass HD with very
little processing, if at all. Their working native format will be SD. The top-
down report suggests the following choice of native formats:

1. Analog NTSC 170M
2. The format of 480i 4:3 or 16:9 (125M) carried by SMPTE 259 Rec. 601
3. 1080i or 720p using SMPTE 292M
4. 480p using SMPTE 293 or 4:3
5. Intermediate compressed formats (SDTI)

The report also points out that the digital component has different production
(SMPTE 125/259) and emission parameters. The production format (pixels/lines)
is 720 × 480, while ATSC encoders produce 704 × 480 (as per table 3 in A/53).

Audio will not have the same native format issues, but it will have the native
in-house number of audio channels that will be processed. Actual channels will
not necessarily equate to actual paths for audio as a four-channel Dolby surround signal can be transported through a two-channel physical plant.

The report also addressed four types of data that will pass through the plant:

■ Picture user data (closed captioning) which must be extracted and sent to
the program mux.
■ Program-related data carried on separate PID. If the data must be synchronized to an existing program, they require reference to that program's PTS
and must be muxed with the other elements that make up the program.
■ Non-program-related data carried on separate PID. Nonsynchronized data
that can be sent straight to the emission mux.
■ System data (internal scheduling data, PSIP source data, etc.). These data
could arrive from outside as part of the NTSC video or data residing in an
MPEG-type contribution stream.

The report also reminds the implementor that a metadata path will be needed
initially for the audio level and, eventually, for other levels as well. Even though

there is no technical reason to operate a digital component facility at 59.94 Hz, this frame rate will continue because it is impractical to operate a plant at 59.94 and 60 Hz simultaneously.

Another important aspect for digital television is the maintenance of lip sync. In the baseband arena frame stores and effects devices tend to delay the video behind the audio. Many boxes available today that cause video delays also provide equal delay for the audio accompanying the video. The report stresses that time stamping should occur when the audio and video are known to be in time. Time stamping now takes place at the video and audio encoders, so both should be locked to the same timing reference such as time code. Ultimately, the reference should be GPS. Video encoding takes longer than audio encoding, and HD encoding takes longer than SD. Care will have to be taken to ensure that video and audio timing do not get too far apart so as to overrun buffers in the DTV receiver.

Another consequence of encoding delay will be a problem that broadcasters with satellite remotes have had to contend with, that is, hearing their own delayed voices if off air is used for queuing. Satellite remotes have had this problem because the transit time for the signal from the remote uplink location through the satellite path to the station, not counting the program audio sent back via landline telephone or even return satellite path, could have well over a $\frac{1}{2}$-s delay. To prevent the talent from hearing his or her own delayed voice in the ear piece, the station would send back a mixed-minus, or all or some part of the complete audio mix, but not (or minus) the talent's voice. For remotes nearer the station, such as microwave shots, off-air audio could be used because the delay would not be enough to irritate the talent. Those who have done these types of remotes know how irritating it is during a standup to hear one's voice, delayed, in the ear piece. If you have ever watched a reporter doing a standup and pull the ear piece out during it, it is because someone has messed up the mix-minus. ATSC encoding delays will now mean that off-air DTV audio will not be usable for microwave remotes.

12.1.2 Throughput scenarios

The paths that a facility installs to handle the new formats will be determined, to a great degree, by the source of the programming the station airs. The simplest situation is a plant that airs its programming straight off tape such as an independent. The tape material is most likely a mixture of programming owned by the station and physically shipped in and of locally produced material. In this case the format will most likely evolve around the tape and storage format selected by the station or, more likely, the format settled on by the program producers. In this day of satellite fiber paths some material probably arrives via that route, which can greatly complicate the choice of formats. Almost all stations, especially network affiliates, face this challenge.

Some programs can be sent through multiple transmission paths before they reach the end viewer:

Acquisition at the venue

Contribution feed to the network production center

Distribution to the local TV stations

Transport and transmission to the home

The ATSC standard only covers transport and transmission to the home. However, due to changes in image and audio formats the impact of ATSC extends to all segments of the program path. The various networks are addressing the contribution and distribution with different approaches. Fox is looking at 8 PSK (68-Mbit capacity) and 16 PSK (90-Mbit capacity) for distribution. Fox plans to send multiple SDs (possibly only for news feeds and the like) and only use 720p30 for special events. It is thought that most Fox affiliates will not initially process the 720p feed but just pass it through. Fox plans to move all feeds to affiliates over to digital.

CBS's SD feeds will be 480i60 at 4:3. The 1080i distribution will be at 45 Mbits/s, and SD will be at 15 Mbits/s. CBS will have separate digital and NTSC distribution feeds. For HD CBS will feed six channels of either uncompressed (three AES pairs) or lightly compressed audio. CBS will provide the QPSK satellite data modem and the HL decoder to affiliates which will have HD out and three AES audio outputs.

This leads to a number of possible station scenarios. Many of these will fall out as viable options soon, but they will be enlightening as to the possibilities that were originally envisioned:

1. Network sends ATSC-ready signal, simple passthrough, no manipulation.

2. Same as item 1, but station adds datacasting. Requires remultiplexing of the network signal.

3. Network sends ATSC, station does bit-stream splicing (switching in compressed domain). Everything in station must reach the splicer as ATSC. Allows network delay. NTSC signal can be upconverted to ATSC and fed to bit splicer.

4. Same as item 3, except network is sending multiple programs, and station can bit-splice all of them. Station must demux (but not decompress) transport stream from net. Transport in the station would most likely be SDTI so that the various programs could be routed. Mux would be needed to put ATSC transport stream back together. A few manufacturers have indicated that they plan to provide bit-splicing capability in the mux. Mux will need carefully controlled data rates from the local encoders, which means a program could be selectively decompressed into baseband so that a key, etc., could be added without affecting the other programs. Seamless switching would be required though. Also, a delay would have to be added to the original signal to account for the delay through the keying process.

5. Network sends HD signal that station decodes to baseband HD. Station must provide HD ATSC encoder. This is an expensive option, as everything else must be HD or upconverted to HD.

6. Same as item 5, except station downconverts to SD. Station starts as 259M plant. Does ATSC MPEG encoding at output of station. Station can still be 16:9. The 19.34-Mbit/s ATSC bit stream will not be fully utilized unless additional program streams are added.

7. Network sends compressed HD that is uncompressed to HD, then either goes to HD switcher or straight to an HD ATSC encoder. The output of the encoder goes to a bit-stream mux with dynamic selection. The local SD program(s), which have gone through an SD switcher (optional) and then an SD ATSC encoder(s), also arrive as inputs to the mux. This allows the station to switch between HD and multiple SD transmissions.

12.2   Engineering Considerations

To successfully complete any project, you must first clearly define the problem to solve or the opportunity to seize. Some will look at DTV as a problem to solve, others as an opportunity to exploit. Depending on the situation and the goals of the company, either may be the case. A number of steps must be taken to ensure that everyone knows and understands the goal. As we will see in Sec. 12.3, the project must be defined and its objectives and the strategy to accomplish it must be spelled out.

12.2.1   System engineering

There is an interrelationship triangle that is often trotted out by project managers (Fig. 12.1). This triangle illustrates that only two of the points on the triangle can be accomplished at one time. A project's outcome generally has three attributes. Is it done on time and within budget and did it accomplish the stated goals? Part of engineering a project is to ensure that these three points are doable. The triangle in Fig. 12.1 indicates that you can obtain two points on the triangle, but not all three. Project costs increase as the time frame is shortened. If the goal is to keep costs down, then it is unlikely that the project can be accomplished quickly. In either case the project needs to be done correctly so if it is to be done quickly, the costs will increase.

In order to meet the requirements of the triangle from the engineer's perspective, it is necessary to define the operating requirements and make sure the operating philosophy is understood.

12.2.2   Engineering economy

Here is where most engineers have been weak. It's convenient and shortsighted to look at engineering as simply an expense. Engineering in a television facility is the equivalent to manufacturing in other industries. It consists of line positions, as opposed to staff positions (accounting, legal, human resources, etc.), and it is actually a cost of sales. However, this doesn't get the engineer off the hook when it comes to looking out for the company's interest. It is everyone's responsibility, especially a manager's, to make a company profitable. Profitability is not

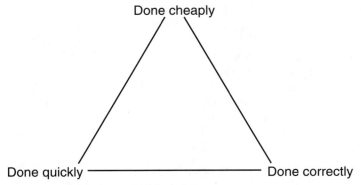

Figure 12.1   Interrelationship triangle.

a matter of spending but one of investing the company's money wisely and in a manner that meets the company's desired return on investment (ROI).

The systematic evaluation of the costs involved in and the benefits gained from a potential project is a discipline known as *engineering economy.* The tools provided by engineering economy are used for selecting between alternative designs by analyzing the economic consequences of each approach. Engineering economy lets you determine if a new approach or technology should be adopted, or if the old system is still viable.

A project usually is conceived when a problem or opportunity is recognized. This leads to a search for feasible approaches. Once the approaches that will possibly solve the problem are identified, economic analysis is applied to each to determine which provides the best return on investment. The approaches are usually mutually exclusive.

Engineering is heavily involved with transforming resources into goods and services. Good engineering and economic analysis allow invested capital to create wealth.

There are numerous types of costs that must be considered and dealt with when a proposed project is analyzed. *Disposal cost* is the expense incurred when a piece of equipment is retired or a system is shut down. *First cost,* also known as *investment cost,* is the money invested to purchase or obtain an asset. *Fixed costs* are costs that will occur whether your project moves forward or not. For example, there are expenses such as utility and janitorial that will be incurred whether or not you build a new control room.

Life-cycle cost is all the money spent on a piece of equipment, including first cost, reoccurring maintenance and operational expenses over its life, and, at the end of the life cycle, disposal cost.

Opportunity cost is the income that is not earned because capital is invested in something else. If your expenditure returns less than if the same money were invested somewhere else, you have an opportunity cost.

Standard cost represents the expense associated with a unit of output. The cost of producing a particular program is considered a standard cost. Each time the KU truck is sent out on location, standard cost is expended.

Sunk cost is money already spent that can't be recovered. The difference between what something costs new and what it is worth today is sunk cost. *Total cost* is the sum of fixed costs and variable costs.

Variable costs are the increases in costs because you move forward with your project. In the case of a control room, overall power consumption might increase and your janitorial service might want more money to clean an additional room. The additional expenses would be variable costs.

Costs can be considered overhead or direct. *Overhead costs* are expenses that can't be attributed to a particular item or activity. The cost of utilities is usually considered an overhead cost because the power consumed by each piece of equipment is usually not broken out. But a service contract on a particular piece of equipment is considered a *direct cost*.

A number of ways are available to evaluate the investment potential of a project as structured. This analysis uses mathematical series to produce scalars to determine financial value. The three scalars are:

1. *Future worth.* Takes all cash in/outflows and compounds them forward to a point in time. The future worth is the amount of money you would have in a bank account at a given interest rate for a period of time and a starting amount:

$$\text{Future worth} = (\text{starting amount} \times \text{interest rate})/\text{period of time}$$
$$= (\text{present worth} \times \text{interest rate})/\text{period}$$
$$= \text{present worth} \times (1 + i)^N$$

where i = interest rate and N = number of months.

2. *Present worth.* Equivalent worth of all cashflows relative to some base or beginning point in time. The present worth is what something is worth today, based on what it can earn in the future:

$$\text{Present worth} = \text{future worth} \times \frac{1}{(1 + i)^N}$$

Here we can see that the inverse of the interest rate is used.

3. *Annuity.* A uniform series of cash payments. This allows you to know the sum of money that will be in a savings account if a given amount is deposited each period at a given interest rate. Or, conversely, how much must be saved to reach a given amount in a prescribed amount of time. It also allows for the calculation of "annuities." Given an amount of money invested in an annuity over a prescribed period of time the payout can be determined.

To calculate a future account balance based on money deposited:

$$\text{Account balance} = \text{money deposited per time period}$$
$$\times \text{compound amount uniform series}$$
$$= \text{money deposited} \times \frac{(1 + i)^N - 1}{i}$$

To determine the account balance required for a desired income:

$$\text{Account balance} = \text{desired annual income} \\ \times \text{ present worth uniform series}$$

$$= \text{annual income} \times \frac{(1 + i)^N - 1}{i(1 + i)^N}$$

To calculate the required deposit to reach a desired account balance:

$$\text{Deposit} = \text{desired account balance} \times \text{sinking fund uniform series}$$

$$= \text{account balance} \times \frac{i}{(1 + i)^N - 1}$$

To calculate a loan payment:

$$\text{Payment} = \text{loan principal} \times \text{uniform series capital recovery}$$

$$= \text{loan amount} \times \frac{(1 + i)^N}{(1 + i)^N - 1}$$

As we can see, these scalars allow the computation of many basic time values of money amounts. In turn, these amounts can be plugged into more complex formulas. The annual worth of an asset can be determined using the capital recovery and sinking fund uniform series:

$$\text{Annual worth} = \text{annual revenues} - \text{annual expenses} - \text{capital recovery}$$

where

$$\text{Capital recovery} = \text{cost of initial investment per year} \\ - \text{salvage value per year}$$

$$\text{Cost of initial investment per year} \\ = \text{initial investment} \times \text{capital recovery uniform series}$$

$$= \text{initial investment} \times \frac{(1 + i)^N}{(1 + i)^N - 1}$$

$$\text{Salvage value per year} = \text{salvage value} \times \text{sinking fund uniform series}$$

$$= \text{salvage value} \times \frac{i}{(1 + i)^N - 1}$$

If the annual worth is greater than or equal to zero, then the project is economically attractive; i would be the minimum rate of return expected by the company.

Another method commonly used to determine a project's attractiveness is called the internal rate of return (IRR) method. Here a series of interest rates are applied to a formula with a given number of time periods. If the value of

IRR is at or above zero at the companies minimum attractive rate of return (MARR), then the project is attractive:

$$\text{IRR} = \frac{(i/100)\,(1 + i/100)^N}{(1 + i/100)^N - 1}$$

As was previously mentioned, it is everyone's job to ensure the financial health of a company. Although this section won't make you a financial expert, it might help you determine if the financial aspects of your project make any sense at all. This section might seem out of place at first but it is an essential engineering step that should be—but isn't always—done by the engineers.

12.2.3   Power and grounding

The electric service that is available should be determined to see how much headroom you have for additional equipment. Remember that not only tech equipment will be added but also HVAC, which will increase power consumption. As more power is drawn through a tech center, any ground-loop problems will be magnified. All technical grounds should be isolated from all other grounds right up to where the power neutral feed for station electric service is tied to the ground. Number 6 wire or larger should be used. Power-supply leakage to the equipment's case (which is usually installed in a rack) is the source of most ground-loop current. Stray capacitance and resistance leakage between the power transformer's primary winding and the frame (typically 1000 pF/1 MΩ) plus inductive leakage will induce unwanted currents into the rack. Most of the induced currents are common mode so the 60-Hz "hum" is induced in both the center conductor and the coax's shield. Equipment with differential inputs will eliminate the common-mode hum. It is the inductive leakage that tends to add the most ground current.

12.2.4   Heating, ventilation, and air conditioning (HVAC)

A simplistic joke that has been kicked around in the industry is that with all the power consumed at a television station the final output is a single video and audio signal sent to the transmitter. It is true that most equipment consumes a lot of power given its final output. Almost all of the power input to a device ends up as heat in the facility. Digital equipment tends to consume more power than the analog counterparts, with HD consuming even more. In many tech areas the ambient noise is greatly increased as digital equipment is added due to all the cooling fans in the equipment. A safe and rough way to determine the amount of HVAC needed is to assume all the power used in a tech facility ends up as heat to be dealt with. One watt of energy requires 3.4 Btu/h of cooling. Therefore, 3.54 kW requires 12,000 Btu/h or 1 ton of cooling.

12.2.5 Auxiliary systems

Auxiliary systems include items such as UPS systems for all the microprocessor-based equipment. A lot of equipment in use today is really just a PC in a case that makes it look like something else. Power hiccups can cause this equipment to crash or reboot, which can take time. Speaking of PCs, it is a wise investment to conduct a PC count. Determine what equipment needs PCs and for what purpose. Much equipment needs a PC for occasional setup, status, and diagnostics, but some equipment needs a dedicated PC or operation stops.

A lot of PCs expect to sit on a network. Determine what flavors of networking the computers expect. Network hubs, repeaters, and bridges might be needed. Computers on networks will most likely use IP addressing (or equivalent), and the vendor of those computers should be responsible for providing the addressing to eliminate conflicts. Additionally, be aware that not all networked equipment will interact well together on the same network. Some vendors' networked equipment won't tolerate another's use of the same network.

The hardest part of dealing with PCs is determining where to put them. A general rule so far has been that rack-ready computers can be harder to work on and are always more expensive to purchase. Non-rack-ready computers tend to use rack space very inefficiently if they can be shoehorned into a rack at all. A temptation for customers who are responsible for buying equipment and who need one or more PCs is to not buy them from the vendor. After all, the vendor usually charges 2 times or more what the nearby computer bargain store charges. The bargain could very well be the vendor's higher price. The analogy is like buying an Ampex VTR and a Sony TBC. When the video out is in trouble, who is responsible to fix it? PC hardware is tough enough to troubleshoot; misbehaving software is even tougher. Don't give the vendor an out by pointing to the hardware as the probable culprit. Anyone who has worked for a vendor who builds both hardware and writes software can tell you about the monumental battles that go on between software and hardware groups in the company pointing fingers at each other over whether it's the software or the hardware at fault. Make the vendor keep those arguments internal and not externalize them with your fate in the balance.

12.3 System Integration

System integration (SI), as it is often called, is the process of designing, installing, and certifying a system or facility that accepts the needed inputs and produces the required outputs in an operational manner consistent with the client's needs. Most projects also must integrate into a larger, most likely preexisting, system. The client can be your own production, operations, or news department or, in the case of an actual systems integrator, your engineering department. As we will see, you have different concerns if you are the client and not acting as the SI. First, though, we will walk through the steps necessary for you to get through the process of systems engineering and integrating a project.

12.3.1 Define the project

First, you must carefully understand what it is you are about to do. This might seem like a trivial step but often it is not clarified and much money and time are wasted before the project finally wanders onto the proper track. Often, this occurs after you have made some career-limiting moves and maybe after you have been kicked off the project. As mentioned already, find out what the client really wants. As an example, take a client who says he or she wants to go digital but really hasn't thought the project through. Facilities that originally didn't go digital now want to because it sounded cool or because digital makes it easier to accomplish what clients want to do. Due to the early successes, digital equipment is what is generally available and, therefore, is the type of equipment used. However, going digital is not a good goal, since you will be forced to go digital eventually whether you want to or not. Find out the real reason for the project. It might be the client wants to improve the on-air look or make the newscast more flexible or simply to cut down on maintenance and operational costs. Going digital is a given, but determine what goal the client wants to achieve.

A number of decision pitfalls occur in getting a group of people to decide on a common goal. One is called *anchoring*, that is, basing a decision on what came first, maintaining the status quo. The act of commission, or doing something, is often punished more than omission, or doing nothing, and making decisions that justify past actions is often the safest thing to do. A common example is audio impedance. Historically, impedance matching in audio paths was paramount because audio devices required power (both voltage and current) to work. The way to transfer the maximum amount of power from one device to another was to match the impedance of the two devices. For the telephone company impedance matching solved a fundamental problem, reflections. Even though audio frequencies are low, the telephone company had paths long enough to notice the reflections. Most broadcast facilities didn't have paths long enough for this to matter. Today, devices don't need much current, just voltage. So if the audio source device is set for low-impedance output and the receiving device is set for high-impedance input, the voltage into the receive device is doubled. This results in doubling the S/N ratio. It also allows audio patch panels to be wired so that you can bridge across sources and not affect the normal path. This is great for troubleshooting. Yes, there are reflections, but the paths are too short to matter. This requires the ability to develop a new mindset and throw conventional wisdom away.

12.3.1.1 Project objective and strategy.
The object of the project should be formally written down so all can read, accept, and sign off on it. Besides stating the scope of the project, it is just important to state how it will be done. Will in-house engineering be used or will an outside SI handle the project? Often a combination of resources is needed. Maybe you will design the system and have outside installers wire the system. Or the opposite could occur—someone else designs and you install.

A word here about professional installers. Anyone who has wired a rack knows that even when every effort is made for a neat installation, wiring can still end up looking "ratty." Good wire installers know how to dress cable harnesses and to facilitate service loops. They are also usually much quicker and less error prone than technicians who install connectors only occasionally.

12.3.1.2 Determine the scope of work. A common problem with any project is something called *scope creep*. A project is started and, as time progresses, other smaller projects and tasks are rolled into the original project. This is usually done without increasing resources or budgets, or at least not enough. To perform the given project well, blinders need to be used to the fullest extent that conditions allow.

12.3.1.3 Determine the required resources. Money obviously is a required resource, but often just as important is staff. How many of the right qualified people can you have access to and for how long? What is this expertise going to cost? Where is the equipment coming from and is there space to stage any of the accumulated equipment? What special tools and test equipment will be needed? In terms of PCs and other office equipment, is it needed and where will it go? If you have an outside SI perform the project, he or she will often require space for a project office.

12.3.1.4 Establish milestones, budgets, and project control. A PERT chart is often helpful to identify and track tasks that must be done concurrently. It also allows for easy identification of milestones and due dates. A *milestone* is when a significant task or event in the project is supposed to occur. Milestones can be delineations between phases of a project or a junction where a number of concurrent paths must be completed to start a single or separate set of paths.

12.3.1.5 Determine work breakdown. Once you have a handle on the scope and amount of what needs to be done, you need to determine who is going to perform the various tasks. If you are using outside contractors, you must clearly define what tasks are for them to do and which ones will be done in house. The delineation between in-house and outside responsibility is a major source of project confusion and delays. Even in-house people must be formally "signed up" to ensure proper execution. *Signing up* is the process of getting members of the project to agree to their part in the project overtly and to be able to articulate what part they are to play in completion of the project.

12.3.1.6 Generate equipment lists. This would seem straightforward on the surface but often it is not. The major "boxes" that are going to make up the system would seem obvious once the project has been properly defined. Most projects don't have the luxury of starting with all new equipment. Often the project is comprised of a mix of existing and new equipment. In many instances that will require additional interface or "glue" pieces. Related to this is a common mistake made with ordering new equipment. Many equipment vendors today sell switchers, DVEs, routers, and even cameras that, in their base form, are nothing more than frames with possibilities. In most cases equipment will have the necessary internal processing capability but no I/O installed. You must *carefully* determine and inventory the types of inputs and outputs to make sure the proper spigots are present. You must make extremely clear to the vendor exactly what inputs (video, audio, machine control, remote control, and time code) and outputs your system needs from their box. A mistake at this step can be expensive or, at the least, irritating to correct.

Along these lines, monitors are especially troublesome when trying to arrive at the correct configuration. How many, what screen size, color/black and white, I/O, and even power consumption need to be determined. Why power consumption? There are many examples of monitor walls that, over time, baked their monitors into the maintenance bench hall of fame. Additionally, you don't want—and probably couldn't afford—all monitors with SDI inputs, just as you wouldn't want all monitors to be color. It is extremely wise to talk to the end users of the project to determine what sources they expect to see and where.

One other thing: Equipment lists tend to be produced early on in the project as they are needed for budgetary planning. However, try to avoid having the early lists cast in stone as the additional engineering required (as we will see shortly) will most likely illustrate areas where the equipment list will have to be tweaked.

12.3.1.7 Investigate equipment attributes. Once the equipment list is initially determined, you need to determine how it will mesh with the rest of the system. You should do this with every major item on the equipment list.

Interconnect requirements. As we just mentioned, in many cases you will need to determine what sources you will feed the box and then ensure that the proper I/O cards are installed to support these inputs. The same will hold true with the outputs. Also, take inventory of what special connectors are involved and what the pin outs for the connectors are. Determine the type of cable that will be needed and how that cable is to be wired. At this time you should begin to draft a set of drawings that will include every type of cable and the wiring diagram of each. This should be included into the list of drawings that you will assemble to support the project.

Space requirements. Although most professional television equipment is designed to be installed in a rack, not all is, mainly equipment that requires PCs as part of the equipment package. Also determine the amount of space any control panels needed to control the equipment will take and determine where they should be located. Determine the rack space required and any additional space that will be needed for cooling. Equipment with air vents on the top or bottom of the unit might need space in the rack, above or below, for air movement. Determine the depth of the equipment. In many cases a single piece of equipment will determine the depth of the rack used (26 and 31 in are common sizes). It is equally important to know whether the equipment needs rack rail support or some other type of support at the back of the rack. If you are ordering new racks, or even if you are inheriting old racks, make sure that the rack has the hardware in the rear to support equipment installation.

Support requirements. Support falls into the category of things like a net connection requirement or a PC for setup and diagnostics. Does the unit require special power, such as 220 V or a "Hubbell" connector, because the unit draws more than 20 A? Does the unit produce excessive heat, requiring special air handling, or is the unit loud (usually due to cooling fans)?

12.3.1.8 Generate conceptual diagrams. After the preliminary homework, you are ready to create a conceptual drawing. In reality a conceptual drawing should have been done at project conception since it would be difficult to create things such as an equipment list without one. A conceptual drawing displays the basic building blocks necessary to satisfy the goals of the project. It includes routers, switchers, DVEs, frame syncs, storage, system I/O, and other major processing units. It will show banks of monitors but not individual monitors. It won't show patching. The purpose of the conceptual drawing is to give a general flow to signal paths in the project. There can be separate conceptual drawings for audio, etc.

12.3.1.9 Generate timing diagrams. Next, the first timing diagram will be generated. This document will be continually refined as the project's design becomes more detailed. In the analog video domain timing was concerned mainly with horizontal timing, ensuring that all signals arrived at one or more reference timing points in exact horizontal phase. Occasionally, a field or frame delay would be encountered because of frame syncs or effects devices. With the advent of digital video signals, horizontal timing is not very important as most digital equipment requires that signals to be within plus or minus one-half line and the equipment will buffer it into exact time. The problem now is equipment that causes one, two, or more line delays and one, two, or more field and frame delays. Thus, vertical time delays (due to line delays) and lip-sync problems due to multiple frame delays must be eliminated from the design of the system.

12.3.1.10 Generate ergonomic and workflow drawings. At this stage, the actual facility being created by the project will take shape. Even when a preexisting building constrains how functions will be organized, you must still lay out exact operating positions and control panel locations. Although not done very often, it is usually beneficial to create a flowchart depicting the functions performed in the facility. Also, the equipment to be used in the rack room should be determined. Many philosophies exist as to what should be in the centralized equipment room and what should not. The more equipment located in one place limits the chances of ground loops, makes timing easier, and aids in troubleshooting. A centralized location tends to limit access, which is generally considered good. Another big plus is that air temperature and purity can be controlled. Systemwide power conditioning and backup are easier. On the negative side power and environmental support are centralized. This often allows for a single point-of-failure scenario. Decentralized equipment placement allows more room for growth because, invariably, no matter how much space is devoted to an equipment room, it will never be enough in the long run. The decentralized approach also allows the people using the equipment easier access, which is not always bad. In this day of microprocessors in almost every piece of equipment, often running software that is prone to crash, it is much more convenient to let the operator reboot the system than have a technician with access to the equipment room do it. One other argument for decentralization is that video/audio to and from remote equipment is usually easier to run than control cables from the equipment room to operator control panels in remote locations.

Floor layout drawings. Once the equipment room scope is determined, along with the other necessary functions, rack and console layout can take place. By this time you should have inventoried how much rack space is required so that you know how many racks are needed. The same is true of operator console space. Many enjoy this phase of moving the furniture around to see how it all fits. A low-tech way that many use is with a basic floor layout drawing and cutouts of the rack and console pieces. If space allows, consider leaving space between racks as this accommodates wiring in the racks. Many installers like to run service loops that go all the way to the floor and back up to the top of the rack; the interstitial space between racks would facilitate that kind of wiring.

Rack and console elevations. Now, the actual location of equipment in the racks and consoles can be determined. The actual location is usually based on operations requirements. Most racked equipment with access needs is at eye level. Obviously, this is true for monitoring equipment. Patch panels should be centered at eye height or slightly lower.

Line-of-sight drawings. These drawings are especially needed for operations positions, mainly in control rooms. Each position is checked for obstacles, such as heads, that block the view of monitors. Also, a feel for the ergonomics of the operation position should be acquired.

12.3.1.11 Generate the document list. At this point, a number of documents should have been created. They are

Conceptual drawings

Equipment lists

Cable schematics

Timing diagrams

Workflow charts

Floor and console/rack layouts

Rack/console elevations

Line-of-sight drawings

To this list add the document list. The following drawings will be in the document list. However, now you should have enough information to determine what detailed drawings are required. These detailed drawings will be grouped into layers or levels, much like router levels: video, audio, time code, and machine control. Additionally, most facilities will need at least one computer network layer. A drawing will be needed in each layer for each control room, editing room, master control, signal routing, signal distribution, etc. All the required documents should be listed. For example:

001	Video conceptual
005	Audio conceptual
010	Control conceptual
020	Workflow chart
030	Video timing diagram
040	Audio timing diagram
050	Equipment list
100	Master floor plan/console layout
101	Master control console elevations
102	Master control line of sight
110	Production control room floor plan/console layout
111	Production control room console elevations
112	Production control line of sight
120	Equipment room floor plan/rack layout
121	Equipment room rack layout
200	Video router synoptic
201	Video distribution/processing synoptic
210	Master control video synoptic
220	Production control video synoptic
300	Audio router synoptic
301	Audio distribution/processing synoptic
310	Master control audio synoptic
320	Production control audio synoptic
400	Machine control synoptic
500	Time code synoptic
600	Network synoptic
700	Cable schematics

You now have determined all the individual pieces of the project that need to be designed in detail. The document list is actually an intricate part of the design process. Like everything else, there are different philosophies as to the exact nature and use of the drawings. Many implementers view the drawings strictly for design and implementation, with little regard as to how those drawings will be used for maintaining and troubleshooting the system. As a result, many more drawings are usually produced with fewer items or smaller pieces of the overall project on each drawing. Some people like to see large areas of the project on each drawing, in extreme cases the whole project on a single drawing. While this might aid troubleshooting a system, assuming you had the wall space to display the drawing, it would make the builder's job much more difficult. A large project often must be spread across many drawings so that many engineers can work on the system at the same time. Many drawings allow better document version control. There's nothing worse than building a system to an old version of the design. Many designers trying to add to a single document won't work. Some consider redoing the drawings once the project is built.

12.3.1.12 Calculate power and heat loads. Actually, this step can be done whenever the equipment list and placement have been settled. As discussed earlier in this chapter, if more power is consumed, it will generally result in increased heat load. You must also determine the amount of power load in each individual rack to ensure that a particular rack has enough electric service.

12.3.1.13 Initial design review. Once the previous steps have been completed, it is time to have all involved parties meet for what is commonly referred to as the *initial design review*. At this meeting, the information gathered, the conceptuals, layouts, project goals, and strategies are reviewed and budgetary details should be approved. This is the time to ensure that all project team members are still signed on to carry out their contribution to the project. This meeting should finalize the conceptual design before proceeding with the detailed design.

12.3.2 Designing the system

Next, the detailed design is carried out. The high-level conceptual design has been completed, so now the nuts and bolts of actually building the project have to be put on paper.

12.3.2.1 Generate mnemonics lists. The first thing that should be done is to generate a mnemonics list. Most neophytes to the process are surprised that this comes early and not later in the project. The main reason why this list comes first is so that the mnemonics used throughout the project are uniform. The engineering group, along with the end users, should generate this list. These mnemonics will be used on drawings, patch panels, and electronic labels in routers, switchers, under monitor displays, etc.

12.3.2.2   Router, audio mixer, and switcher(s) I/O lists. An overall inventory of inputs required for all devices, including all inputs used for monitoring, should be developed at this point. This list will indicate the amount and type of distribution needed. In most systems the design starts with the router. All signals that need to go through the router are determined along with destinations needing router feeds.

12.3.2.3   Detailed design of audio/video/time code/control/auxiliary systems. This is when the individual synoptic drawings, that is, the schematics of the various sections of the project, are created. Each engineer involved will have one or more, probably many, of these drawings to create. The drawings should indicate every I/O port on every device associated with the layer the drawing represents (video, audio, etc.). The location of every box on the drawing should be indicated, often right below the box. Conventionally, the name of the box and actual vendor and product name are located inside and at the top or bottom of the box. At this stage, jackfield locations are determined and included in the drawing. However, the actual jackfield will not be determined until the next step, after which it will be added to the synoptic. Pointers to other drawings will be included for all signals that arrive or leave the drawing.

12.3.2.4   Patch-panel layout. All the jackfields called out on the individual drawings are compiled and linked to physical patch panels. At this time, you must reconcile actual patch panels to signals that must travel through a jackfield. Most systems have all router I/Os on patch panels. Generally, the more patching, the better. Modern patch panels are good for 50,000 or more operations. They seldom fail. Patching allows a facility to have much more flexibility, which allows a facility to perform operations often not envisioned in the design. It also allows reconfiguration in the event of equipment failure and it allows for easy troubleshooting of system problems. These attributes all make the facility, and you, look better.

12.3.2.5   Specify cable use. This is simply a step where you determine path lengths, signal frequencies, and attributes along those paths; determine the number of conductors needed; and go through the cable catalogs selecting the proper cable. Conductor diameter and construction, along with shielding, loss, and jacketing, will usually come into play when selecting cable.

12.3.2.6   Determine cable types and color coding. With all the various flavors of audio, video, and data running around a facility today it is often useful to use different colored cable jackets to signify the various signals that will be found in the rack. For example:

Black   Analog reference

Blue   NTSC video

Yellow   SDI video

Green   AES audio

Red   MPEG video

The disadvantage to this approach is that ongoing discipline must be maintained to abide by and maintain this color-coding scheme. Also at this time, all the needed cable pin-out drawings should be completed.

12.3.2.7 Generate wire lists. If you are using professional installers, this step will be mandatory. If not, it is still a good idea, if only during the installation phase. However, if kept up on an ongoing basis, it has been known to help in troubleshooting a system. Wire lists are just that; each cable is given a number, a source and destination, wire type, length, and entered onto the list. This list is usually kept in a spreadsheet that can be sorted many different ways. When sorted by type, it lets the people building the cables set up and cut and connect all cables of the same type at once. When sorted by source and then destination, bundles from one area to other areas can be built.

If wire lists are only to be used for building, then labels should be put on each cable end that identify where both ends of the cable terminate. These labels can become hard to read if a lot of information is put on them. Another approach is to keep an on-going wire list and only put a single alphanumeric label at each end of the cable. A workable approach, this label can indicate the type of signal through the cable and the source and destination areas of the cable. The detailed connection location would reside in the actual wire list. For example:

S240714   SD video from rack 24 to rack 07. This is the 14th cable between the two areas.

A35MC04   AES audio from rack 35 to master control. This is the fourth cable between the two areas.

This approach requires additional discipline to maintain the list.

12.3.2.8 Final design review. Once the detailed design steps have been completed, it is time to meet again to review the final design. First, the project manager should restate the project's goals and strategies. Then someone should go over the conceptual drawings that were approved in the initial design review. Next, each person who was involved in the design should present his or her part of the design to the group. Deviations from the initial design should be pointed out and discussed. Once again, a consensus should be reached among the group as all need to sign on a final time to complete the last stages of the project.

12.3.3 Building the system

At this stage, cables have actually been cut, connectors installed, racks placed and stuffed. This process is often complicated when an existing plant, already in place, must be integrated with the new equipment but kept on the air at the same time. However, like the design, the router is usually installed first along with the domain change and other processing equipment. This allows signals in the legacy system to slowly be moved over and used in the new system. As the system is being built, the synoptics are marked off and modified as changes occur. The objective is to use the marked-up drawings to create a final set of "as built" drawings. Wire numbers should be added to these drawings.

12.3.4 System shakeout

Often as the system is being built, testing of the system starts. The system should be thoroughly exercised and wrung out as soon as possible to catch any design flaws or installation mistakes that have occurred.

12.3.4.1 The test plan. Like all other aspects of the project, you must design a plan to ensure that the system thoroughly works, and works as planned. These are two different things; the fact that each box, connector, and patch bay works does not necessarily mean that the system does what it was intended to do. As a result, the test plan should have two goals: (1) to test for path continuity and equipment operation and (2) to see if the intended functionality has been achieved. Encourage end users to look over the system and ask questions as it comes to life. Admittedly, there is a fine line between simply "shooing" interested parties away and letting their questions, criticisms, suggestions, and even just their presence slow or stop the project. If handled properly, these people might stop a mistake before it gets too embarrassing or expensive to correct. Many people who couldn't offer much input in the conceptual stage will offer a great deal when they can see and touch the system. Part of your job is to filter out the casual observations while latching on to a nugget of inspiration.

12.3.4.2 Proofing the system. This is another name for ringing out the system. As we have said so often earlier, digital systems do not generally fail gracefully; they work well up to a certain point and simply fail. You need to be able to determine how far you are from the error cliff. In Chap. 4 we discussed using the third harmonic of any digital signal to gauge error-cliff distance. There is another indication that you are very near, but not at the error cliff. This situation can be found using what have come to be known as *pathological test signals*. The receiver circuitry in a serial digital receiver must regenerate the clock signal. To help it do that, most ASICs devoted to receiving serial digital signals equalize the incoming signal to boost the high frequencies so that clock regeneration and data value determination are easier. Pathological signals produce bit streams that stress these circuits. Many devices produce these tests signals, including some actual production equipment, like Sony's digital Betacam VTRs. One common pathological signal stresses both the clock regeneration and the equalizing circuitry by producing values for C and Y that force the bit-scrambling circuits to produce a run of 19 zeros and a single one approximately every frame. Because of the nature of NRZI, the single one ensures a phase reversal for the next run of 19 zeros. This stresses the equalizer circuitry in the receiver by providing a large dc component "blast" every so often. The fundamental of this signal is at 13.5 MHz, which adds to the low-frequency energy component.

Other common types of pathological signals have C and Y values that produce runs of 20 ones, followed by 20 zeros periodically. This produces edges at only a 13.5-MHz rate, which is one-twentieth of the optimum zero crossings. This rate stresses a receiver's clock recovery circuitry by making the receiver circuit's PLL "coast" for long periods of time. It should be understood that there are literally thousands of possible Y and C combinations that can produce pathological bit streams.

Experimentally, it has been found that a path will fail with a pathological signal 2 dB above the point where a nonpathological signal will fail. Therefore, if there is any doubt about a signal being near the error cliff, this signal will help to confirm that.

Another killer of digital signals is jitter. Although the use of a time domain reflectometer is the best tool for seeking impedance mismatches along a path, a digital oscilloscope set to a long persistence will give hints that reflections are occurring. Figure 12.2 shows a path that is double terminated. The receiver looking at this data stream will probably report many errors, if it is able to recover data at all. Reflected energy, not only out of phase, but possibly many bit cells earlier than the present data, is creating a ringing effect at transitions.

Also, in the real world impedance is not constant as a function of frequency. Impedance is a complex number (or value). A *complex number* is a number that not only has a magnitude but also direction (or phase). It is the ratio of resistance and inductive and capacitive reactance that determines the magnitude and direction, or the phase angle, of the resulting impedance. Figure 12.3 is a polar plot, produced by a network analyzer, of the reactance versus resistance through a patch jackfield as frequency is swept from 300 kHz to 1000 MHz. The X axis represents the resistance of this path, while the Y axis is the reactance.

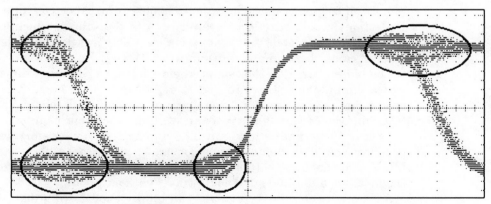

Figure 12.2 Double terminated path.

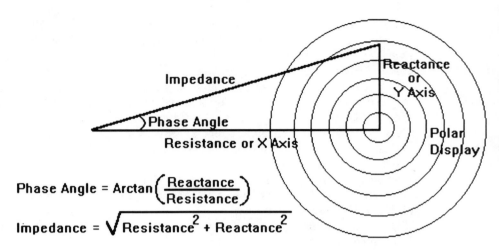

Figure 12.3 Real/imaginary triangle that is the base of polar plot.

The positive Y axis represents inductive reactance, while the negative Y axis represents capacitive reactance. The origin (center) of the X axis represents the nominal resistance of the path. Look at the right triangle in Fig. 12.3. You can calculate the impedance and its phase (polar form) by treating the magnitude of the resistance as the base of a right triangle and the reactance as the side opposite (the phase angle). The hypotenuse of this triangle is the magnitude of the impedance, and the arc tangent of the ratio of the side opposite over the base will provide the phase angle.

As can be seen in Fig. 12.4, the resistance and reactance are changing as a function of frequency. When the trace is not on the X axis, the total impedance will not be equal to the resistance. When the impedance in one segment of a digital path is not equal to the impedance of the next segment of that path, reflections result. Any time the plot on a polar chart is not at the origin, reflections will result. Figure 12.4 (SDI through a jackfield) demonstrates that, at almost all frequencies, the impedance is not at the expected value (the origin). Reflections result in reflected power back to the source, which lowers power transmitted, and leads to transmission loss. This jackfield is better than most when it comes to changing impedance and VSWR values.

The plot in Fig. 12.5 shows a jackfield where the reactance function is much worse than the one shown in Fig. 12.4. As can be seen here, at certain points on the plot the reactance starts to become a significant enough value that the total impedance versus the resistance will be such that significant reflections can occur along that path. Figure 12.6 represents a horror story when it comes to reflections. This illustration demonstrates significant reflections by the 270-MHz point (near the bottom Y intercept). This jackfield would greatly reduce the distance a digital signal could travel without errors.

As can now be seen, reflections can totally shut down a digital path, even when all the other attributes of the path are normal and healthy. In fact, a

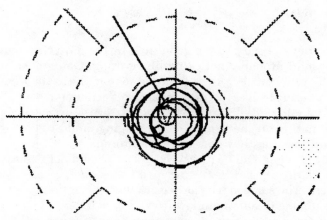

Figure 12.4   Polar plot of normal path.

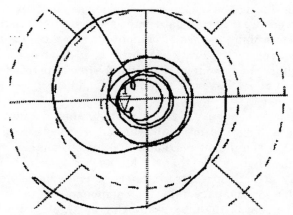

Figure 12.5   Polar plot with increased impedance versus frequency variation.

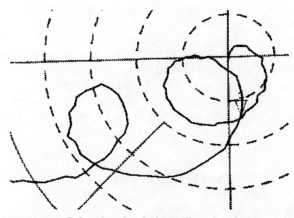

Figure 12.6   Polar plot of path that will produce bit-stream errors.

373

casual observer of the eye pattern generated by the data through the jackfield in Fig. 12.6 would probably notice nothing wrong. Going back to Fig 12.2, if you are having problems in a digital path and don't have the luxury of owning a network analyzer or even a time domain reflectometer, you should remain critical of the analog attributes of the eye pattern. Just because we are now "digital" does not mean that group delay, ringing, and pre/postovershoots are still not important, and they should be monitored.

Additionally, jitter should be measured. Jitter should not become troublesome because most paths are nowhere near involved enough to have sufficient non-clocking devices. The same is true for reclock devices that remove high-frequency jitter but are unable to eliminate low-frequency wander. A lot of equipment, as we will see in the next section, disassociates input timing with output timing. This starts the jitter budget over from scratch. Excessive jitter is usually caused by a box with a problem in its digital transmitter (output) circuitry.

Good engineering during the design and implementation stages of a digital video project should ensure that your data paths are some distance from the error cliff. In addition, determining the signal's amplitude above the noise floor (especially at higher frequencies), ensuring that jitter is as low as possible and that there are no reflections along your paths, should provide proof of performance and documentation that you have installed a well-engineered digital system.

12.3.4.3 System troubleshooting. It has become common today to refer to digital video as data. There is no doubt that in a video stream defined by SMPTE 259 or 292, a significant portion of the bit stream can be devoted to ancillary data. While slightly less than 10 Mbits/s are available in composite bit streams (barely enough for four channels of embedded audio), over 55 Mbits/s are available for ancillary data in a component digital stream. That's approximately 20 percent of the bit stream. Currently, almost all of the data present are embedded audio.

But back to the video—is it data or just discrete video samples? Well, in the *serial digital transport interface* (SDTI) bit stream all bits are thought of as data. What is SDTI? It is a serial digital signal conforming to SMPTE 259M, at 270 Mbits/s, where the active video portion of the bit stream is replaced by data. This data could be any type of data, but at this time the data are usually MPEG video and audio data. Since 422MP@ML has a data rate of approximately 18 Mbits/s and component digital video has approximately 207 Mbits/s of capacity, either multiple streams of MPEG data or MPEG data at faster than real time can be sent. The SDTI bit streams can pass through an SDI (serial digital interface) component video path and can even be displayed on an SDI monitor. Although you won't see recognizable video, you will see the data stream array on the raster.

Therefore, the point is that even when baseband serial digital component video is the bit stream, it is often helpful to think of this stream as all data. In the DSP realm, where video spends more and more of its time, this data stream would often be referred to as a number sequence. Each sample of luminance or

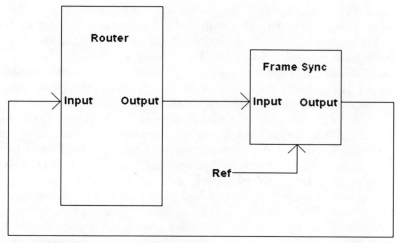

Figure 12.7   An example of video feedback.

chrominance information is just a number representing the original intensity value of a picture in spatial space. The beauty of digital video is that once the video signal is in the digital domain, these number values should never change, not by one literal "bit." You should be able to send this stream of "numbers" or values on a never-ending journey and not have any change in their value.

Is that really true? Like most sweeping statements, it depends.... First, let's describe a scenario that would be a mistake in an analog plant. Take one analog frame store (yes, we all know that internally it's a digital TBC on steroids, but it still has analog ins and outs). Feed a signal to the frame sync from the output of a router. Next take the frame store's output and run it to an input of an analog router. Finally, switch the crosspoint feeding the router to the output of the frame store (Fig. 12.7). What happens? Of course, "video feedback." A coherent picture quickly dissolves into a blur and usually pulsates or undulates at some rate, based on the overall resonance of the path. Not useful. Let's try the same thing, but with an entirely digital path. What happens when the router crosspoint feeding the frame sync is switched so that the frame sync sees its output? Nothing. A still frame occurs. The output of the frame sync circulates through the router/frame sync path forever. In fact, you can add many other digital boxes to this path and have the same result. Why? Because our "number sequence" is faithfully being reproduced by each box in the path. In fact, you can keep this bit stream recursively transversing through the frame sync for many days (in a normal "healthy" path) and not see any errors. How would you know if you picked up an error? Well, if it is in the active portion of the video, you will see a "dot" appear in the picture. Once that error is generated, it too will continuously circulate with the "good" data. The longer the bit stream circulates through the path under test, the more transparent that path is.

Now this doesn't mean that you could send this digital bit stream through just any infinite path. It's generally believed that as long as a box reclocks the bit stream at its input or output (or both), the bit stream has a totally new "lease on life." Not true. Reclocking eliminates the higher-frequency accumulated jitter but not the low-frequency jitter (also called *wander*). The wander jitter component slowly continues to accumulate until it finds a reclock circuit it can swamp. It usually takes more than 30 reclocks before this has much of a chance of happening. Now you're probably wondering why this didn't happen with our "frame store" test. Let's remember what a frame store does—it disassociates input timing and bit rates (along with the accompanying jitter) from the output. A simple reclock circuit PLL is actually genlocked to the recovered clock information in the incoming bit stream. Although a frame sync uses PLL circuitry locked to incoming video to write to RAM, and PLL circuitry locked to local reference to read video back out of RAM, the local reference should have very little accumulated jitter, while incoming video might have a great deal. Hopefully, no one ever experienced this, but imagine what would happen if a genlocked sync generator fed a second genlocked sync generator, that fed a third, and so forth. Imagine what the tenth or so sync generator's reaction would be if the first one hiccuped. Thankfully, in the late 1970s the potential for this cascaded genlock scenario was ended by the introduction of the frame sync.

Now that we have demonstrated how you can prove that most digital paths are transparent, let's look at some of the reasons why a path might not be transparent. Let's cover the most obvious and talked about first—compression. This topic is discussed at length almost every month in most professional television periodicals, and we won't dwell on it here. Let it suffice to say that intraframe (JPEG) and interframe (MPEG) compression schemes are labeled as lossy compression schemes for good reason. Anytime your digital video bit stream is converted from the time domain to the frequency domain (which is what the discrete cosine transform does) and then scaled, you have changed the data that made up your original picture. The actual DCT process is not lossy in itself, as you can end up with more data than you started with; hence the frequency coefficients are scaled to make the small-valued ones go to zero. Go back to the time domain and then back into the frequency domain of the compressed world, and you change your video data some more. If you continually do this, the amount of data change will diminish exponentially, meaning that the fifth or sixth change will be very small compared with the first change, with one caveat. If any part of the picture is changed, a key laid on the existing compressed video stream, a wipe added, a mix, whatever, and the exponential degrading for that area of the video starts over. However, this is only true if you stay in the digital domain.

If you are going between the analog and digital domains, all bets are off. No D/A or A/D is completely linear or perfectly accurate. Nonlinearity breeds unwanted harmonics. Each step of the way in the analog domain adds noise and quantization noise coming back into the digital domain. Techniques such as 1-bit D/As to improve accuracy in audio can't be used in the digital video domain due to the speeds required.

However, compressed video is not the only choke point for digital video. It turns out that any domain change introduces video data change. Going from digital component (SMPTE 125M) to digital composite (SMPTE 244M) or vice versa is not transparent. It has some cost. Just because you are now in the digital domain doesn't mean that what Fourier had to say is no longer true. Most approaches to these domain changes involve subsampling and, thus, low-pass filtering and, in some cases, interpolation. But it's digital; it shouldn't do that you say. Yes, DSP techniques, namely, when applied to filtering, have made many DVE/DMEs quite remarkable when it comes to video quality during complex effects. DSP has made 28K and now 56K modems possible, but even they can't approximate enough poles when it comes to filtering to make the process completely transparent.

Additionally, if you traverse into the composite domain from the component domain, you no longer have separate luminance and chroma information. Upon reentering the component realm, you will have some of the luminance in the chroma and vice versa—what is traditionally known as *cross modulation*. In situations such as this one, the analog reference fed to these digital boxes can become important when it comes to minimizing artifacts. All boxes should also have the same reference. What should be stressed here is that once you are in a particular domain, stay there. As long as nothing in the path is broken, the quality should not diminish. The main beneficiary of digital operations up to now has been the VTR. The effects of multiple generations of recordings are no longer a factor. Even VTRs that are doing mild compression, say 2:1 or 4:1, can go dozens of generations with no noticeable artifacts.

You should also be aware that there are items in the serial digital stream outside of the actual video data that can impact the handling of the video information. Many pieces of equipment, such as video switchers, strip off timing signals. In digital component video, there are no sync pulses anymore. There are now what are called timing reference signals (TRSs)—one at the beginning of the active video for each line, called start of active video, and one at the end of active video called end of active video. Each is comprised of four words. The first three are unique values, 3FF (all ones) and two sets of 000. These first 2 bytes are a big reason that some equipment strips off these signals. Once these serial bit streams are converted to parallel data, which most equipment does for processing, they can play havoc if many data lines of video signals arriving in a given box are going from all on to all off at once. The third word indicates whether the TRS is an EAV or an SAV, which field of the frame you're in, and whether or not you are in the vertical interval. You obviously want the video to have the same TRS values out of a box that you put in.

In addition, during the vertical interval, some equipment uses the video index data on line 14, in the vertical blanking, to carry signal legacy information that helps some equipment process the signal (SMPTE RP-186). These data are hard to see on a traditional waveform display because only the chrominance data are used, and each word used represents 1 bit of an 89-byte stream. A 0 bit is represented by hex value 200 for the whole word, and a 1-bit hex value is 204. This will show up as a minute change in value in the Cr and Cb signals. This information is needed to pass the data through intact.

Finally, another system can be used to ensure your video data values are not changing between devices. This is the error-detection and handling device (SMPTE RP165). Devices that conform to this recommended standard place active and full-field check words in the horizontal blanking portions of line 9, which is in the vertical blanking. Included are two words that contain active and full-field error flags. These flags indicate whether an error has been discovered at the input of a particular box or upstream from that box. Since line 9 is out of the optional video area described in SMPTE 125M, a box that strips and reinserts TRS data and does not support EDH will remove the history or errors that the EDH data stream contained (Fig. 12.8). Digital video has proved to be very robust and, once placed in a particular domain, will retain the state and quality it had when it got there. Each successive domain change, even as a digital signal, will introduce some, although often small, degradation in quality.

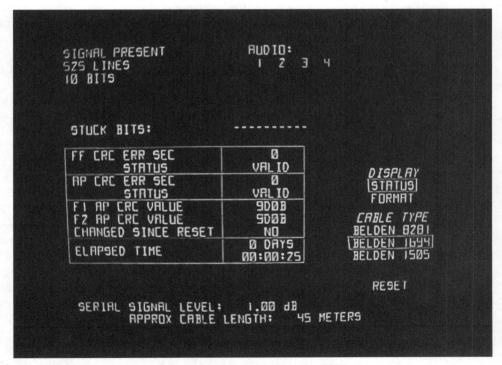

Figure 12.8 Display on a piece of test equipment that provides EDH status.

12.3.5 Final project review

This stage is also known as the postmortem. Here the result or output of the project is compared to the initial spec or input of the project—as sort of a project transfer function. At its worst, this review can turn into a blame and finger-pointing session. The intention is to catalog what worked and what didn't. It's supposed to be a learning experience, but sadly this usually depends on the personalities involved.

12.4 System Integrators

A word on system integrators. First, the obvious: There are good ones and bad ones. Slightly less obvious is that there are usually good project managers, engineers, and installers in any given SI company, but there are also bad ones lurking in the same company. If you are considering an SI to help you with a project, don't assume you'll only have people working on your project that have worked on the company's marquee projects. Ask for resumes of the people involved. Also be aware that some integrators are inclined to push certain products or approaches, based on internal and external arrangements that the SI has. This doesn't rule out the use of any of those as they are generally extremely competent and experienced people.

If you enter into an agreement with an SI, you need to be aware of the following. [*This is not legal advice so get the company lawyer(s) involved.*] Make sure that the entire scope of the project is spelled out. If the SI has not described an aspect of the project, assume it is not included. Determine what the SI and your company are responsible for and ensure it is listed. Along this line, project management, engineering, installation, and testing should be clearly listed as to which party is responsible. Primary contacts between the SI and your company should be identified.

The project schedule should be spelled out, including substantial completion dates. This is one of three normal paydays for the integrator. The SI will expect some amount (25 to 50 percent) upon signing the contract and an equal amount when the substantial completion milestone has been met. The rest will be due when the SI has completed the "punch-down" list. The punch-down list consists of small items that don't have a major effect on the operation of the new facility but do not meet some requirement of the project, missing items in the equipment list, or resolution of some box not performing as expected.

Material and labor costs should be spelled out, as well as who pays for shipping, which can be a considerable expense. Labor costs are often higher than the actual time needed because this is the main insurance the integrator has if things don't go well with the project. It is best to consider the padding as an insurance policy. It also inspires the SI to complete the project in a timely manner since the profit will be higher. Finally, it allows the integrator more flexibility to gracefully accommodate your viewpoint in disputes that might occur (Fig. 12.9).

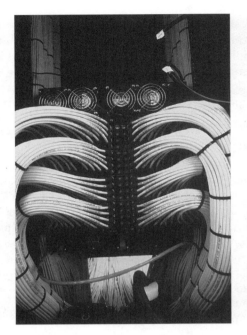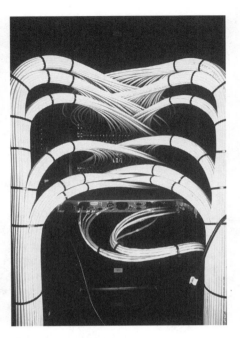

Figure 12.9 Examples of professionally wired racks (*courtesy of Sony Systems Integration Center: Benson & Rice*).

One of the attributes the SIs brings to the game is extremely skilled installers. These people usually take great pride in their work. They not only produce extremely neat installations, but they also usually wire racks so that equipment can easily be serviced and the cables of interest quickly identified. The downside is that you will expend additional time to maintain their craftsmanship. Breaking in to their harnesses, neatly adding additional cables, and rebundling the harness makes for extra work and increased discipline among the engineering staff.

12.5 Bibliography

Anderson, Kare, "Fooling Ourselves into Making Decisions," *Broadcast Engineering,* December 1998, p. 76.

Boston, Jim, "Building HD Remote Trucks," *Broadcast Engineering*, June 1998, p. 78.

Boston, Jim, "Keeping Digital Transparent," *Broadcast Engineering*, July 1997, p. 54.

Robin, Michael, "Grounding Considerations for Analog and Digital Facilities," *Broadcast Engineering,* November 1998, p. 36.

Chapter
13

Transmission

13.1 DTV Transmission

This chapter is what ATSC, in the purest sense, is all about. ATSC uses 8-VSB as a modulation scheme. This modulation scheme, like QAM, COFDM, QPSK, and others, allows transmission of digital data. The modulation schemes are not digital in nature; they just lend themselves to propagating digital information.

13.1.1 DTV channel assignments

The FCC's sixth report and order was issued on April 3, 1997. It was the initial DTV channel allocation table. Less than 10 percent of the DTV stations were issued VHF channels. This report also mandated that network affiliates in the top 30 markets must be on the air by May 1999. Six months after, all the affiliates in the top 30 markets must be signed on, and by May 2002 all commercial stations must be up and running. Noncommercial stations got an extra year to implement their DTV plans. The sixth report and order initially mandated that NTSC transmissions cease in 2006, but the Budget Reconciliation Act of 1997 modified that rule to allow for NTSC transmissions to continue until 85 percent of the television households in a given market could receive DTV signals. The order also stipulated that by April 2003 DTV stations must air 50 percent of the same programming as the NTSC channel. The simulcast required goes up to 75 percent the next year, and all NTSC programming must be available on the DTV channel by May 2005.

The station assignments are as follows: 943 UHF moving to UHF, 630 VHF moving to UHF, and 49 UHF moving to VHF.

The FCC used the grand alliance guidelines in assigning DTV power for each station 12 dB below the station's NTSC power level for UHF to UHF transitions. The FCC didn't use this guideline for conversion of VHF to VHF. The difference in power between NTSC and DTV is due to the signal-to-noise ratio between NTSC (-28 dB) and DTV (-15 dB), which is about 13 dB.

Moving from VHF to UHF requires great increases in radiated power because the FCC has increased its time factor from FCC (50,50) to FCC (50, 90) for replicating their grade B coverage. Low-frequency NTSC has large grade B zones due to the large diffraction zone at low frequencies. UHFs tend to have slightly larger grade As due to much higher power but significantly smaller grade Bs.

$$\text{Channel 4 grade A} \geq 68 \text{ dB}\mu$$

$$\text{Grade B} \geq 47 \text{ dB}\mu$$

$$\text{Channel 26 grade A} \geq 74 \text{ dB}\mu$$

$$\text{Grade B} \geq 64 \text{ dB}\mu$$

VHF signals can extend somewhat beyond the horizon because the VHF signal is diffracted along the curvature of the earth. The diffraction zone for UHF is about one-third that of VHF. VHF has good penetration into houses and buildings. The VHF signal propagates well beyond the optical horizon. Height is still important. At UHF frequencies the signal's optical and radio horizons are nearly the same. This means that UHF requires lots of power and antennas perched at high elevations to match VHF propagation. Channel 7 is probably the best rf channel. Low channels like channel 2 see a lot of skip, which causes cochannel interference. These low channels are also susceptible to RFI, EMI, and motor noise. Also, the channel bandwidth–to–actual frequency ratio is much wider; therefore, it is harder to set up good bandpass characteristics.

The core DTV spectrum is 2 to 51; any stations with assignments outside the core will have to vacate the new DTV assignment and go back to their original NTSC channel when NTSC goes dark. A few NTSC stations that are now outside the core have been assigned DTV channels that are also outside the core. It is yet to be determined what channel reassignment they will undergo when the transition is over.

DTV coverage is not protected beyond the existing NTSC grade B contour. The DTV service contours are as follows:

Channels 2 to 6 — 28 dBμ

Channels 7 to 13 — 36 dBμ

Channels 14 to 69 — based on $\{41 + 20 \log[615/(\text{channel midfrequency})]\}$

The major concern is DTV's interference with NTSC stations. DTV is expected to be robust enough to withstand NTSC interference, but DTV interference into NTSC will mostly appear as increased noise in the video. As such, although new DTV channels are placed nearer existing NTSC cochannel stations than a new NTSC would be, some channel DTV stations are placed much closer to each other. NTSC interference into DTV and DTV interference into DTV have no protection.

Back when television receiver technology was young, many taboos existed as to the assignment and spacing of UHF channels. The restrictions were a result of limitations in the rejection of unwanted signals in the receiver. The UHF NTSC taboo channels were:

N ± 2/3/4/5 has a minimum NTSC to NTSC spacing of 20 miles because of intermod interference.

N ± 7 has a minimum spacing of 60 miles because of if beat and local oscillation radiation.

N ± 8 has a minimum spacing of 20 miles because of if beat interference.

N ± 14 has a minimum spacing of 60 miles because of interference between the visual and aural carrier.

N ± 15 has a minimum spacing of 75 miles because of interference between the visual and aural carrier.

DTV interference into NTSC has a much simpler set of rules:

Channel $-2 = -24$ dB desired-to-undesired ratio

$$+2 = -28$$
$$-3 = -30$$
$$+3 = -34$$
$$-4 = -34$$
$$+4 = -25$$
$$-7 = -35$$
$$+7 = -43$$
$$-8 = -32$$
$$+8 = -43$$
$$+14 = -33$$
$$+15 = -31$$

To ensure that DTV does not interfere with NTSC, the FCC has prescribed an extremely sharp bandpass filter or channel mask: ± 2.5 MHz $= -35$ dB, and -110 dB at the far side of an adjacent channel.

In some areas channels 16 and 17 are shared with land mobile. A few other channels may have adjacencies as well such as with radio astronomy. Adjacent DTV stations may have special out-of-channel emission limits imposed.

Many stations have DTV channel assignments adjacent to their existing NTSC assignment. The channel above the current NTSC channel is known as $N + 1$ and an adjacent assignment below is known as $N - 1$. An $N + 1$ channel assignment is more problematic because the DTV signal provides interference with the NTSC's aural carrier. The DTV pilot and the NTSC aural carrier are fairly close together. The filter transition region for $N + 1$ between the upper sidebands of the NTSC aural carrier and DTV energy is approximately 460 kHz apart. For $N - 1$ there is approximately 800 kHz of separation.

13.1.2   The modulation debate

The "which modulation approach is best" argument is another issue that plagues the initial DTV roll-out. Many are still lobbying for the ATSC or FCC to scrap the 8-VSB system. The 8-VSB road has been traveled too far already to return to the start, though. However, we will look briefly at the major contender and why many like it. Lately, some have advocated allowing more than one modulation scheme to be used.

Coded orthogonal frequency division multiplexing, or COFDM, is sort of a parallel path between the transmitter and receiver, as opposed to the usual serial path of most transmissions. European, Japanese, and Australian systems use OFDM. This parallel approach is accomplished by dividing transmitted data among hundreds of narrowband low-speed carriers located side by side, instead of the single wideband carrier normally used. The frequency division multiplexing part of COFDM alludes to the use of many carriers. The "C" in the acronym stands for coded, which means forward error correction is added. The use of many carriers means that the individual symbol time of each carrier can be fairly long, which minimizes multipath interference. The carriers overlap one another, but since they are orthogonal, that is, perpendicular, or quadrature to adjacent carriers, the product of two adjacent carriers integrates over a symbol period to equal zero.

However, OFDM is more computationally intensive than 8-VSB. It uses thousands of quadrature-modulated carriers to transmit large quantities of data at a low symbol rate. A major factor in the ATSC's choice of 8-VSB was the need to keep DTV signals from interfering with NTSC signals.

13.1.3   8-VSB modulation

Two flavors of VSB are available. The first is 16-VSB, which is intended for use by cable systems. It is less robust against white noise than the terrestrial 8-VSB version. One segment error per second exists when the 8-VSB S/N ratio approaches 15 dB. If 16-VSB is used, the S/N ratio must be about 28 dB for the same error rate. For comparison, an NTSC signal is considered to have marginal quality with a S/N ratio of 34 dB, which means that an 8-VSB signal has 19 dB more S/N ratio headroom before it is no longer viewable. As a result, assigned DTV power levels are generally 12 dB less than their NTSC counterpart. This still allows DTV a 7-dB margin over NTSC.

An 8-VSB transmitter consists of 10 blocks. The first four are data processing blocks and final six are for signal processing. The first block is known as the "data randomizer," and this block enables the 8-VSB modulation process to produce a flat spectrum across the channel. This also minimizes DTV interference into NTSC receivers by making the DTV receiver signal appear as white noise. It only randomizes MPEG data. Parity bytes, data field sync, and data segment sync are added later.

The next block is the Reed-Solomon encoder. Twenty Reed-Solomon parity bytes are added to the end of each 187-byte MPEG data packet (the single byte of data sync at the start of each MPEG-2 data packet is not coded), which has been randomized. The result is 207-byte packets. This scheme can correct up to 10 byte errors per packet.

The next block is the trellis encoder. This forward error corrector adds 1 FEC bit for every 2 data bits. Therefore, each ATSC symbol carries two randomized MPEG packet, or Reed-Solomon bits, and 1 FEC bit.

Next comes the multiplexer. This block adds segment and frame sync. Segment sync has no FEC bit added. Since it is 1 byte long, it can easily be encoded across four symbols. Segment sync occurs every segment. Counting the four symbols used as segment sync, each segment consists of 336 symbols. Thus

$$207 \text{ bytes/segment} \times 8 \text{ bits/byte} = 1656 \text{ bits/packet}$$

$$1656 \text{ packet bits} + 828 \text{ FEC bits} = 2484 \text{ packet and FEC bits/segment}$$

$$\frac{2484 \text{ bits/s}}{3 \text{ bits/symbol}} = 828 \text{ symbols/packet}$$

$$828 \text{ symbols/packet} + 4 \text{ segment sync symbols}$$
$$= 832 \text{ symbols/segment}$$

A frame sync is inserted every 313 segments. Frame sync is 832 symbols long. The first four symbols are for sync, and the 511 symbols are a pseudorandom sequence of data to produce a flat spectrum that is analyzed by the receiver. These symbols are used to determine linear distortion in the signal, which the receiver then corrects for. Next, a pseudorandom sequence of 63 symbols is sent three times. The middle 63-bit sequence is inverted every other frame sync. Twenty-four symbols indicate which VSB mode is present. Although the ITU has five modes (2-/4-/8-/16-VSB or 8T-VSB), ATSC only has two modes (8-/16-VSB). Mode 8-VSB is the terrestrial broadcast mode, while 16-VSB is intended for cable transmission. It should be stressed that ATSC timing bears no reference to NTSC. Each segment is 77.3 µs long, and frame sync occurs every 24.2 ms, resulting in a frame rate of 41.322.

The digital VSB transmission system uses three supplementary signals for synchronization:

A low-level pilot is employed for carrier acquisition.

A data-segment sync is employed for synchronizing the data clock in both frequency and phase.

A data-frame sync is employed for data framing and equalizer training.

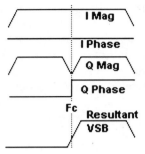

Figure 13.1   Hilbert transform.

This is far more robust than a regular data modem that uses the data signal alone to achieve synchronization. A regular modem will lose all synchronization if the received signal drops below the received threshold (Fig. 13.1). The 8-VSB modulation scheme is a quadrature-modulated system, with I and Q components just like NTSC's color subcarrier. Like the color system, a baseband I and an orthogonal Q modulate a carrier. However, the purposes of the two ATSC components are radically different than their NTSC color equivalents. The I actually conveys the ATSC data values by assuming one of eight values of magnitude (Fig. 13.6) along the I axis. The Q signal is used to minimize bandwidth by canceling the lower rf sideband. The Q channel has a fixed relationship to the I channel and can be described as a pseudo–Hilbert transform. The Hilbert transform has lower sideband power with a phase relationship 180° apart from upper sideband power. The Q channel is constructed so that it has no dc component. Since the Q channel's lower sideband has a phase opposite that of I, most of the lower sideband is eliminated when I and Q are combined (see Fig. 13.1). In a perfect DTV rf system, at the midpoint of the symbol period, the I signal value will be one of eight data amplitude levels. Any linear or nonlinear distortion in the transmission process will cause the I signal amplitude at the center of the symbol time to be incorrect. The Q signal lacks the discrete levels that the I has and contains no digital data information for the receiver. Each I symbol value can be thought of as an impulse, which causes considerable pre/postringing. However, the ringing passes through 0 at preceding and succeeding symbol sample times. Therefore, this ringing does not add to other I samples (that is, no intersymbol interference). The final 8-VSB signal bears no resemblance to the I signal vector that created it. The extent to which the actual I component sample point departs from the ideal sample point is called the *error vector magnitude* (EVM; Fig. 13.3). The difference between one of eight I values along the I axis (−100 to +100 percent) is 28.6 percent (200/seven spaces between I values). Therefore, when the EVM value exceeds one-half of 28.6 percent (14.3 percent), slicing errors will result.

A simplified method of developing 8-VSB modulation follows. Start with QAM (I and Q), cut the Q data rate in half, double the I data rate, cut one sideband to create a single sideband, and then suppress the carrier. The result is amplitude components, without reliance on phase components.

The pilot signal in the ATSC channel is a continuous wave (CW) signal, while the rest of the spectrum is uniformly filled with noiselike information. The low-level rf pilot is a constant rf level, and adds only 0.3 dB to the total signal power (although the voltage level of the pilot is 3 dB above average power). However, this low-level pilot aids carrier recovery independent of data. This provides reliable carrier recovery down to S/N ratios of 0 dB, well below the data-error threshold, which is ≥ 10 dB. The low-level pilot is created by adding a dc value to the baseband data, which have a zero mean because all data levels are equally probable. It should be noted that the apparent level of the CW pilot signal with respect to the rest of the noiselike data signal is a function of the spectrum analyzer's resolution bandwidth filter setting. To check the DTV signal, the analyzer's resolution bandwidth should be set to 30 kHz and heavily averaged.

Some exciters use DSP filters with many taps; one uses 72, for EVM precorrection. The tap coefficients are derived using an HP89441A vector signal analyzer (VSA) to determine the values required. These are then loaded into the exciter from the VSA. Power out of the exciter is usually around 5W, but some are capable of as much as 250 W out (Figs. 13.2 and 13.3). EVM is a parameter that many seem to pay attention to. However, it turns out that the channel signal-to-noise ratio is a direct indication of EVM. In the testing protocol used by the Model Station Group for early reception testing, S/N ratio was used instead of EVM. It's interesting to note that the eyes can be completely closed, with no errors occurring in the receiver due to the heavy error correction employed. Errors start to occur when EVM approaches 15 percent. Vectors at the corners of the I/Q quadrant represent maximum power (Figs. 13.3 and 13.4). Any group delay in the system will skew the dots off the vertical access. EVM of 2.3 to 3 percent out of an exciter is equal to approximately a 31-dB S/N ratio. This is generally the best specification claimed. An EVM of 4 percent equates to an S/N ratio of 28 dB. An exciter output with 2.4 percent EVM has an S/N ratio of 31 dB. Thus

$$\text{S/N ratio} = 20 \log \frac{1}{\text{EVM}}$$

The DTV transmitter total power out (TPO) rating is a peak value because the rf peak envelope excursion must transverse the linear operating range of the transmitter on a peak basis to avoid high levels of IMD spectral spreading. The DTV peak envelope power (PEP) is similar to the NTSC peak of sync rating for setting the transmitter power level by noting that the NTSC linearizes the PEP envelope from sync tip to zero carrier for best performance. ATSC has no peak power, but its average power is very stable. Typically, 99.9 percent of the transmitted digital VSB signal peaks are within 6.3 dB of its average power. Vectors at the corners of the I/Q quadrant represent maximum power (Fig. 13.4).

While many UHF transmitters use a pulsed system where the major portion of envelope linearization only extends from black level to maximum white, the DTV signal has no repetitive peak to apply a pulsing system. DTV power is always stated as average (rms) because this is the only consistent parameter of an otherwise pseudorandom signal. Four times average power occurs about

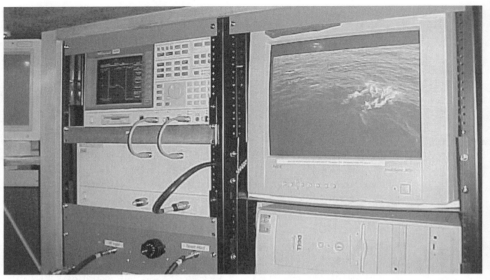

Figure 13.2 VSA (left rack) installed in an ATSC reception test vehicle. Notice DTV display on PC monitor (RGB) in the right rack.

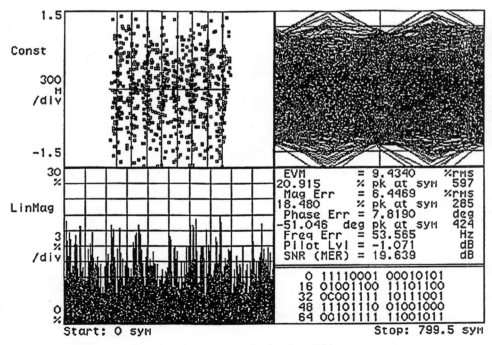

Figure 13.3 8-VSB constellation and eye pattern display from VSA.

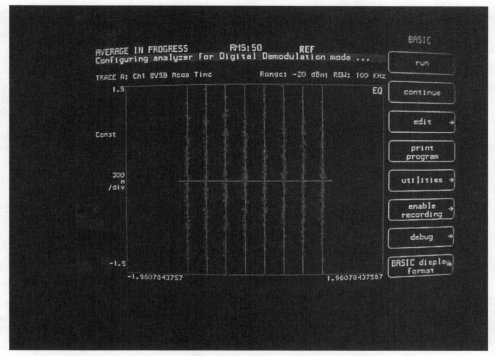

Figure 13.4 Samples in the corners require more power than samples near the center.

2 to 3 percent of the time, and 6 to 6.3 dB above average occurs about 0.1 percent of the time (Fig. 13.5).

It is usually under continuous wave that an antenna's radiation pattern and gain are defined. NTSC is considered to be narrow enough to essentially be considered CW, but DTV occupies the entire band and is considered wideband. DTV occupies the middle 5.38 MHz of its 6-MHz space. NTSC channel energy consists of three carriers—visual, chroma, and aural. The signal energy rapidly falls away from these carriers. NTSC maximum power is generally 2 MHz above the visual carrier. However, the DTV spectrum is flat across the whole 6-MHz channel, except for the last 0.3 MHz. The effective NTSC bandwidth is less than 4 MHz, but the entire DTV channel is of equal importance.

There is nothing digital about the transmission of a DTV rf signal. The channel, like NTSC, is 6 MHz wide. But the entire channel is occupied with energy. The DTV rf signal is essentially a single-sideband suppressed-carrier transmission. DTV transmission has better bandwidth and power efficiency than AM. With 100 percent AM sinusoidal modulation the overall power increases to 150 percent, with 100 percent still in the carrier and 25 percent in each of the upper and lower sidebands. With 8-VSB, all the available transmit power is used to convey the information contained in only the upper sideband. There is a pilot 0.31 MHz for the lower end of the channel.

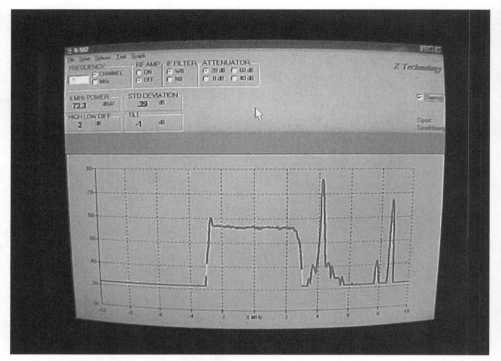

Figure 13.5 DTV signal adjacent to NTSC signal.

The 8-VSB signal has eight possible levels in the *I* component: +7, +5, +3, +1, −1, −3, −5, and −7 (Fig. 13.6). At eight levels, it means each symbol carries 3 bits. The symbol rate for the ATSC channel is 10,762,238, which is exactly double the 5,381,119-MHz, 3-dB bandwidth for the DTV channel. The symbols are exactly double the channel bandwidth because two symbols butted together, each with a different value, can be thought of as one complete cycle of a waveform. That waveform would have a fundamental frequency based on the whole waveform period, which is 2 times the period (or one-half the frequency) of each symbol. At 3 bits/ symbol, this means that the DTV channel carries 32.28 Mbits/s of data. However, more than one-third of this data is used for data synchronization and FEC. The actual data payload is 19.28 Mbits/s.

13.1.4   Channel management

DTV does not mean that your analog skills will go away. You must stay cognizant of the fact that processing does not end at your ATSC transmitter's encoder. A lot of what makes ATSC so robust is the high-powered VSB signal processing that takes place in the consumer's receiver, whether an out-and-out television or just a set-top box.

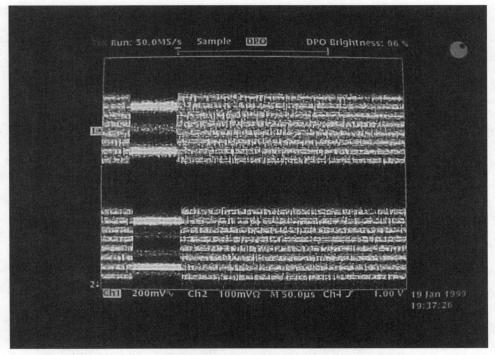

Figure 13.6 An *I* signal can assume one of eight values. This component carries the information content of the 8-VSB signal. The top signal is before processing; the bottom is after.

Once again, it is being driven home to us that just as analog audio and video will be around, your analog skills need to stay intact as well. Noise used to be the limiting factor in terms of how far out a signal could be viewed. With ATSC, it is not only noise but also transmit linearity. As nonlinearity gets worse, out-of-band performance gets worse (remember the FCC channel mask) and the in-band signal-to-noise ratio gets worse. Passband amplitude distortion and group delay affect the S/N ratio.

Although transmitter manufacturers are playing "specmenship" when it comes to S/N ratios out of their boxes, Zenith claims that 27 dB is sufficient. Some manufacturers claim values well into the 30-dB range. One has to wonder if this is necessary or just desirable. At 33 dB, the 8-VSB noise threshold is 15.1 dB. At 27 dB, it is 15.25 dB, which is a worst-case scenario if all the noise is uncorrelated. To keep this value as high as possible, all the analog specs remain important. Also, the signal-to-noise ratio relates directly to EVM. This is the value often reported as an indication of 8-VSB's health. Again, manufacturers' "specmenship" has values between 2 and 5 percent. Those values can be seen out of the transmitter but not at most receive sites. Values approaching 15 percent can be seen as the error cliff is approached. Linearity is extremely important. The coverage area will shrink as the system becomes nonlinear. Group delay will start to rotate the *I/Q* 8-VSB constellation.

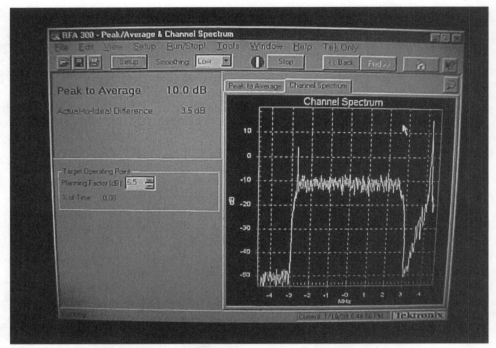

Figure 13.7 DTV channel spectrum. The spike at the lower (left) end of the channel is the pilot signal.

Intermodulation distortion in the rf amp of the DTV transmitter will produce emissions in adjacent channels. Any nonlinear amp will produce IMOD products. The shoulders at either end of the channel rise because of intermod. Each amplifier in the rf chain not only amplifies the band of frequencies in but also harmonic images of the original signal. The second harmonic band will be twice as wide as the original. The output of the second amp would create an 18-MHz image under the desired 6-MHz-wide signal because of additional heterodyning between the undesired image and the desired signal. Low-pass filtering can minimize the upper sideband images, but the lower sideband images are hard to eliminate (Fig. 13.7).

All DTV devices are run class AB. Devices are more efficient as they are driven harder because the signal approaches the power supply rails so there is less wasted power in the amp. Also, more signal drive means more power variation. Only deltas in power can be coupled out of the final.

13.1.5   Channel conversion

Most DTV stations will face the decision on which channel to continue transmission on at the end of the DTV transition. For those stations that are assigned out-of-band channels (>51) there will be no choice; they will have to go back to the in-band channel that was previously occupied by their NTSC channel. For many UHFs that moved to UHF the decision will probably solely rely on economic issues as to whether they should stay on the new channel or move back. But for VHFs that have new UHF assignments, and the few UHFs that find themselves in the VHF band, there will be issues other than the economics of maintaining a transmitter.

It is suspected that many VHF NTSCs with UHF DTVs will opt to go back to VHF when the transition is over. Why not? Coverage is easier and transmitters are generally tamer beasts that use less space and power. But does it actually make economic sense? Let's look at the engineering needed to convert a transmitter from one channel to another. The channel conversion for an IOT transmitter is

Exciter. Retune or swap the output filter. Generally, there are three different filters to cover the UHF band and one for VHF. Regardless of the filter required, new channels can usually be selected with dip switches.

IPA. Replace circulators. There are four different circulators that cover the UHF band. Adjust the feed-forward delay lines.

HPA. Change the cavity. The UHF band has two different cavities based on whether you are at the high or low half of the UHF band. The IOT will also need to be retuned and metering will have to be recalibrated.

RF. Magic tee and channel masking filter will have to be replaced. The total conversion cost for the IOT transmitter could be above $400,000.

The channel conversion for a solid-state transmitter is as follows. Amplifier modules are each wideband enough to cover one-fourth of the UHF band. If the channel being moved to is in a different UHF quadrant, the modules will have to be replaced. Modules represent about 80 percent of the total transmitter cost. Today, they cost around $800 each. As in the IOT box, circulators will have to be replaced if the new channel is in a different quadrant than the old channel. Combiners and dividers will be replaced or retuned. Additionally, the same exciter and rf plumbing requirements exist in the solid-state unit as in the IOT unit. Total conversion costs for a solid-state transmitter could be over $500,000.

13.2 Television Transmitters

There are approximately nine transmitter vendors selling in the United States. Power ranges from 200 W to 300 kW. Where an average 20- to 40-kW VHF transmitter would cost approximately $2700/month to operate, the average UHF transmitter can cost over $10,000/month in operation costs. Most VHF stations can afford full redundancy in their transmitters, but full backup for UHF is costly. UHF transmitters can require 30 to 50 tons of air conditioning. (*Note:* 1 ton of air conditioning can handle approximately 4 kW; 12,000 Btu = 1 ton of air conditioning = 3516.85 W.) A typical tube DTV transmitter consists of 8-VSB mod > shoulder corrector > rf converter (0.5 W out) > class A amp > (15 W out) > class AB amp (800 W out) > class AB HPA (25 kW out).

13.2.1   Intermediate power amplifiers (IPAs)

Most IPAs are solid state. The difference between a tube and a solid-state transmitter is mostly in the final amplifier. A tube uses either a triode, tetrode, klystron, diacrode, or an IOT. A solid state is just that; solid-state power amp modules comprise the high-power rf generation stage. It is common to have four 250-W IPA modules to drive a tube final. The transmitter can continue to be used if one of those modules fails, albeit at reduced drive to the final.

The solid-state IPA accepts 5 W of power from the exciter and passes it through three cascaded class A amps before the first class AB amp is encountered. In some systems at the higher UHF frequencies two IPA assemblies are required. The class AB IPA amp feeds four additional class AB final amps in parallel. Class AB allows for a simpler design than class A. Also, there is less heat and power consumption. The four amps are combined in pairs through two circulators, and these are the only components in the IPA which are frequency sensitive. The two resultant signals are then combined in a star point combiner, which creates a 125-W signal out of the IPA (Fig. 13.8).

13.2.2   Power amps

Many transmitters were configured to run final or power amps in parallel. If two amps were used, each would be run at full power and would be combined to produce the full required power. However, a 3-dB loss penalty is generally encountered in the process of combining the two. If one final was lost, the output power is initially at one-fourth power, which could be brought back up to one-half power by reconfiguring the combiner to pass only power from the amp that was still good. No additional power was gained by using two final amps this way, but it is possible to continue at one-half power with one final failed. The NTSC signal would be noisier to fringe viewers but overall coverage would not suffer much. For DTV it is not desirable to run two half-power finals in parallel because of the 3-dB loss encountered when combining them. This reduced power would greatly limit the coverage area as the S/N ratio would drop below the required value (15 dB) at outlying areas. The DTV signal wouldn't degrade; it would simply go away. With DTV it is better to run a full-power main and a full-power backup final. Here one final has the full load, with a standby ready to replace the main if required. Therefore, a main/alternate configuration where there is no combining penalty is advisable, but magic tee doesn't have that high a loss. In addition, amplifiers run more efficiently at full power. They are less efficient if drive to the amplifier is decreased.

The three major power amp technologies for DTV are: IOT, diacrode, and solid state. An IOT transmitter requires more maintenance, coolant replacement every 2 years, cleaning of high-voltage items monthly, and usually quarterly

Figure 13.8   IPA replacement for older transmitter. These racks feed the original transmitter final amp.

tuning along with IOT and thyratron replacement. In addition, all the pumps and fans/blowers will require some maintenance effort. Solid-state transmitter maintenance is typically one-tenth that of IOTs. There is no routine tuning. Mainly, routine maintenance consists of air-system maintenance, that is, the replacement of filters. It has been said that solid-state transmitters turn transmitter engineers into filter replacers. However, it should be noted that because of the high currents found in solid-state transmitters, connectors should be inspected regularly.

Historically, klystrons are most likely to be used for high-power UHF NTSC stations, while diacrodes and tetrodes might be used for low-power or staging. The problem with UHF is that the short wavelengths mean things must be smaller, but since high power is needed, things also need to be larger. Tetrodes (and triodes) were initially used for VHF but almost all new VHF transmitters are solid state. Tetrode tube life is 10,000 to 25,000 h, while the diacrode tube life is 12,000 to 20,000 h.

13.2.2.1 Adjacent channels. Currently, only the tetrode and diacrode cavities can be tuned wide enough to accommodate side-by-side channels. Diacrodes can have a 1-dB bandwidth of 14.6 MHz, which means that the channel edges are far enough apart so that there are no group delay problems. Diacrodes are capable of up to 104 kW of unsaturated peak envelope power, giving them a rating of 60-kW peak of sync along with simultaneous provision for 6 kW of aural power. If higher power is needed, diacrodes can be run in parallel. When adjacent NTSC and DTV channels are to be driven through a common diacrode, separate IPAs and exciters are needed for each channel.

13.2.2.2 Tetrode. The *tetrode* is a refinement of the triode. The triode has three elements:

1. A cathode that emits electrons (not counted is the filament that heats the cathode and produces thermionic emission of electrons from the cathode) (Fig. 13.9).
2. A positively charged (with respect to the cathode) plate that receives the electrons.
3. A negatively charged (again with respect to the cathode) control grid between the cathode and plate or anode that controls the amount of electrons that flow between two.

The output of the IPA would be fed to the control grid. The output to the antenna would be connected to the plate. Small currents from the IPA would control large currents from the power amp's power supply. The changing final current and, thus, power would be capacitively coupled out of the transmitter.

A fourth element in the form of another grid was added for the tetrode. This additional grid is called a *screen grid,* and it was added to increase the ability of the tube to amplify without increasing the plate voltage power supply. Some electrons arriving at the plate would essentially bounce off the plate, which would lower the amplification factor of the tube. The screen grid, which was placed between the control grid and plate, prevented that. The screen grid has a positive potential but not as high as the plate's. Its effect was to reduce the interelectrode capacitance between the control grid and the plate. A major advantage of the tetrode plate voltage is that it is generally around 8 kV. This

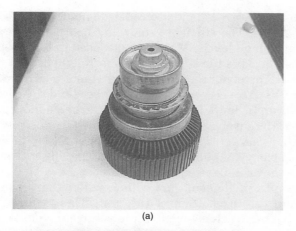

(a)

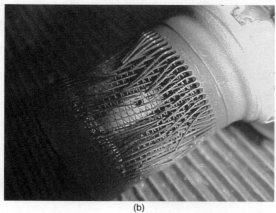

(b)

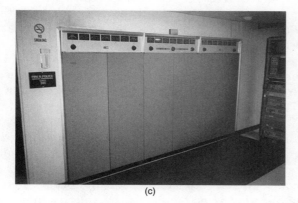

(c)

Figure 13.9 (*a*) Low-power triode. (*b*) Cathode and grid of a failed device. (*c*) Vintage 1970s VHF transmitter that used either triodes or tetrodes.

is much lower than the klystron's, plus tetrodes do not require focusing magnets as the klystron does. The efficient removal of heat is the key to making the tetrode practical at high levels. Most tetrodes used in television transmitters are air-cooled. A tetrode's life is about 8000 to 15,000 h.

Tube design for UHF presents conflicting requirements. UHF requires more power for equivalent VHF coverage. As a result, tubes tend to be larger to help dissipate heat. However, higher UHF frequencies call for smaller tubes because of the short wavelengths involved. Tetrodes were used extensively in the low- to medium-power UHF market. These tubes were often used in common-mode amplification of both the visual and aural signals. The linearity of tetrodes is good and, therefore, they can be used in DTV service.

In many tetrodes all the elements are cylindrical or coaxial to each other and connected to pins at the bottom of the unit. This means that current decreases as you travel up each element, while voltage increases. As a result, the center of the tube has the maximum ability to make power. Thus

$$\text{Typical tetrode power} = 8.8 \text{ kV} \times 5.5 \text{ A} = 30 \text{ kW}$$

13.2.2.3 Klystrons. The *klystron* is a linear device that overcomes the transit time limitations of a grid-controlled tube by accelerating an electron beam to a high velocity before that beam is modulated (Fig. 13.10). The electron beam modulation is accomplished by varying the velocity of the beam, which causes the drifting of electrons into bunches to produce rf space current. One or more cavities in the klystron reenforce this bunching action at the operating frequency. The klystron's output cavity acts as a transformer to couple the high-impedance beam to a low-impedance transmission line. The spent electrons land on a collector. The collector is generally water-cooled (coolant). A large percentage of a transmitter's wasted heat is carried off as steam from the klystrons. Some actually convert the coolant to steam as moving water to the vapor phase consumes large amounts of heat. The problem is that the heat exchangers at the other end give off equal amounts of heat as the vapor condenses back to water.

The klystron's electron gun and heater usually operate at -15 to -26 kV relative to ground or chassis potential. Electrons emitted from the cathode are accelerated through the rf cavities and drift tubes to the collector, which is at ground potential. The beam is focused by internal and external electromagnetic assemblies (Figs. 13.11 and 13.12). The resultant electron beam is tightly focused, and has uniform density before rf drive is applied. The grid assembly controls the amount of beam current. Sync pulsing schemes usually control this grid to increase beam current during sync pulse time. The grid voltage is generally biased between 0 and -10 kV to set the quiescent beam current. In the klystron rf amplification is accomplished by velocity modulation of the electron beam.

The rf drive is coupled to the beam via the input cavity, which includes a pair of capacitive rings that form a structure called a *gap*. Electrons passing through

(a)

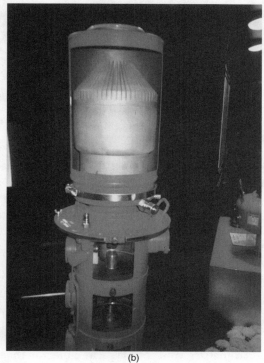

(b)

Figure 13.10 (*a*) Two visual cabinets of a UHF transmitter. (*b*) Klystron with collector exposed.

Figure 13.11   External focusing cavity for klystron.

the gap are velocity modulated by the field across the gap. The electrons drift by in what is called a *drift tube* toward the collector. The "bunched" electrons pass additional gaps where they impart some of their energy. These additional gaps are part of additional cavities that are resonant at the desired frequency. The "ringing" that occurs in these cavities further amplifies the bunched electron phenomena. The final gap is the output gap and cavity where rf power is coupled. After the electrons travel past the output gap, they continue on to the collector. It is here that many of the klystron's nonlinearities arise. Electrostatic repulsion of the electrons to each other as they arrive at the collector tends to eliminate bunching at higher rf levels.

The frequency response of a klystron is limited by the impedance/bandwidth product of the cavities, which may be extended by stagger tuning or by the use of multiple resonance filter cavities. Klystron output power ranges up to 60 kW. The klystron has high gain and little external support circuitry. The klystron is a class A device. Tetrodes and diacrodes can be either class A or AB. Since the klystron is a class A device, the average dc input power does not vary significantly with picture content. Typical collector (beam) current is 4 A in a final for visual and 1 A for aural finals. This current is tightly focused by focus coils around the body of the tube. Current that strays from the beam and does not end up in the collector but in the body of the tube is known as "body" cur-

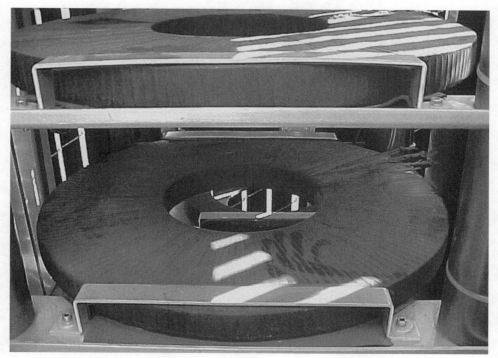

Figure 13.12 Close-up of a focus coil.

rent. High body current is an indication that a tube has a leak and is getting gaseous because electrons in the beam are hitting gas molecules and being scattered. High body current could also mean that the drift tubes are damaged, the cathode is about used up, or the focus coils are misaligned. Typical body currents are under 100 mA. Focus current is generally around 10 A.

A typical klystron transmitter might consume 360 kVA to produce 60 kW of TPO. Common input power would be 480 Vac/3 phase. The high voltage would be 24 kV. Klystron drive from the IPA is usually around 50 μA (Fig. 13.13). Thus

$$\text{Figure of merit (FOM)} = \frac{\text{rf peak power output}}{\text{average dc power input at 50\% APL}}$$

Early klystrons had FOMs of 0.30 to 0.40. With the introduction of modulated-anode pulsing (increased power at NTSC horizontal sync time), FOM increased to above 0.40. The latest generation of external cavity klystrons has achieved FOMs of 0.50, which, when pulsed, may be raised to between 0.60 and 0.70.

Since DTV has unpredictable signal peaks, klystron pulsing is not viable. In addition, since the klystron is operated as a class A device, the beam current has to be biased for full power all the time. However, to stay under the compression

Figure 13.13 Transformers used to create high voltage in a klystron transmitter.

point in the klystron transfer characteristic curve with an extremely small percentage of DTV peaks 7 to 8 dB above average power, the DTV signal drive would have to be backed off. It's conceivable that a 60-kW klystron would only make 5 kW of output power. This is an efficiency of under 10 percent, which makes use of klystrons an expensive option. Another shortfall klystrons have in the DTV realm is bandwidth. DTV requires a wider bandwidth because the data envelope occupies the whole channel, right out to the channel edges. This wider bandwidth requirement would further reduce the klystron's gain. Klystron bandpass is either tilted or bowed, and generally there are no ripples across the bandpass. Any ripples would be from the IPA or before.

Some have proposed that the possibility of modifying existing klystron transmitters to operate the visual klystron's combined visual/aural NTSC service and to operate the aural NTSC amp for DTV service. DTV service can be supported by biasing the aural klystron to the original 20 percent aural beam power, tuning the aural klystron for 6-MHz bandwidth, adding a DTV exciter with the proper predistortion, and adding a masking filter on the DTV section's output.

13.2.2.4 Diacrode. The *diacrode* is a derivative of the tetrode (Fig. 13.14). The diacrode UHF amplification tube was introduced in 1995, and it is basically a

(a)

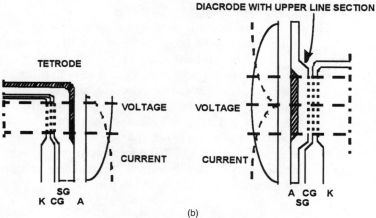

(b)

Figure 13.14 (*a*) High-power 60-kW diacrode (*courtesy of Thomson Tubes Electroniques*). (*b*) The upper shorted quarter-wave line section above the diacrode causes two current maximums instead of the single one in the tetrode.

double-power tetrode. An external cavity is placed on top of the tube and connected across the plate and screen grid. In a diacrode the top of grid G2 is shorted one-quarter wavelength away from the center of the tube. This creates a current null and a peak in voltage one-half the way up the tube, which means that there will be two power peaks, at one-fourth and three-fourths up the tube, instead of one, which is usually found in a normal tetrode. This means power out of a diacrode is double that of a tetrode. While ≈1600 W into a tetrode produces 25 kW out, ≈800 W is needed for the diacrode, a gain of 15 dB. This cavity is actually a two-conductor concentric cylinder, that is, in effect, a one-quarter wavelength-shorted transmission line measured from the top of the cavity to the tube's vertical center. Its purpose is to reflect an open circuit into the vertical center horizontal plane of the tube. The result is that the rf current between the cathode and the plate in this plane is nearly zero. This creates two horizontal planes of maximum current flow, one at the base of the tube and one above the tube at the cavity short circuit. The diacrode uses the same plate voltage as a basic tetrode. In the diacrode the anode current and the rf power output capability are effectively doubled over that provided by the basic tetrode.

In a ground grid-amp design an rf drive of 800 W is applied between the cathode and ground. The rf output power is then available between the plate and ground. The diacrode uses a plate voltage of 8500 V, which is about the same as a conventional tetrode. A diacrode rated at 25-kW DTV average power is approximately 7 in tall and 8 in in diameter, it weighs 14.3 lb, and it is water-cooled. Lower-power diacrodes are air-cooled. Diacrodes cost about $26,500, and have a life of +20,000 h.

These tubes offer sufficient linear performance for amplifying combined visual and aural signals. Intermodulation and cross-modulation correction circuits are used to prevent one carrier from contaminating another. The diacrode is the only current device capable of handling adjacent channel operation because it has low-input impedance, which eliminates the need for an external channel combiner. It still requires separate paths up to the final. There are 128 $N + 1$ adjacent channel assignments. The ratio of DTV peak to NTSC power must be one-tenth to one-fortieth. If other technologies are used, $N + 1$ channel assignments need two separate antennas because of intermod in the combiner. $N - 1$ assignments can use one combiner and one antenna.

Several items have made the diacrode possible:

Hypervaportron cooling, that is, the vapor state right at the tube surface.

The vapor into the tube is around 15°C above ambient at 18 gal/min.

Some of the cooling water goes through a filter to remove impurities.

If the impurities get too high (≈100 kΩ), the water becomes too conductive. It will carry too much heat away, and the transmitter will shut down.

Pyrolytic graphic grids, which have a zero temperature coefficient of expansion. They stay extremely elastic, even at high temperatures.

Improved ceramic to metal bonding.

Improved cathode emissivity for high power.

13.2.2.5 Induction output tube (IOT).   The IOT is a variant of the klystron (Figs. 13.15 and 13.16). While the klystron modulates the velocity of the electron beam, the IOT modulates the density of the beam as the grid of a tetrode would. The advantage of an IOT over a conventional klystron is the same as a tetrode has over a klystron—high efficiency without the need for a long electron beam drift space. However, the IOT also has the same advantage as the klystron has over the tetrode—electron beam collection takes place with a collector and not at a plate, which is part of the rf circuitry, as in a tetrode. The IOT uses a high beam-supply voltage to reduce beam current, which increases the efficiency of the tube. Power is extracted from the IOT in the same manner as a klystron.

The IOT comes in multiple sizes. Average power out of the devices ranges from 9.2 to 23 kW. IOTs have beam supplies that range from 25 to 35 kV. The

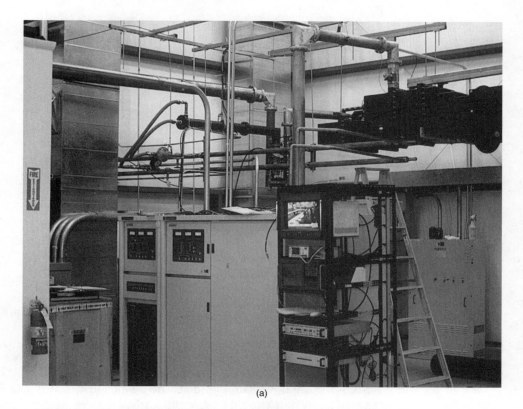

(a)

Figure 13.15 (*a*) IOT transmitters. Transmitter on left has DTV channel mask above it. (*b*) Close-up of transmitter.

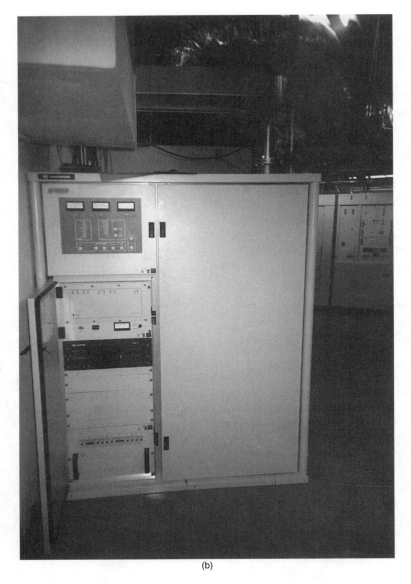

(b)

Figure 13.15 (*Continued*)

larger IOTs need approximately 35 kV of high voltage. This requires a separate oil-filled power supply. Most purchasers of transmitters that use IOTs are
opting for power levels at the high end. IOT has gain of around 22 to 24 dB.
Although absolutely not recommended, the output of the IOT can be pushed
past its rated limit, but the S/N ratio will suffer.

A common IOT transmitter can be configured for 10- to 90-kW average power.
The transmitter would achieve this power range using one to four amplifier

Figure 13.16   IOT installed in transmitter.

cabinets. The outputs of the cabinets are combined with a magic tee, so that they can be switched in and out hot. Most designs have no high voltage in the front of the PA cabinets; it is all at the back.

IOTs need high-speed thyratrons, which act as a "crowbar" to protect the IOT grid when high voltage is removed. In addition, young IOT tubes tend to cause thyratron arcing more than older IOTs. The thyratron must crowbar approximately 4 J of energy in about 1 ms. This is done by sensing a rapid raise in beam current. If the thyratron filament is set incorrectly or if the tube is gaseous, it will fire or crowbar on its own. The thyratron is not normally needed on start-up because the control circuitry brings the beam current up slowly using a simple step-start circuit. Thyratrons can fire tens of thousands of times. They are used in radar systems. A thyratron is faster to fire than a spark gap. They should be replaced when the IOT is replaced.

IOTs have mean time between failures (MTBFs) of more than 10,000 h. Solid state has MTBFs of 300,000 h. Some claim average IOT life is greater than 25,000 h. Although most tubes are prorated for 10,000 h, tubes have remained in NTSC service for over 50,000 h. Some IOT finals operate class A for NTSC and class AB for DTV. A general rule is that the cost of ownership is less for solid state than IOT transmitters over 10 years when dealing with power levels under 10 kW. The opposite is generally true when the power level exceeds 15 kW. Redundancy is inherently built into solid-state transmitters because one or more power modules can fail and the transmitter will continue to operate but at a reduced power level. Modules can be replaced hot. Redundancy in IOTs requires additional cabinets, which decreases the efficiency of the entire system because the standby IOT consumes some power. In addition, full filament heat counts against warranty hours, while black heat operation (reduced filament voltage) does not. The rms efficiency of solid state is 18 to 20 percent and 25 to 30 percent for IOTs. Ongoing maintenance for solid state is one-tenth that required for IOTs. However, system cost per watt at 10 kW of average power is approximately 40 percent higher for solid state.

Another big advantage to a solid-state transmitter is that it can go from a cold start to transmit in less than 3 s. The IOT has a 5-min warm-up, but the system warm-up time is actually 10 min because that is the time it takes for the thyratron to warm up. Also, if a power interruption longer than 20 s occurs, the full 10-min warm-up period must be endured. Most transmitters today have power-fail memory, where the transmitter will return to the state it was in (after any required warm-up time) after the fault is removed. Solid state is usually the choice for VHF. Since the gain of an IOT decreases by approximately 1 dB as it warms up over 20 min, an AGC system is required.

13.2.2.6 Solid state. When new VHF transmitters are purchased, solid-state transmitters have been steadily replacing the tetrode (Fig. 13.17). The same is true for low-power UHF. Bipolar and metal oxide silicon FET (MOSFET) are the two basic technologies used, but the use of silicon carbide transistors is currently under development. The actual power line ac to rf efficiency of a solid-state transmitter may not be any greater than a tube transmitter. A considerable amount of rf efficiency is lost in the output combining process from the individual solid-state modules (Fig. 13.18).

A power loss of 0.25 dB is typical of each level of power combining. It is not uncommon to find nine such levels in a solid-state transmitter. This would amount to 2.25 dB or 40 percent wasted power. A typical solid-state rf module is capable of 250 to 300 W of average power. A single cabinet may contain from four to eight such modules. If more power is needed, additional cabinets must

(a)

Figure 13.17 (*a*) Acrodyne TRU/5-km solid-state transmitter. (*b*) Basic block diagram of solid-state transmitter (*courtesy of Acrodyne*).

be added. Eighteen percent of the power feed into a solid-state transmitter goes up the waveguide.

A tube amplifier is simpler in design, and it requires fewer parts. Fewer components means less chance of failure. However, tubes use high voltages, which increase the chances for disaster, and a tube slowly consumes its filament and cathode.

A typical solid-state transmitter is capable of 5 kW per PA cabinet. Up to five cabinets could be tied together. Overall power levels of 1.25 to 30 kW are possible. The system is usually fairly silent, with an acoustic noise level of −65 dBA. The transmitter should meet ANSI C62.41 transient voltage tests with no damage. It should also comply with IEC-215 safety standards.

Typically, each amplifier module outputs 400 W. Although that means 16 modules in a cabinet would total 6400 W, 1400 W is lost in the combining process, for a total of 5000 W/cabinet. There is generally a separate power supply in the cabinet for every two modules. The power factor for the transmitter is 0.99 (Fig. 13.19). The modules are 25 percent efficient, although the whole

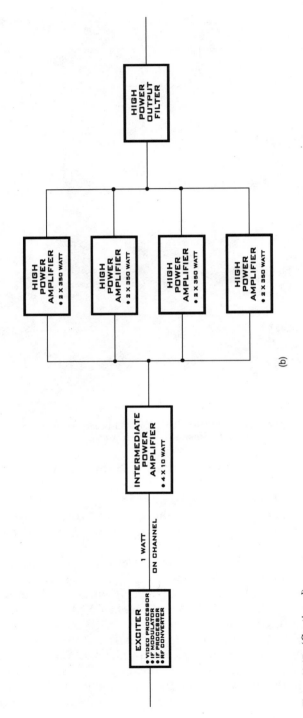

Figure 13.17 (*Continued*)

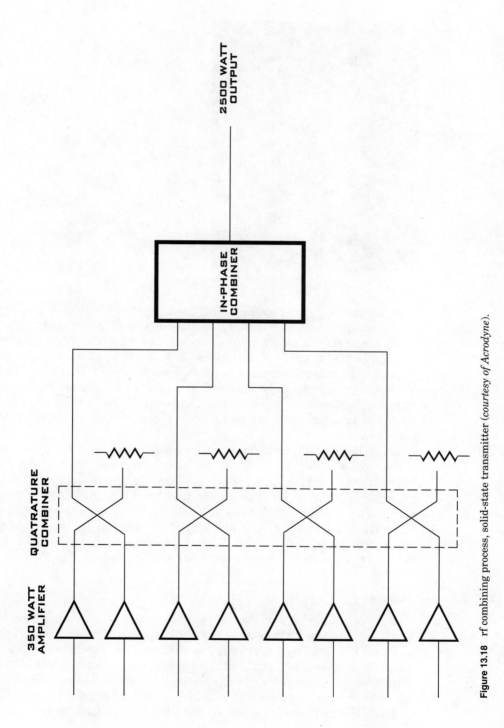

Figure 13.18 rf combining process, solid-state transmitter (*courtesy of Acrodyne*).

413

Figure 13.19 Solid-state or power amplifier module (*courtesy of Acrodyne*).

transmitter is 20 percent efficient. Since each module is drawing 50 A, each power supply, which is supplying two modules, generates 3 kW of power. Each amp module which runs class AB has 41 dB of gain. The highest rail voltage in the amplifier modules is 32 V. No feed-forward correction system is needed because the modules are extremely linear. The future use of silicon carbide technology might lead to higher power handling, which would result in less splitting and recombining. This would lead to higher module efficiencies. The problem with the technology at this time is that the yields are still very low.

There is phasing circuitry for each module so that the rf's out of the various modules are all in phase. Some losses are from phasing problems due to board traces and combining. Each module usually has a class A stage followed by a class AB stage. Next there are parallel AB amps. Each of these amps is made of two sets of two push-pull FETs. The typical exciter output for a solid-state transmitter is in milliwatts. The exciter output would go to a driver which creates a 10-W output. This 10-W signal is distributed to each cabinet via a power splitter. Each cabinet sees approximately 1.75 W.

13.2.3 Power supplies

The power supplies are the transmitter subsystems that provide the high power used in the final or power amp to create the transmitted rf power. With IOT and klystron transmitters, these supplies are generally known as beam supplies. High-voltage supplies that are outside are generally oil filled, which eliminates corona effects and insulation breakdowns. Dry supplies need to be larger to stand everything off. In addition, to cool dry supplies air has to move through them. Beam supplies have to float since they have to supply negative beam current. Plate supplies can be tied to the ground because the negative potential is at ground. Usually, for safety all control and status are brought out via fiber optics.

Some IOTs have switching power supplies. These active component supplies are hard to repair, but a dry supply is fairly light. They are used because some installations in high-rise buildings have weight limitations (limited by elevators) and restriction on oil-filled devices. Since the diacrode high-voltage supply is low compared to IOTs, the diacrode high-voltage power supply can be made smaller and be a dry supply. As long as you don't run out of headroom, triodes and diacrodes stay linear as plate voltage changes. Therefore, you can have high-voltage power supplies that do not regulate.

13.2.4 Cooling

Cooling is a greater concern at higher altitudes. High voltage tends to behave differently as well. Typical IOT transmitter cooling requirements are

The control cabinet gives off 300 W or 1020 Btu/h of heat.

The IPA cabinet has 320 ft^3/min of airflow. It exhausts 2500 W or 8530 Btu/h.

The IOT cabinet has 72 ft^3/min of airflow. There is ≈15° temperature rise over ambient. This is 650 W or 2220 Btu/h.

Water cooling an IOT consists of using a 50/50 glycol mixture with a 13.8-gal/min flowrate. The coolant flows back out of the amplifier cabinet 12° hotter. Overall, the ac load is between 13.6 and 14.3 kW (36,640 and 48,840 Btu/h).

13.2.5 Efficiency

Typical IOT performance. 66 kW ac in, 17.5 kW out (TPO), 26.5 percent efficiency

Typical solid-state performance. 76.6 kW ac in, 15 kW out (TPO), 19.6 percent efficiency

It is more efficient to run a single HPA at full speed than to run two in parallel.

13.3 rf Plumbing

The rf power generated in each PA (or HPA) cabinet must be totaled or combined. Final channel mask filtering must be applied as well.

13.3.1 Combiners and filters

The path that rf takes through the rf system after the transmitter involves some losses that are a function of frequency. The loss through the output rf system for channel 14 = 3 kW, for channel 36 = 3.3 kW, and for channel 52 = 4.3 kW.

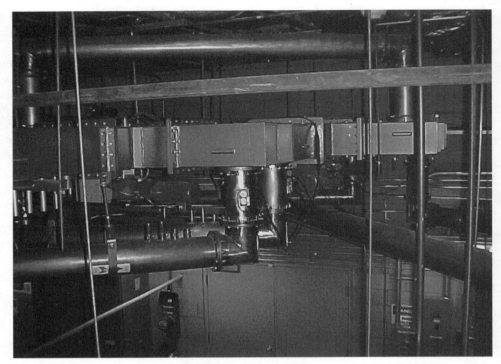

Figure 13.20 RF plumbing combining two visual, one aural, and one standby cabinets in a high-power UHF transmitter.

The first thing the rf system must do is combine the rf power from the various HPAs. This can be done with either a coax switch, which selects an HPA for use, or with a combiner known as a magic tee. With a magic tee you can switch HPAs hot; you can't do that with a coax switch. However, coax switches have less loss than a magic tee does (Fig. 13.20).

In the case of DTV the rf power must also go through an extremely sharp channel mask. Any energy just outside your channel must be −47 dB down from in-band power. At the far side of adjacent channels, though, the power must be −110 dB down. Filters will drift as the temperature changes (Figs. 13.21, 13.22, and 13.23). In addition to the channel mask filter, the rf system consists of reject and dummy loads. Some systems use the same coolant that flows through the tube to also flow through any system loads. Some systems use resistorless loads in the rf plumbing. Water is used to actually absorb the rf. Early examples of this approach had varying impedance values. Modern cooling systems use a glycol mixture which has a rust inhibitor in it so that brass or bronze pumps can be used instead of only stainless steel pumps. There is a separate pump cabinet in most transmitter systems. The pump cabinet produces a large percentage of overall transmitter system noise.

Figure 13.21 DTV channel mask.

Figure 13.22 Air-cooled reject load (on top of waveguide).

Figure 13.23   Water-cooled dummy load (at center, to left of coax patch).

13.3.2   Transmission lines and waveguides

The transmission lines and waveguides provide the path from the output of the rf combining and filtering system and the antenna. These paths use either transmission line, which is often referred to as rigid coax, or waveguide, which has no center conductor and often appears to be water pipe.

13.3.2.1   Transmission lines (rigid coax).   Transmission line efficiency is usually around 70 to 80 percent. Transmission line selection is typically based on frequency of operation, power handling, attenuation or efficiency, characteristic Z, and tower loading (size and weight) (Fig. 13.24). Common transmission line size for rigid coax is $3^1/_8$, $6^1/_8$, and $8^3/_{16}$ in. Larger rigid coax can handle more power but it has a lower cutoff frequency. Coax cutoff frequency is based on the dielectric constant or relative permittivity of dielectric to air and the diameters of the inner and outer conductors. For rigid coax lines the dielectric losses are considered negligible. Dielectric spacers or pegs are placed at intervals along the length of the coax. With age, oxidation will increase the conductor losses. Conductivity varies with temperature, with tests done at 20°C. The inner conductor runs hotter than the outer conductor, often by as much as 100°C.

Group delay is now a concern in some lines. However, group delay is not an important factor in coaxial lines that are homogenous throughout their length.

Transmission lines have greater wind load than does the antenna. A given transmission line's power-handling capability can be increased by using a

Figure 13.24 Disassembled rigid coax. Smaller-diameter copper is center conductor with dielectric spacers installed.

more inert gas or the same gas under higher pressure. Here, average power, and not peak power, is the major concern. A flexible transmission line has a lower wind load than does rigid line. There is no bullet VSWR (see Fig. 13.26), installation is faster, and its cost is less than rigid line. Flexible line also has long life but only handles low to medium power, and therefore, is only used for VHF and low UHF. Solid transmission line has higher windloads, its bullets cause VSWR, and it is costly to purchase and install. But it has less loss, a very long life, and a wide frequency and power range. Transmission lines of 50 Ω are good up to about 45 kW, and 75-Ω lines are good up to about 50 kW. Transmission lines cost less than one-half of what waveguide does. Also, coax is nowhere near as frequency dependent as waveguide.

13.3.2.2 Waveguide. Waveguide attenuation is inversely proportional to the frequency (Fig. 13.25). Both waveguide and coaxial lines exhibit very little change in attenuation values across a 6- or 8-MHz channel. The variation is typically less than 0.05 dB. However, unlike coax, waveguide has both lower and upper cutoff frequencies. The upper frequency limit is where undesirable propagation modes occur. The lower cutoff frequency is where true wave propagation begins. The velocity of propagation is different for waveguide due to the lower cutoff frequency. There is also a slight amount of group delay in waveguide because the lower edge of a channel has a slightly slower velocity of propagation than the upper end of the channel. For channel 30 that difference

Figure 13.25   Square waveguide.

is 35 ns. The velocity of propagation is based on the ratio between the lower cutoff frequency and the actual frequency. A typical lower cutoff frequency would be around 460 MHz. A $1^5/_8$-in waveguide has a loss of 2.3 dB/300 ft.

The disadvantage of waveguide is that it is generally wider than coax, which increases wind loading. Also, waveguide has limited bandwidth. The group delay is greater in waveguide than coax. The disadvantage of coax is that the bullets need to be replaced every 10 years, and there is also a greater chance of burnout (Fig. 13.26).

In a comparison of rectangular to circular waveguide, rectangular waveguide has higher wind loading, more group delay, and is more expensive. However, when compared to circular waveguide, it has lower loss and can handle higher power. Circular waveguide has slightly less wind load than rectangular waveguide and is less expensive, but transitions to other lines can be tricky.

13.3.3   Antennas

Like everything else, antenna selection involves a number of trade-offs. High-gain antennas (rf propagation limited to desired directions only) allow less transmit power. However, the higher the gain of an antenna, the larger its size and the narrower its vertical pattern.

Figure 13.26 Bullet used for mating sections of rigid coax.

13.3.3.1 Propagation. In rf transmission systems two terms are often used to describe power. The first is TPO, that is, the total power out of the transmitter into the rf plumbing system. Between the transmitter and the antenna the rf power is attenuated by combining, filtering, and transmission/waveguide losses. These losses are known as the *line efficiency*. The power that is actually radiated by the antenna is known as the *effective radiated power* (ERP). Thus.

ERP = transmitter power out × line efficiency × antenna gain

The *directivity* of an antenna is its ability to concentrate radiated power in a particular direction. The *gain* of an antenna indicates its directivity. Antenna gain considerations are transmitter size and expense versus good coverage. Low gain means better coverage, but the transmitter needs more power out.

Tower-top pole-type antennas can be classified into two categories:

1. Resonant dipoles and slots

2. Multiwavelength traveling wave elements

RF waves are comprised of transverse electromagnetic (TEM) waves. A TEM wave has E (electrostatic) and H (magnetic) waves transverse (or normal, 90°) to the direction of propagation. Maxwell's equations indicate that changing electrostatic fields create magnetic fields and vice versa. Once created, they therefore sustain each other. E fields are much stronger than H. E fields generally induce currents into receive antennas.

With regard to antennas, near fields versus far fields are often mentioned. In a near field E is based on a Poynting vector and a dielectric constant, along with distance. The Poynting vector is a power density vector associated with an electromagnetic field. The Poynting vector points towards the source. The surface integral (an envelope encompassing all the radiating surface of the antenna) of the Poynting vectors over a closed surface equals the power leaving the enclosed volume. This is referred to as Poynting's theorem. In far fields H and E are mainly based on the antenna current and the distance. The ratio between E and H becomes a constant. Near fields are where the distance is much less than the wavelength. Far fields are where the distance is much more than the wavelength.

When computing the coverage properties of an antenna, the distance to the radio horizon is usually done using a chart with the earth's radius that is four-thirds the actual size to account for atmospheric refraction effects.

13.3.3.2 Radiation resistance. The *radiation resistance* of an antenna is the ratio of voltage at any point on an antenna to the current flowing at that point ($R = E/I$). Since current decreases toward the ends of the antenna and voltage increases, R increases toward the antenna's ends.

13.3.3.3 Polarization. The orientation of the E field can be in one of three directions: (1) vertically polarized (AM transmission is vertically polarized), (2) horizontally polarized (such as FM and most of television), or (3) circularly polarized (used by some FM and UHF TV).

Many natural, along with a wide selection of synthetic, obstacles produce shadows for broadcast signals. This situation has sometimes caused degradation of NTSC coverage. As has been widely publicized with the advent of DTV, gradual degradation of the signal should not occur in the digital domain. The baseband, modulated, and transmitted digital signal should be decoded perfectly by a receiver right up to the point where noise, reflections, and other factors swamp the error-correction system's ability to recover the data, that is, the well-known cliff effect. It has been determined from testing that error-free DTV reception can plummet to an almost one in two chance for errors with less than a 2-dB change in the signal-to-noise ratio.

Signal reflection or multipath problems, while annoying in NTSC, have been rumored to be lethal to ATSC signals. CP propagation has been employed by some broadcasters since the late 1970s to minimize ghosting. This is largely attributable to the fact that reflections off buildings and other objects tend to have the opposite polarization than what was transmitted. It has also been found that signal components in the H and V fields fade to a great extent, independent of each other. H seems to fade or be defracted off its axis to a greater degree over water than does V. Indeed, defraction off the

H axis has been evident since television's inception. Early CATV systems often mounted receive antennas in vertical orientations rather then horizontal for the strongest receive signal. Horizontally polarized *E* fields sometimes appear to be defracted by obstacles along the path and end up with vertical orientation.

A number of DTV receive attribute tests were conducted. One of the measured attributes concerned *H* versus CP polarization. Although the tests were inconclusive, hints that CP might add a degree of robustness to the propagated signal were apparent. One hint came from the DTV receiver. A parameter called *tap energy,* which is a measure of how hard the receiver's equalizer is working, was slightly lower with CP broadcasts than with *H*. A second hint was that the received signal strength indoors actually increased slightly with CP.

There are a number of disadvantages to using CP. Most center on cost. Although true circular polarization is nearly impossible, to approach it will mean doubling radiated power, with nearly equal vertical and horizontal *E* fields. Fairly high assigned radiated power, along with the desire for using a fairly low-gain antenna for a fat vertical pattern, means that it would take more rf cabinets if CP instead of *H* polarization were implemented. This equates into increased up-front costs and increased monthly costs, mainly for power. Doug Lung has stated that he believes only 25 percent vertical component might be necessary. This would obviously lower the power required. Some have stated that noteworthy diversity improvement can be obtained without doubling the power.

Another problem arising from using CP is that more tower reflections are possible from the vertical component. One of the advantages of using a panel antenna is that it produces minimal reflections from its support tower. Finally, at the receive end, antennas with poorly shielded downleads can have reflections induced into the received signal from the CP's *V* component.

For circular polarization the sense of rotation is determined using the right-hand rule. Point the thumb on the right hand toward the radiator and curl the fingers. Looking at the curled fingers from the index figure end of the hand represents the direction of polarization as seen by the receiver. The vertical radiation should not exceed the horizontal component. On side-mounted CP antennas, tower interaction, or reflections, will occur mainly in the *V* plane.

Some early CATV systems experimented with *H*- and *V*-mounted receive antennas which were combined. This was done long before there were any CP transmit stations. The vertical component of the signal was apparently generated through a diffraction process over obstacles along the path, and, as a result, the polarization diversity was site-specific. Due to reflection from non-vertical objects or to ionosphere contour, the polarization of a ground or sky wave may be twisted out of its original polarization. It has been found that *H* and *V* signals fade independently to a large extent. The main attribute of CP operation is ghost cancellation. Inherent CP reflection cancellation can reduce the equalization effort required of the ATV receiver's adaptive equalizer.

A common technique for producing circular polarization has been to place two linear dipoles at right angles in front of a reflecting screen and feed them with equal voltage magnitudes but with a 90° phase difference. However, the azimuth beam width for horizontal and vertical polarization is about 60° and 120°, respectively. Thus, the power ratio between the two planes is low only for directions near the normal to the reflective screen. Flat-crossed dipoles enclosed in a circular cavity are one technique for equalizing the azimuth beam's width for vertical and horizontal polarization.

13.3.3.4 Gain. The gain of an antenna is always specified relative to a one-half-wavelength dipole. The gain–beam width product of antennas is essentially a constant. The product for practical antennas varies from 50.8 to 68. This illustrates that as the gain increases, the beam width decreases.

13.3.3.5 Beam width. The approximate beam width of an antenna corresponds to 70.7 percent of the value of the field (50 percent of the power), or 3 dB below the maximum. Beam width is determined by the number of radiators, the distance between the radiators, and the frequency.

13.3.3.6 Beam tilt. The phasing of cables feeding various antenna elements will introduce beam tilt. Any specified beam tilting of the beam below the horizontal is easily achieved by progressively phase shifting the currents in each panel. Antenna beam steering changes with frequency.

13.3.3.7 Turnstile antennas. Turnstile, or bat-wing, antennas are the oldest type. Turnstiles are made of four bat-wing–shaped elements mounted on a vertical pole. The bat wings are, in effect, two dipoles that are fed in quadrature phase. The azimuthal field pattern is a function of the diameter of the support mask. Usually, there are six stacked turnstiles for channels 2 to 6 and 12 for channels 7 to 13.

Most existing VHF antennas are tuned to operate on specific channels and, therefore, are not suitable for multiplex operation. Bat-wing antennas can be used for $N + 1$ operation, and can be designed for multiplex VHF operation.

13.3.3.8 Slot antennas. The vast majority of UHF antennas currently used in NTSC service are slotted cylinder designs. They have excellent omni-directional azimuth patterns, low wind loads, and smooth null fill. Most slotted cylinder antennas have diameters between 8 and 14 in. In addition to reducing wind loading, this also translates into a small electric radius (radius/λ) for the slot radiators, resulting in excellent circularity or azimuth pattern. Multislot antennas are available for channels 7 to 13, and these antennas are comprised of an array of axial slots on the outer surface of a coaxial transmission line. The slots are excited by an exponentially decaying traveling wave inside a slotted pole. The azimuthal pattern deviation is less than 5 percent. The antenna is generally approximately 15 wavelengths long.

The slotted cyclinder antenna, commonly referred to as the pylon antenna, is the most popular top-mounted antenna for UHF applications. Horizontally polarized radiation is achieved using axial resonant slots on a cylinder to generate circumferential current around the outer surface of the cylinder. A good azimuthal pattern is achieved by exciting four columns of slots around the circumference of the cylinder, which is a structurally rigid coaxial transmission line. Slots along the pole are spaced approximately one wavelength per layer. Typical gains range from 20 to 40.

Helical and slotted traveling wave antennas have low wind loads, but they only radiate on a single channel. They have a limited number of azimuthal patterns, but can be used for ellipitcal polarization.

Branch feed–type antennas (such as turnstiles or panels) are assumed to have equal path lengths to each element, so that all elements receive the signal from the feed system at the same instant. In contrast, slots in a traveling wave antenna receive the input signal sequentially, not simultaneously. A traveling wave–type antenna is essentially a leaky transmission line. The traveling wave antenna has less gain than the branch feed antenna because the signal must propagate up the antenna to each radiating element. Slew rates are longer than branch feed types, since the first element is fed (gain of 1), then the second element (gain of 2), etc.

Bottom-fed slotted cylinder antennas have slots one wavelength apart (at a center frequency of interest). Therefore, the beam tilt of a slotted cylinder antenna varies with frequency, which means that, at the lower band edge, beam tilt is higher than desired and lower than expected at the upper edge of the band beam tilt. Beam tilt can be as high as ±2.5° (−0.75° is common).

An alternative feed design is to electrically center-feed the slotted antenna. A "T" is used on a side-mounted antenna, and a triaxial configuration is used for the bottom half of the antenna on a top-mounted antenna. Center feeding will cause the bottom and top halves of the antenna to have electric tilts in the opposite directions. This produces a constant electric beam tilt for the entire antenna.

Slotted antennas cannot be made to have low gain. These antennas are generally about 16 wavelengths long. Each slot, which is spaced about one wavelength apart, effects a gain of 4. For NTSC visual and aural carriers the wavelength is slightly different. Therefore, the amount of beam steering is different for each carrier. To minimize this difference, the antenna's slot spacing is usually a compromise between the two.

For UHF the antenna height is everything. UHF power cannot travel over the horizon; UHF signals cannot go around objects. Height is more important than power when dealing with UHF. The exception to this rule is if the audience is close in, then lower the gain of the antenna and increase power to better penetrate buildings.

13.3.3.9 Panel antennas. Although slotted antennas are not wideband devices, panels are, and DTV needs wideband antennas. Although panel antennas are fairly light, they produce much wind loading. Panels can't take high power; they are generally restricted to about 4 kW/panel, although some are now rated up to 8 to 10 kW.

Panel antennas are primarily used to control or minimize the reflections from the supporting structure. The simplest panel is either a horizontal dipole or, in the case of a CP antenna, two crossed dipoles in front of a reflector. The reflector is usually a wire grid for VHF and a solid sheet for UHF. For good omnidirectional patterns the tower width should not be greater than one wavelength.

Panel antennas can be used for multiplex operations, but are not suitable for $N + 1$ operation. Panels can achieve excellent pattern circularity.

13.3.4   Towers

Tower structural standards are 1949 RETMA TR-116, 1959 EIA RS-222, 1980 ANSI A58.1, and 1996 RS-222-F (also known as TIA/EIA RS-222-F). Electromagnetic radiation rules are covered in OET 65 (www.fcc.gov/oet/rfsafety).

An *appurtenance* is a thing added to a more important thing. This term is applied to things hung on a tower.

Terms important with tower loads are

Dead load.   Antennas, transmission lines, waveguides, lighting systems, conduits and junction boxes, ladders, safety climbs, work platforms, guy wires, and elevators.

Wind load.   Round members are more aerodynamic. Wind load increases with height.

Ice load.   Ice is a leading cause of tower failure.

Seismic load.

Other tower loads.   Bending moments, shear force, compression, tension, and axial load.

Several tower upgrade methods are

Leg member upgrades.   Reduce the unbraced length of legs with added leg bracing. Fill legs with high-strength concrete or grout. Weld half-round sections of pipe to legs. Weld or bolt plates, angles, or channels to legs. Install additional horizontal bracing (Fig. 13.27).

Guy wire or anchor upgrades.   Add additional guy level between existing guys. Add a star frame with torsional guys.

Reducing wind loads.   Add antenna radomes, which reduces wind loads by as much as 45 percent. Hide coax and transmission lines.

Although towers can inflict heavy damage if they fail, tower insurance ranges from $5 to $10 million for general liability, as compared to $1 million for

Figure 13.27 Solid leg for tower.

personal auto liability. Major causes of tower failures are weather and human error (such as replacement of tower members and top-mounted antennas).

13.4   DTV Reception

To many, DTV reception might be the Achilles heel of the whole DTV scheme. At least to begin with reception will be via antenna, either set-top or roof-top. Cable systems will initially not carry DTV signals. Much of what will make the DTV transmission scheme a viable replacement for NTSC is the processing that will take place in the receiver. Massive DSP activities surround channel equalization, and multipath elimination at the receiver end kills a lot of the demons in the transmit path.

Many facilities will need to conduct DTV reception testing. To understand the protocol and what a DTV reception test indicates, the methodology used by the Model Station Group is described here.

13.4.1   Test vehicles

The objective of a test vehicle is to detect and measure signals that range from over 100 dBμV/m down to what the FCC considers the edge of a UHF DTV

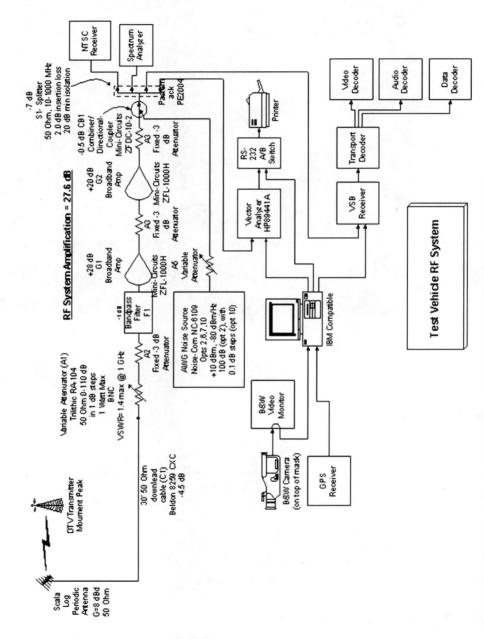

Figure 13.28 Test vehicle block diagram based on model station specifications.

428

service area (41 dBμV/m plus the FCC's dipole factor, which for channel 52 is around 43.5; Fig. 13.28). To be able to measure and recover signals at fringe locations, the test truck's design is based on parameters found there.

Therefore, the truck's receive signal path must be such that the signal fed to the VSB receiver must be at least 15 dB above the noise floor (value for the edge of error-free operation) at fringe locations. The amplification must be high enough to boost the received signal so that it is not only higher than the vector analyzer's noise floor ($\sim$ −80 dBm/6 MHz), which is needed as part of the test, but enough that the signal into the VSB receiver is higher than the noise floor by at least 15 dBm. This must be the case even as the field strength of the signal falls to under 45 dBμV/m.

A noise floor increase by the system's amplification above −80 dBm is desired so that the HP vector analyzer can display the actual system noise floor, and not its own. This is critical if the received signal's S/N ratio is to be accurately measured. The noise floor in our vehicle turned out to be approximately −65 dB. This is also a needed control, so that we can judge receive system health as the test proceeds. The preamps must have a gain large enough to ensure that the receive test system's overall noise value is determined by the amps, and not any of the test equipment (Fig. 13.29).

The truck's rf system should be robust enough to handle large NTSC and DTV interfering signals and have a large dynamic range. To simplify testing of the entire coverage area, it is necessary to leave the rf system intact throughout the testing period. It is not desirable to have to switch out the preamps at strong signal level sights. This will cause calibrated gain changes and also affect the truck's noise floor. As a control, it is necessary to have a known measurable noise floor which is eternal to the VSB receiver on which to evaluate system performance. If the limiting noise floor is within the VSB receiver, which is determined internally by its tuner, there is no simple way to determine the noise floor other than to use a noise figure meter, which is not practical in the field.

A practical way to handle a large dynamic range of received signal strengths over the entire testing area is to use a manually set attenuator to prevent preamp overloading. The attenuator is adjusted so that the rf system output is at a predetermined level, such as −30 dBm, which is the value that was used for testing. The rf attenuator's value must be documented at every test site so that the site's field strength and site margin can be calculated.

To restate, the preamp's large gain ensures that the overall truck noise value is determined by the preamp, and not any downstream test equipment. The FCC's planing factor calls for a 10-dB noise value in the receiver. The truck's first preamp has almost the same noise value, so that the collected data closely correlate.

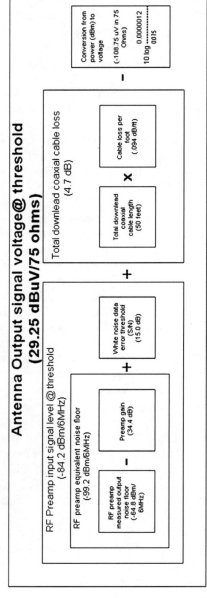

Figure 13.29 Algorithm for determining minimum received field strength for test vehicle.

13.4.2   Site tests

The following steps are performed at each site:

1. The time, date, geodetic location, distance and bearing from the transmitter, and weather are recorded.

2. Next, the rf system attenuation is set for -30 dBm on the vector signal analyzer. The attenuation is recorded. This attenuation value, along with the rf system gain measured during the AM calibration, is used to determine the DTV field strength at that site.

3. The rf feed is removed, and the 6-MHz noise floor is measured. This value is used in the DTV SNR calculations.

4. Next, the DTV pilot power is measured. It generally ranges between -11 and -13 dB below DTV channel power.

5. Next, plots are taken of the DTV channel passband and tilt using the VSA. Ripple and tilt are often caused by multipath. Extreme close-in ghosts (less than one symbol time) that are out of phase will boost frequencies at the high end. This creates positive passband tilt. Close-in ghosts that are in phase create negative passband tilt. NTSC video appears "enhanced" or peaked in the first case and "rolled off" in the second case. A single ghost that is not close-in will create ripple across the passband with a frequency separation equal to the inverse of its period in the time domain. As an example, a 1-μs ghost will create ripples across the channel with a spacing of $(1/1E - 6)$ 1 MHz (Figs. 13.30 and 13.31).

6. Then a test receiver that can record tap energy is given a command to record the equalizer's input and output S/N ratio tap energy and all the individual coefficients along with a summation value of all the coefficients. The S/N ratio into and out of the receiver's equalizer system is also recorded. Many receivers use the 256 tap coefficients displayed here (Figs. 13.32 to 13.35) to cancel reflections to maximize reception. All DTV receivers must do this, but the test requires a demod that also displays these values as a measure of multipath at a given site. The magnitudes of all the coefficients, except the main one, are squared and then summed. Then the logarithmic ratio between the summed squares and the squared main tap is displayed in the upper left corner as tap energy.

At 0 dB, the energy in all coefficients is equal in magnitude to the main tap. Sites with no reflections will have tap energy values below -18 dB. Decoding may no longer be possible when the tap energy approaches -3 dB.

The signal-to-noise values reported by the test demod represent the 8-VSB constellation landing errors that result from noise and intersymbol interference. In other words, they indicate how far off from a valid sample level the vector has landed or deviated from its ideal value along the I axis. Intersymbol interference due to multipath can be thought of as noise since the data are random (noiselike). The equalizer input S/N can be less than 15 dB and still be decodable. The reason why the S/N value in the equalizer can be less than 15 dB and still be recovered is because this value may be partially comprised

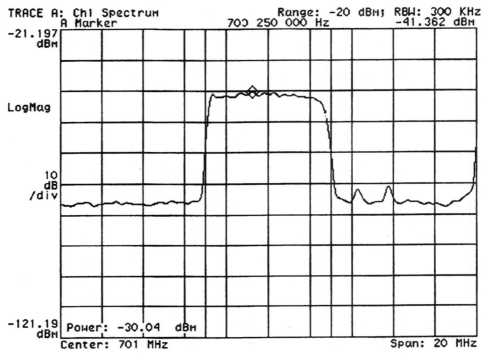

Figure 13.30 DTV passband.

of correlated reflections. If it is pure white noise, the value remains 15 dB, which means there is now an ambiguity as to the proper I value and errors will occur. The equalizer output S/N value in Fig. 13.32 illustrates multipath error improvement. However, it should also be noted that if many taps were on, in order to cancel the correlated ghost(s), the uncorrelated noise would increase as more taps are used. This is known as *white noise enhancement.* As tap energy approaches -5 dB, an additional 1.25 dB of noise will be introduced. This means that as multipath and hence tap energy increase, the 15-dB white noise threshold increases, decreasing headroom and pushing you closer to the error cliff. The difference between equalizer input S/N and equalizer output S/N is an indication of how much "work" the demodulator is doing to eliminate the effects of multipath. Phase output S/N is the S/N value due to uncorrelated noise after tuner phase noise has been corrected. A 511-symbol training string in the frame sync of the ATSC transmission data frame is used by the equalizer as a reference to create base tap coefficients that are applied to the data frame. It is interesting to note that if these time domain plots are taken to the

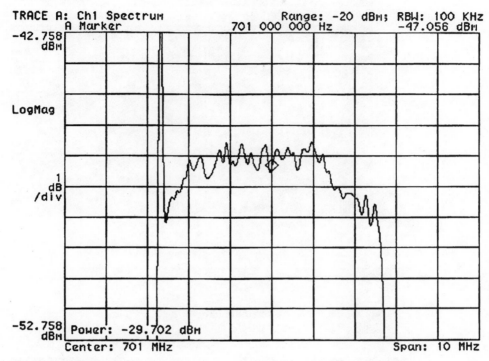

Date: 11-11-98 Time: 02:06 PM
pr29236h. 1:

TRACE A: Ch1 Spectrum   Range: -20 dBm; RBW: 100 KHz
       A Marker   701 000 000 Hz   -47.056 dBm
-42.758
 dBm

LogMag

  1
 dB
/div

-52.758
 dBm   Power: -29.702 dBm
       Center: 701 MHz   Span: 10 MHz

Figure 13.31 Channel tilt.

frequency domain, the result is a channel passband characteristic that cancels the errors in the actual passband.

7. Noise is then added to the system until the error threshold is reached. The DTV channel power and the noise power (rf input signal removed) are measured to determine the site margin and as a sanity check to confirm that the DTV signal is still recoverable with an S/N ratio of approximately 15 dB. Sure enough, when a power level of -30 dBm is input into the VSA, errors begin when the noise floor is brought up to -45 dBm.

There are two ways to determine the white noise margin of the received signal at a test site. The first way is to attenuate the receive signal until errors occur. However, any local EMI/RFI in the truck or test site can make this approach inaccurate. The preferred way is to add gaussian noise.

8. The test receiver is given the command to make the same equalizer and tap measurements again in this simulated near-threshold condition.

9. Peak sync power and C/N are measured for the NTSC channel as a basis for comparison. The CCIR impairment rating, a subjective five-point quality rating, is recorded for the NTSC signal.

Figure 13.32   Software program that indicates the DSP efforts required in a test receiver to ensure proper decoding of the received DTV signal.

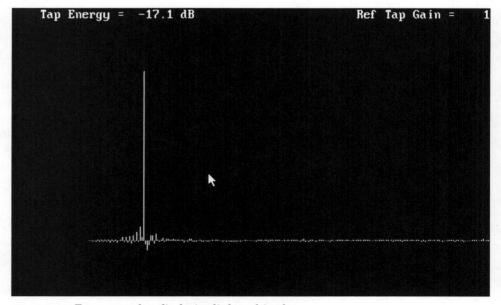

Figure 13.33   Tap energy plots displaying little multipath.

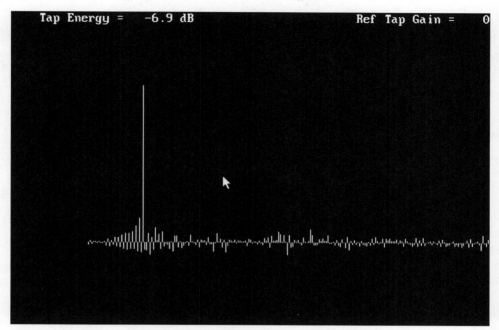

Figure 13.34 Tap energy plots displaying a medium amount of multipath.

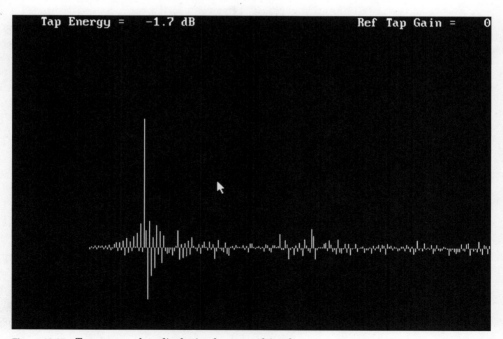

Figure 13.35 Tap energy plots displaying heavy multipath.

13.4.3 Reflections

Sections 13.4.1 and 13.4.2 illustrate that signal strength and reflections are the two big limiting factors in determining DTV coverage. Tap energy and S/N are parameters measured mostly as an indication of 8-VSB health. With NTSC, reflections simply degrade the picture. The more reflections, the greater the degradation. With NTSC, reception reflections delayed by more than 0.25 μs could be perceived by most viewers as "ghosts." A reflection level of 3 percent could be seen by critical NTSC viewers.

Receiver tap energy represents the coefficients needed in the DSP circuitry to eliminate multipath. It has been found that the signal can be recovered even if the coefficients representing reflections total one-half the main signal level. However, the DSP wizardry comes at a price, known as white noise enhancement. Besides amplifying the time-domain components in such a way to eliminate the ghosting, random white noise become amplified as well. So when lots of tap energy is present, more noise is present. If significant ghosting exists, for example 50 percent, the signal-to-noise ratio required to recover the DTV signal is increased from approximately 15 to 16 dB due to the noise added by the ghost cancellation circuitry.

13.4.4 NTSC versus DTV coverage

Many have openly expressed concern about how DTV coverage will compare with current NTSC coverage. Additionally, because of the noiselike nature of the ATSC channel, NTSC channels will be protected from DTV interference while DTV channels will receive no protection from NTSC interference. While noise levels will slowly rise as NTSC signal strength drops, DTV reception will work right up until the carrier-to-noise (C/N) reaches the critical level. As is widely known, the signal is recoverable until the C/N has dropped to a mere 15 dB. The cliff is steep as signal recovery goes from no errors to unrecoverable in just 0.6 dB. For a UHF field strength of about 43.5 dBμV/m has been determined by the ACATS to be the noise limiting value for DTV reception.

Band	dBμV/m			
	City	Grade A	Grade B	DTV
Low VHF	74	68	47	28
High VHF	77	71	56	36
UHF	80	74	64	43

As you can see, lower signal strengths have been mandated for DTV.

13.4.4.1 Overlaying DTV stations.   For DTV the FCC used essentially the same height above average terrain (HAAT) antenna pattern as for existing NTSC. The FCC used a coverage algorithm known as Longley-Rice (L-R) to allow DTV signals into areas not actually covered by NTSC so that DTV stations could be located closer. In some instances L-R predicts coverage into the same areas, which results in interference. Interference is limited to 10 percent of the actual NTSC service area.

The L-R input includes frequency, H or V pol, HAAT, transmit power, transmit antenna pattern and gain, receive antenna height [−30 ft (9.1 m)], receive antenna pattern, and gain. L-R calculates propagation input parameters (ground conductivity, atmospheric refractivity, and climate descriptor code) and propagation loss variability factors (long-term time variability, location variability, and confidence, which accounts for other effects). L-R gets terrain between TX and RX and processes terrain to each RX site using the propagation model. This process is repeated for 500,000 to 3 million locations for each station.

13.4.4.2 Determining DTV coverage. Coverage is based on a function of location (L) minus the percentage of the location's receiving this signal and time (T) minus the percentage of time received. From this, a set of curves is plotted, using various times. The curves are a function of location and time and are, thus, known as $f(L,T)$ curves. These curves have caused a fair amount of confusion and consternation. The curves are based on signal strength at a particular location over the prescribed time. As an example, common sets of curves are $f(50,50)$ and $f(50,90)$. The $f(50,50)$ curves mean that, at a particular distance from the transmitter, 50 percent of the locations tested received the average signal strength for that distance 50 percent of the time. The $f(50,90)$ curves are the same, except that the locations must receive the average signal 90 percent of the time. Thus, $f(50,90)$ is a more stringent test than $f(50,50)$.

Height has always been important for providing good coverage, and that definitely doesn't change with DTV. The higher the antenna, the better. A standard measure of antenna height is known as the height above average terrain. The HAAT elevation is determined using terrain between 2 and 10 mi of the transmit site and measuring elevations at 1-mi intervals (nine measurements per radial). Eight radials are used. The elevation at the actual tower site is used as well. The result is HAAT. HAAT affects the location where the radio horizon is met. Thus

$$\text{Optical horizon} = 1.22 \times \text{square root of height}$$

$$\text{Radio horizon} = 1.44 \times \text{square root of height}$$

A simple way to show this is with two circles tangent to each other at their tops. The inside one is equal to the earth's diameter, while the outside one is equal to one and one-third times the earth's diameter.

A HAAT of only 300 ft and an ERP of 316 kW (25 dBk) for UHF will cause the radio horizon to be hit before the signal has dropped in field strength to a grade B signal (64 dBμ). A HAAT of 300 ft has a horizon about 25 mi out, a 600-ft HAAT has a horizon about 35 mi out, a 900-ft HAAT has a horizon about 43 mi out, a 1200-ft HAAT has a horizon about 49 mi out, a 1500-ft HAAT has a horizon about 55 mi out, a 2000-ft HAAT has a horizon about 63 mi out, a 5000-ft HAAT has a horizon about 77 mi out, and a 6000-ft HAAT has a horizon about 89 mi out. Thus, a 600-ft HAAT gets about 4 mi into grade B, while a 2000-ft HAAT gets 16 mi into grade B, and a 4000-ft HAAT gets 30 mi into grade B. Field strength at any given distance is higher with greater HAAT. Thus, for a given frequency and ERP, you will have a family of $f(50,50)$ curves (field strength versus distance) for various HAATs. A low VHF with an ERP of 100 kW (20 dBk) will hit the radio horizons at the same point as the 316-kW UHF, but the grade B contours are at 47 dBμ versus 64 dBμ. Thus, VHF stations tend to have much larger grade A coverage and often very little grade B coverage. Only a HAAT of 4000 ft will come close to hitting the grade B field (about 4 mi short). A high VHF whose grade B point is at 56 dBμ will make it about 4 mi into the B zone with the same HAAT.

The FCC used the existing NTSC facility database to determine existing coverage (current NTSC power, current antenna pattern, and current HAAT for eight radials). It applied those parameters to the new DTV (FCC used 360 radials interpolated from the original eight), determined power at $f(50,90)$ to replicate NTSC grade B, and adjusted antenna pattern to produce specific output to more closely duplicate coverage, based on HAAT and $f(50,90)$. An existing directional pattern is exaggerated by the technique. FCC then used Longley-Rice to find interference conditions.

Due to the cliff-edge effect of DTV, a given signal level is needed for a higher percentage of the time. The $f(50,90)$ curve provides higher availability of the signal. The same difference for $f(50,50)$ and $f(50,10)$ is used for interference predictions. The difference is roughly 6 dB. While $f(50,50)$ produces curves that extend farther out in distance, $f(50,90)$ produces curves that fall much more quickly with distance.

13.4.5   Receive antennas

The FCC relies on receive antenna gain to provide gain to the desired station and rejection to the undesired stations outside of the protected grade B contours. As a result, antenna selection by you and your viewers is important.

13.4.5.1   Dipoles.   The dipole is the basic antenna. Due to capacitive fringing, a one-half-wavelength dipole is usually only (0.95) one-half wavelength long. The beam width (3-dB rolloff points) for a dipole is 78°. As such, the basic dipole doesn't have much directivity.

A folded dipole's input I will be one-half that of the one-half-wave dipole with the same E (voltage). Therefore, input Z is 4 times that of the regular dipole ($4 \times 75 = 300$).

The dipole bandwidth is approximately 8 to 16 percent of its resonant frequency. The bandwidth of a folded dipole is typically 10 percent greater than that of the equivalent dipole and it exhibits two peaks, much like a double-tuned circuit. The bandwidths of resonant antennas, like dipoles, depend primarily on the wire or tube diameter. The larger the diameter, the lower the reactance, and therefore the lower the Q. Lower Qs mean wider bandwidth but with less gain.

13.4.5.2 Yagis. The basic yagi consists of a folded dipole at its core, and this folded dipole is the active element. "Directors" are located in front of the active element and "reflectors" are located behind the folded dipole. This arrangement is known as a *linear array*. The directors and reflectors are known as *parasitic elements*. The reflector is an aluminum rod cut approximately 5 percent larger than the dipole, and the director is 5 percent shorter than the dipole. Spacing between the dipole and other elements is 0.15 to 0.25 wavelength, and is determined for maximum directivity. For the yagi this is usually about 9 dB.

Signal energy coming from the director strikes all three conductive elements and excites them into resonance. Reradiation of the energy from the parasitic elements adds constructively at the dipole for an overall power gain. Additional reflectors add little gain, but two additional directors add +2.5 dB. An additional six directors (for a total of eight) add only 1.5 dB more gain. Front-to-back ratios for a yagi can be as high as 30 dB. Yagis can be stacked vertically one wavelength apart to increase gain and directivity.

Some popular yagis have multiple 60° swept angle dipoles for VHF and an eight-dipole yagi with a corner reflector (>) for UHF. The parasitic elements of a corner reflector structure perform the same functions as the reflector elements in a linear yagi.

13.4.5.3 Log periodic. The log periodic dipole array has moderate gain, and is a highly directional broadband antenna. It consists of a horizontal array of one-half a dipole. Each dipole is cross-wired to the ones next to it. They become successively shorter from the back element to the front, along with successively shorter spacing between elements. The downlead is connected to the front, and smallest, dipole. Broad bandwidth is obtained with log antennas because multiple dipoles are resonating at various frequencies. When a particular dipole is resonant, the ones in front of it act as directors and the ones behind it act as reflectors.

13.4.5.4 Loop antennas. Loop antennas have wide bandwidths. The loop diameter should be less than wavelength/16. The shape of the loop doesn't matter. It can be circular or square shaped. It can even be coiled multiple times around a ferrite core. The loop antenna has a toroidial or doughnut-shaped pattern. The doughnut-shaped pattern envelopes the coil, with a null in sensitivity as you face either side of the loop.

13.4.6 DTV decoders

Most early DTV receivers will actually consist of separate demodulator/decoder boxes, or set-top boxes, feeding separate displays. Most chip sets that perform these functions will come from only a half dozen manufacturers to start.

DTV decoder chip sets generally consist of ATV demod/FEC, MPEG-2 system layer demux, ATV video decoder, Dolby AC-3 decoder, ATV display processor, and a line doubler/tripler for HD monitor (Figs. 13.36 to 13.39). Features include demodulation 8-VSB terrestrial 10-bit sampled signal input to the

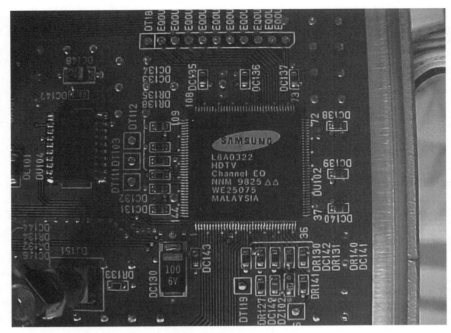

Figure 13.36 Channel equalizer IC for DTV chip set.

demod and PCR recovery and decoder timing for A/V synchronization. It converts all formats to either 1280 × 720p or 1920 × 1080i, and receives VSB or QAM terrestrial and cable broadcasting. It should be capable of decoding video in MP@HL of ISO/IEC 13818-2 MPEG-2, and is compliant with 13818-1.

Most receivers will decode all 18 ATSC formats for display, and will decode 5.1 sound. All are 16:9 (Figs. 13.40 and 13.41). At the receive end inside a DTV receiver, or a set-top box, the VSB is demodulated and fed to a transport demultiplex. The DTV receiver uses the low-power pilot to frequency-lock to the signal. The video and audio channels are decoded and fed to a display. Data will eventually be available out of most STBs.

13.4.7 DTV displays

Many will view their first DTV on computer-type VGA monitors, which are all component video (RGB) and progressively scanned (Fig. 13.42). These displays can either be actual computer monitors or displays that might look like a conventional consumer television but which, internally, are more like a computer than an NTSC device. Almost all DTV displays will have a native scan format. Regardless of the format they receive, it will be converted to a format that the display uses. Displays will use a single-presentation or native format to simplify deflection circuitry. Many will upconvert SD since 720p and 1080i native displays

Figure 13.37   Channel decoder IC for DTV chip set.

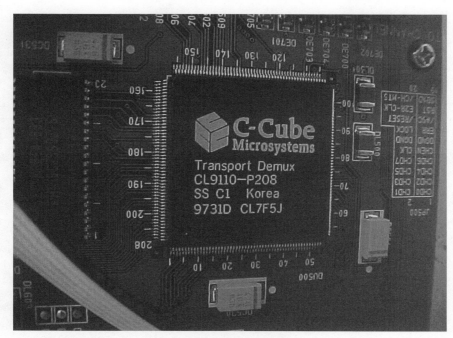

Figure 13.38   MPEG transport stream demux IC for DTV chip set.

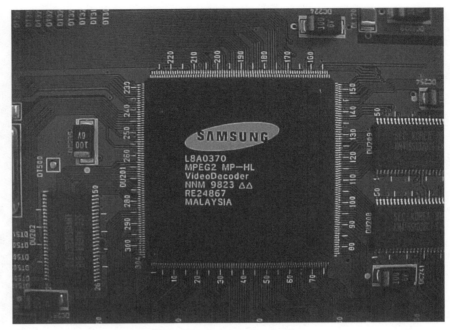

Figure 13.39   IC that decodes MPEG back to baseband video.

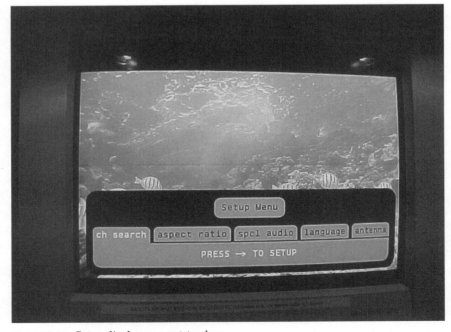

Figure 13.40   Setup display on a set-top box.

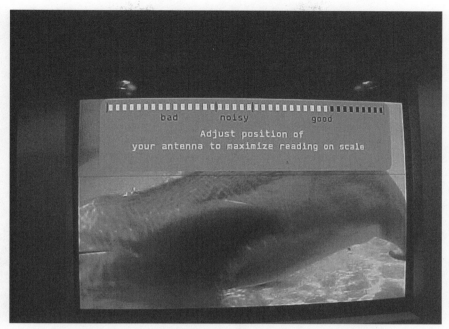

Figure 13.41   Display that helps user aim antenna for maximum received signal strength.

appear to be becoming prevalent. Sony uses a special circuitry to upconvert all DTV to 1080i using something called "digital reality creation." They claim to double SD H and V resolution. Common V scan rates are 60 and 72.

Common display attributes are as follows:

61-in diagonal projection display, 480p native scan.

55-in rear screen, 800 horizontal lines native scan.

34-in flat CRT, 1080i native scan.

55-in rear projection screen, native 800 × 600 pixels, uses a set of Texas Instruments digital micromirror devices (DMDs)—one is for green and the other is for red/blue by using a color wheel.

65-in 1080i display.

73-in 1080i display projector, 1280 × 720 native display, full 1920 × 1080i display.

DMD technology is almost 20 years old, and DMDs have been demonstrated using 5000-W xenon bulbs with a light output of 10,000 lm onto a 47-ft screen. The DMDs now available have 600 × 800, 1024 × 750, and 1280 × 1024 resolutions. Each DMD cell controls light output by how long it is open. To display extremely low light levels, it must be able to operate at extremely high light levels.

Figure 13.42 VGA monitor displaying video for a DTV set-top box.

Inputs available on various receivers are VGA, smart-card slot that will take DirecTV card, S-video, SXGA (1280 × 1024), and firewire.

At this time, plasma displays don't have the resolution to be used to display HD. Their resolution tops out at 800 × 450.

13.5 Digital Satellite Transmission

Two additional areas important to broadcast facilities will be reviewed before concluding this chapter on rf. The first is digital satellite transmission. The satellite transmission will prevent the signals received from degrading as they cascade through various equipment encountered along the path traversed. Since news departments have been early adopters of digital satellite transmission, we will focus on that application. However, can digital satellite news gathering (DSNG) beat analog SNG in solving the obstacles encountered? It can, and in this section we will discuss why and how that is so.

A new mindset is required to understand what is happening in the digital satellite transmission process to move to this new technology. As has been often mentioned before, it is helpful to think of a digital video stream as data. In a transmission path this type of thinking becomes imperative. A digital signal's baseband path is no where near as strenuous as its trek through an rf satellite path.

Satellite transmission can often be compared to telephone modems. Modems come with a slew of standards, with labels such as v.27, v.29, v.32, v.33, etc. These labels specify quadrature phase-shift keying, or quadrature amplitude modulation, and baud rates, among other things. Baud rate is also referred to as the symbol rate. A common misunderstanding is that baud rates and bits per second are synonymous. The baud rate must be less than the channel bandwidth. Telephone modems can't have baud rates higher than approximately 3300, as that is the bandwidth of the average telephone line. Modulation techniques allow the bits per second rate to be higher than that. The same is true in a satellite channel. We will review these modulation methods shortly.

13.5.1 Link budgets

Many more things can ruin a satellite transmission than a telephone transmission (Fig. 13.43). A measure of the health of a satellite link is the parameter called *received carrier-to-noise* (C/N) ratio. In the case of digital satellite transmission another specification is often stated, energy per bit versus noise (Eb/N). These two ratios have a one-to-one relationship. The difference between the transmitted Eb/N and the received Eb/N is known as the *link margin*. Satellite transmission paths commonly have losses in the 200-dB range. With uplink and satellite amps having gains in the range of 50 to 70 dB, and send and receive dishes having gains of 40 to 50 dB, not much room is available for unplanned degradation. Ku-band rain fades alone can subtract an additional 10 dB on the uplink or downlink side. Obviously, this means that you need to start with as high a transmission effective isotropic radiated power (EIRP) as possible and as high a receive signal and as low noise power as possible. Working in conjunction with the received signal and noise is a parameter called antenna gain–to–system noise temperature ratio. A common value for this is >30 dB/K. Noise will increase in northern latitudes as the dish must be pointed closer to the horizon, which means increased atmospheric absorption and increased noise radiated from the earth. The second reason Cassagrain feed systems in dishes are used is so that the feed horn points at the sky, not at the earth. Lower link margins can occur at the edges of the country as satellite transmit/receive antennas commonly have more gain at the center of the mainland than near its borders or coasts—as much as a 6- to 8-dB difference.

Let's state some obvious facts. The creation of a component digital bit stream from an analog one means that what previously fit in a 6-MHz terrestrial AM channel or an 18-MHz FM half-satellite transponder channel now occupies many hundreds of megahertz of baseband bandwidth—at least 540 MHz. What's even worse is that the introduction of HD in the form of SMPTE 292M will result in bandwidths of at least 3 GHz as it has a 1.5-Gbit/s clock rate. Sounds like the term "dc to light" had HD in mind. As we know, MPEG was conceived to solve this problem.

Figure 13.43 Small satellite "farm."

Various MPEG-II profiles and levels have been derived to allow video, audio, and other data into the available bandwidth. The level aspect of MPEG-II specifies the horizontal, vertical, and temporal resolution, along with the resulting bit rate of the data stream. The profile aspect of MPEG-II describes the features available to keep the data rate correct. Profiles allow for the decoding of I (interframe), P (predictive), or B (bidirectional predictive) frames and the luminance-to-chrominance ratio (4:2:2, 4:2:0, etc.). Some of the most common profiles and levels of the MPEG-II compression algorithm are contribution quality 4:2:2 profile @ main level (4:2:2 sampling with data rates that can be varied from 1.5 to 50 Mbits/s, with 18 Mbits/s commonly used) and main level and profile (4:2:0 sampling with MPEG data rates of 1.5 to 15 Mbits/s).

13.5.2 Transponder use

The previous bit rates mean that "digital" signals can be made to fit within today's satellite transponder widths. Common transponder widths tend to vary from 24 to 110 MHz wide. Transponders of 110 MHz are really two transponders with contiguous feeder link frequencies feeding a similar or the same antenna. Common transponder bandwidths are 24, 36, and 54 MHz. However, as any SNG engineer knows, using a whole transponder is considered wasteful for almost all news operations. There just wouldn't be enough transponder space to go around if everyone did that. Therefore, half-transponder usage is the norm. Analog video is frequency modulated for its trip through the satellite path. Universal deviation rules never really evolved for bandwidth use of those half transponders. It depended on the individual satellite operator. Some tended to spec both normal and peak deviation rates, while others only specified peak deviation. Peak deviation of ±7.50 MHz seems to be common, with normal white levels around ±6.85.

Some satellite operators will now let you buy slices of a transponder even smaller than one-half. "Spectral occupancy" can be based on where the skirts at the edges of a signal fall. One satellite operator considers occupancy to exist between the −26-dBc falloff points. Satellites are a precious commodity. The cost of launching a satellite is in the $50 to $100 million range. The cost of the satellite itself can rival the cost of its ride into space. They are fragile devices that consume approximately 1.2 to 5 kW of total power, including all receivers and transmitters and the control systems. The source of this power is generally solar. Efficient use of these resources has been, and will continue to be, required by the marketplace. MPEG compression does not have to stop once the elementary stream (video and audio data) fits into the desired bandwidth. We can continue to compress it even further. We can "shrink" the data rate until many programs can be frequency multiplexed into a single transponder. Or we can take the time multiplex approach and allow a single MPEG data stream that occupies an entire half transponder and combine multiple elementary streams into one program stream. For a two-camera newscast, you could send the video from both cameras back to the station and have it treated as two separate sources instead of trying to preswitch it in the truck. Or, instead, don't send two different sources; send a single source at faster than real time. This would solve a 10 min of raw tape, 5-min satellite window dilemma. Finally, you could be a glutton and use the entire bandwidth to send studio-quality digital video.

13.5.3 Modulation techniques

Digital SNG does not use frequency modulation but quadrature phase-shift keying or the newer 8PSK eight-quadrant phase-shift keying modulation such as BPSK, QPSK, or 8PSK. Each cycle of these modulation types would be considered one symbol. BPSK signifies a high or low bit by 180° phase shift from one symbol to the next. Therefore, it is able to send one bit per symbol. QPSK has four different phase states, that is, each symbol, or cycle, can convey one of four states. In a binary system four states would indicate 2 bits in use. That means QPSK can transmit 2 bits/symbol and 8PSK 3 bits/symbol. QPSK is twice as efficient as BPSK. But what's the catch you ask. Very little between BPSK and QPSK. That's why QPSK is often used and BPSK is not. The two types have the same power efficiency but QPSK has better bandwidth efficiency. Power efficiency is the bit error rate that occurs with a given Eb/N. Both of these signals have no carrier component in the spectrum; therefore, local carriers must be derived at the receiver to recover the information. This last fact usually means that most Ku DSNG systems must use digital LNBs at the receive end. What's a digital LNB? Analog LNBs can have local oscillators that drift as much as 2 to 3 MHz. Some digital decoders need 70 MHz if or, in some cases, L band outputs that drift no more than 100 kHz, although some newer ones can tolerate drifts of hundreds of megahertz. Good LNBs will use temperature-compensated crystal oscillators in their PLL circuits. LNB LO's stability is measured with a parameter known as phase noise, a measurement to determine how much energy is found at various frequencies away from the desired LO frequency. An example of a good phase noise measurement would be −65 dB @ 1K.

Another consideration in the digital realm is spectrum inversion. The signal, just like analog signals, is upconverted at least once on the uplink side, downconverted in the satellite, and downconverted at least once on the receive side. These conversions are usually accomplished through heterodyning. Basic communications theory tells us that the product of this process is the original signal, the local oscillator sine wave, and the sum and differences of the first two. The sum product is a replica of the original signal but at a new frequency. However, the difference signal has a spectrum at its new frequency that is a mirror image of the original. Some frequency converters use the sum product (filtering out all the other products), while others use the difference product. If an even number of these difference products is used in the path, there is no problem, since the double inversion cancels out. If an odd number is encountered, it can cause problems. Digital receivers must be able to cope with this situation. Analog FM satellite signals do not seem to be affected by this because once the signal is discriminated, a simple inversion of the baseband signal is all that is needed. Most newer receivers sense the inversion and correct for it automatically. Older ones had switches to perform this manually. When QPSK receivers are used for digital signals, the I and Q signals must be able to be inverted to solve this problem.

Another modulation method is quadrature amplitude modulation. In the United States, QAM is generally used in terrestrial microwave links and not in domestic satellites. The satellite transponders that are generally available for demand usage have traveling wave tube (TWT)–type power output transponders that are inherently nonlinear in amplitude transfer characteristics. Amplitude modulation predistortion has been tried but not successfully, so QAM has been left to the Canadians. Most Canadian transponders are FET solid-state power output type and amplitude linear. The phase pattern that QAM generates is referred to as a *constellation*. You can think of the dots on a vectorscope due to color bars as a constellation. Just as NTSC/PAL color modulation uses phase and amplitude, so does QAM. The number of points in the constellation, such as 16, determines the QAM type. Sixteen points would indicate a 16-QAM signal. Each point in the constellation would signify a state. 16-QAM has 16 states. It would take 4 bits to specify one of 16 states. Therefore, every symbol, or cycle, of 16-QAM would each convey 4 bits. This means that 16-QAM has 2 times the spectral efficiency of QPSK. For 16-QAM to maintain the same average transmit power as QPSK, the constellation would have to be packed more tightly. However, as the space between the points in the constellation diminishes, the error probability goes up. 16-QAM needs a higher S/N ratio for the same error performance as QPSK. One trick used to minimize errors in QAM is to use "gray" coding for mapping points in the constellation. This means that the value of any point in the constellation is only 1 bit different from any adjacent point. As we will see shortly, when we look at link budgets and error detection and correction, generally, error immunity decreases as data payloads increase. More spectrum-efficient modulation schemes require higher S/N ratios, which means increased satellite link budgets.

13.6   DSTL

The same bandwidth demons that haunt digital pipes haunt the path between the studio and the transmitter. Most facilities encode and create their ATSC transport stream at the studio and ship it to the transmit site for 8-VSB modulation and transmission. Adding more microwave hops in congested areas is becoming very problematic. Thus, there is much incentive to share the existing paths with the additional DTV paths that are needed.

Most microwave hops are 25 MHz wide. With current analog transceivers, the modulation method is FM. The wider the channel, the higher the index of modulation, which equates to a high S/N ratio. So although narrowing the channel for the NTSC signal is an option, video quality will suffer. A 25-MHz-wide channel can handle approximately 22 Mbaud of phase-modulated traffic. If QPSK is used, that is 2 bits/baud for 44 Mbits/s of traffic, which is essentially a DS-3 path. Depending on the ECC inner code (R) ratio, the following net traffic could be sustained:

$$R = \frac{1}{2}\ 20.3 \text{ Mbits/s} \qquad R = \frac{2}{3}\ 27.0 \text{ Mbits/s} \qquad R = \frac{3}{4}\ 30.4 \text{ Mbits/s}$$
$$R = \frac{5}{6}\ 33.8 \text{ Mbits/s} \qquad R = \frac{7}{8}\ 35.5 \text{ Mbits/s}$$

This means that a 19.392658-Mbit/s ATSC signal would still leave upwards of 16 Mbits/s to carry the existing NTSC signal in the DS-3 path. Hence, MPEG compression would also be performed on the NTSC signal. Both the ATSC and the NTSC data streams would be sent to the transmitter over the same DS-3 microwave path. At the transmitter, the NTSC bit stream would be converted back to NTSC via an STB. A drawback to this approach is the latency of the MPEG encoding/decoding of the NTSC signal. If the on-air NTSC signal is used for interruptible feedback (IFB) in ENG operations, the delay to the talent will seem even worse than most satellite hops.

Another approach is to combine the analog NTSC with the digital ATSC bit stream and transmit them on two separate rf carriers within the microwave channel. The NTSC video and audio are FM modulated and the ATSC signal is digitally modulated using 16-QAM or 8-VSB modulation schemes. The ATSC would occupy approximately 6 MHz of the channel, while the NTSC would occupy the rest. This narrowing of available bandwidth for the NTSC would slightly lower the modulation index, slightly increasing the S/N ratio.

13.7   Bibliography

Bloomfield, Larry, "Some HDTV Receivers to Display Low Pixel Counts," *Broadcast Engineering,* December 1998, p. 12.

Boston, Jim, "Engineering KICU's Path to DTV," *Broadcast Engineering*, April 1999, p. 202.

Penhume, James, "Digital TV: Where's the Consumer?" *Broadcast Engineering,* December 1998, p. 76.

Index

ABOUT THE AUTHOR

Jim Boston is Senior Project Engineer for The E. W. Scripps Company. He has written numerous articles for *Broadcast Engineering* on digital audio and video signal transmission.